Beyond the Paths of Heaven

The Emergence of Space Power Thought

A Comprehensive Anthology of Space-Related Master's Research Produced by the

School of Advanced Airpower Studies

Edited by

Bruce M. DeBlois, Colonel, USAF
Professor of Air and Space Technology

Air University Press
Maxwell Air Force Base, Alabama

September 1999

Library of Congress Cataloging-in-Publication Data

Beyond the paths of heaven : the emergence of space power thought : a comprehensive anthology of space-related master's research / edited by Bruce M. DeBlois.
p. cm.
Includes bibliographical references and index.
1. Astronautics, Military. 2. Astronautics, Military—United States. 3. Space Warfare. 4. Air University (U.S.). Air Command and Staff College. School of Advanced Airpower Studies-
-Dissertations. I. Deblois, Bruce M., 1957-
UG1520.B48 1999 99-35729
358' .8—dc21 CIP

ISBN 1-58566-067-1

First Printing September 1999
Second Printing September 2002

Disclaimer

Contents

PART IV

High-Ground Perspectives

Illustrations

Table		Page

Table *Page*

Overview

Bruce M. DeBlois

Major issues have plagued the US military space community for years. Foremost among these issues is the relationship between air and space. At a recent airpower conference, military leaders from the western powers presented discussions of airpower and space issues with a pervasive underlying assumption: that the next logical step from the exploitation of airpower and space capabilities was the merging of the two environments toward the exploitation of "aerospace" power.[1] The current distinction between air and space rests on the fiscal and technical inability to merge them—an inability that is soon to be overcome. Conferees dismissed environmental distinctions between the two on the grounds that there is no absolute boundary between air and space.[2] In *Paths of Heaven*, the chapter titled "Ascendant Realms: Characteristics of Air and Space Power," I examine this assumption from the perspective of 21 different military characteristics and conclude it to be invalid. The reasons extend well beyond an inability—fiscally and technically—to merge the two realms.

Similarities based upon functions and the lack of a distinct boundary are offset by distinctions in the physical environments. The physical laws of air and space are profoundly different. A vehicle flying on a cushion of air is not equivalent to a vehicle in free-fall orbit. Aside from the issue of access due to huge differences in energy requirements, the airborne vehicle is maneuverable and allows for flexible operations while the space-borne platform is fixed to a high-velocity orbital path. The latter expends little energy to stay in a fixed orbital position, allowing it a duration capability well beyond airborne vehicles. The issue is not whether the two environments can be merged technically, but given that they can be merged, should they be merged. An analogy is useful to illustrate the argument.

Land and sea forces maintain a two-dimensional perspective and relatively slow pace of operations. The amphibious mission certainly illustrates the fact that there is no absolute boundary between land and sea for military purposes. Fiscal and technical capability to merge the two environments in an attempt to exploit surface power exists. In spite of these similarities, land power and sea power have not been merged as surface power because of environmental differences. The question is not whether to make a land/sea capable vehicle or system, but whether they should be the mainstay of a military surface capability. The answer is a resounding no. Given limited fiscal resources, the choice between making either 1,000 land/sea vehicles or making 490 land vehicles, 490 sea vehicles, and 20 land/sea vehicles is trivial. A land vehicle will out-perform a land/sea vehicle on land, and a sea vehicle will out-perform a land/sea vehicle at sea. Most missions are either at land or at sea; only a few cross the hazy boundaries. It makes sense to invest in the best capability for the environment in which the mission will be performed. Doctrine, organization, and strategies flow from the environments and the systems employed to exploit those environments. Hence land power is distinct from sea power. Surface power would be a less optimal approach.

The same argument holds true for air and space power. Air and space forces maintain a three-dimensional perspective and relatively fast pace of operations. The similarities end there. Although there is no absolute boundary between air and space, no physicist would refute the fact that once the fuzzy boundary is transcended, the nature of the environment changes radically. Fiscal and technical capability to merge the two environments in an attempt to exploit aerospace power is emerging, but should it be pursued? Again, environmental differences drive the answer. The question is not whether to make an aerospace capable vehicle/system, but whether we should make many as the mainstay of a military aerospace capability. The answer, again, is a resounding no. A space vehicle will out-perform an aerospace vehicle in space: A typical aerospace vehicle will carry the baggage of air capability, such as wings, into space. An air vehicle will out-perform an aerospace vehicle in the air: A typical aerospace vehicle will carry the baggage of space capability, such as radiation shielding, in the air. Most missions are either in the air or in space, and only a few missions are performed at the boundary. As was the case with land and sea, it makes sense to invest in the best capability for the environment in which the mission will be performed. Hence, airpower is distinct from space power. Aerospace power, like surface power, would be less than an optimal approach. The crux of the argument rests on the distinction in physical environments, which may not be obvious to a society raised with science fiction presenting maneuverable, flying space fighters. The fact that the environments and related physics are drastically different is above reproach. The chapters in this book embody independent graduate research on space-related issues, and all assume the distinction between air and space.

Many of the chapters are products of one of several schools of space power thought. From a theoretical perspective, the seminal work by David Lupton sorts the "how-to-approach-space" controversy into four categories.[3] The *sanctuary school* views space as a realm free of military weapons, but allows for military-related systems providing such functions as treaty verification and intelligence activities. Advocates maintain the only way to ensure the legal overflight aspect of current space treaties is to declare space as a war-free zone or sanctuary. This school calls for virtually no funding of military space programs involving weapons in space. The sanctuary school has a substantial following in the domestic and international populace, though many in the military see it as a "head-in-the-sand" approach to national security. This military perspective is unfortunate, since the strong case in favor of the military advantages of a space sanctuary posture warrants objective consideration.[4]

The *survivability school* argues that military forces should deemphasize space access, but for less idealistic reasons—the assumption that space forces are inherently exposed and vulnerable. Survivability adherents assert that the probability of using nuclear weapons in the remoteness of space is higher. This, the fact that weapons effects have longer ranges outside of an inhibiting atmosphere, and the vulnerability associated with predictable orbit locations support the survivability position. Remoteness also allows for plausible deniability, thus making the decision to attack more likely. The survivability school calls for the recognition that space forces are not dependable in crisis situations. They are critical systems openly exposed and make for likely targets. Military space missions should thus be limited to communications, surveillance, reconnaissance, and weather reporting. From this perspective, investment strategies ought to fund those missions, along with redundant space-terrestrial programs, and perhaps ground-based antisatellite (ASAT) systems.

The *space control school* recognizes the importance of space as coequal with air, land, and sea power. The result is that military space policy must balance investments in space, air, sea, and land power to meet the anticipated threat. Of the four schools, space control is the face worn by the Department of Defense (DOD) and the Air Force since the 1980s. Current political emphasis on jointness prompts a space control approach as evidenced in Air Force Manual (AFM) 1-1, *Basic Aerospace Doctrine of the United States Air Force;* Air Force Doctrine Document (AFDD) 1, *Air Force Basic Doctrine*; AFDD-4, *Space Operations Doctrine;* Field Manual (FM) 100-5, *Operations;* and Joint Doctrine, Tactics, Techniques, and Procedures (JDTTP) 3-14, *Space Operations.*[5]

The *high-ground school* advocates space as the location from which future wars will be won or lost. The view of using space-based ballistic missile defense (BMD) to convert the current offensive stalemate of mutually assured destruction to mutually assured survival has some appeal. The growing number of supporters of this school advocate expanded militarization of space and the adoption of a corresponding policy. In their view, investments ought to focus on both offensive and defensive space systems at the expense of air, land, and sea systems. Funding would include space-based ASAT systems, directed-energy warfare (DEW), and BMD with maneuverable, space-to-space, space-to-air, and space-to-ground capability. Air-to-space (airborne laser or kinetic miniature homing vehicle ASAT) and ground-to-space (direct ascent ASAT) systems would also warrant investment.[6]

These schools of thought often extend beyond the military perspective into the policy arena. Each school has support from a variety of constituencies, and each plays a role in the way the military has approached space as a potential war-fighting realm. Beyond the theoretical controversies, the fundamental problem within the military space community stems from a violation of military principle: unity of command/effort. Former commander in chief for space (CINCSPACE), retired Air Force Gen Charles A. Horner, when asked by the chairman of the Senate Armed Services Committee, Senator Sam Nunn, if he was in charge of space, replied that—it depends because he is the one commander in chief (CINC) that exercises little control over his own command. The National Aeronautics and Space Administration (NASA), the Defense Information Systems Agency (DISA), the Ballistic Missile Defense Office (BMDO), the Central Intelligence Agency (CIA), the Central Imagery Office (CIO), the National Reconnaissance Office (NRO), the National Oceanographic and Atmospheric Administration (NOAA), the departments of Commerce, Transportation and Interior, the National Science Foundation, and the White House Office of Science and Technology Policy all intrude upon CINCSPACE's budget, while many of the same organizations intrude upon his launch, on-orbit control, research and development (R&D), and acquisition authority.[7] In addition to the governmental intrusion into his joint command, CINCSPACE must also deal with service infighting over who should have the dominant role in space.

Military space lift vehicle requirements, space architectures, and ground support infrastructure are more major issues. Graduate students at the School of Advanced Airpower Studies (SAAS) researched and discussed a variety of these issues and their efforts are brought together here as a collection of master's degree research theses. The significance of this book lies in the synergism of the contributions. Although each of the following articles reflects varying, well-documented, independent perspectives with both strengths and weaknesses, in total, the articles give a mature summary of the best available military thought regarding space power. A summary of each thesis follows. The first three papers examine space organization, doctrine, and architecture. The remaining are loosely grouped as predominantly sanctuary/survivability, space control, or high-ground perspectives.

Space Organization, Doctrine, and Architecture

"An Aerospace Strategy for an Aerospace Nation" analyzes the need for a national aerospace strategy that encompasses the linkage of the aerospace industry and military aerospace. Stephen E. Wright's assessment of the US aerospace industry reveals that it provides the kind of high-technology and high-wage jobs necessary to improve the nation's standard of living. Likewise, a vibrant military aerospace is essential to national security. The writer evaluates current military strategies against a set of political imperatives and the reliance each strategy has upon aerospace power. The results of this process show that each military service relies on aerospace power for the success of its strategy. By coupling these facts with the serious problems that exist in the aerospace industry and in military aerospace, the author shows the need for the United States to develop a national aerospace strategy. The final section of the study proposes this goals and objectives of such a strategy and recommends the formation of a national aerospace council to develop and implement a national aerospace strategy.

The strengths of Wright's work lie in his presentation. The critical issue is not how to get to space or what to do when we get there. The issue is, and has always been, support of a flourishing economy and a national security policy that protects it. The commercial and/or military use of space is pertinent only as it supports national interests. Wright recognizes this and establishes that the health of the US aerospace community is in the US national interest. The breadth at which the author examines the issue is evidenced by his nonparochial approach examining the criticality of aerospace from Navy, Marine, Army, and Air Force perspectives. Broaching the topic from this vantage shows several limitations. Although he examines future conflict broadly, he addresses current and emerging political imperatives as they direct current and near-term employment of aerospace forces. This limitation is somewhat excusable, as it would require an extensive futures study to establish future political imperatives, and even then, those future political imperatives would be, at best, educated guesses. As for the emerging political imperatives, each of the services' strategies conveniently supports the imperatives. While the services have produced effective, satisfying strategies for nurturing and employing aerospace power, it is hard to believe that they have produced efficient, optimum strategies. The fact that the services claim that a joint, national strategy for aerospace is a necessity suggests that there must be some redundancy between the separate services' strategies. Further research into how such a joint, national strategy would impact each service is necessary, but was beyond the scope of Wright's work. Finally, lumping of air and space together makes it difficult to cull which of Wright's main points apply to space power. The argument can be made that even if the environments and systems are radically different, air and space capabilities both emerge from the same technical community—the aerospace community. Thus the claim that the United States needs a coherent, national aerospace strategy has merit.

Such a national strategy would, no doubt, have a significant impact on doctrine. The lack of a national aerospace strategy may in part be responsible for the many doctrinal shortcomings cited in this book.

Frank Gallegos' purpose in writing, "After the Gulf War: Balancing Space Power's Development" is to expose such doctrinal shortcomings which caused significant problems in the employment of space power during the Persian Gulf War. Comments like "the Gulf War was the first space war" wreak of revisionist history and seem to indicate that the United States entered the war with a well-thought-out strategy for employing space power. Nothing could be further from the truth. Space technology was certainly exploited, but its effectiveness against a lack-luster adversary tends to overshadow the inefficiency in its employment during Desert Shield/Desert Storm. Ironically, the success of space technology in that war may be the biggest obstacle in correcting significant doctrinal shortcomings.

Gallegos presents many perspectives on the role space played in the Gulf War. Each results in different points of view on space shortfalls, which once brought together, produces a rich pool of recommendations. While United States Space Command (USSPACECOM) recognized the lack of capability (normalized operations and theater missile defense), the war fighter, that is United States Central Command (USCENTCOM), accented a lack of doctrine, training, and support. *The Gulf War Airpower Survey (GWAPS)* emphasized a different set of issues exemplified by a fundamental flaw in space architecture: a cold war mentality which focuses on supporting strategic levels of war and overlooking operational and tactical support. The *Conduct of the Persian Gulf War: Final Report to Congress*, unlike other sources, emphasized technology's shortcomings, particularly space launch and communication satellite vulnerabilities. Gallegos' summation of these shortcomings provides a comprehensive summary of the many limitations space presented to the war fighter in the Persian Gulf War.

The strength of Gallegos' work lies in his clear summation of lessons from the war, many of which boil down to poor doctrinal development, a problem which he claims continues today. One weakness of his analysis is the assumption that lack of doctrine is a problem. A valid counterposition is that the lack of doctrine aimed at weaponizing battlefield space is a well-thought-out, military sanctuary strategy. Gallegos recognizes that the newly formed Fourteenth Air Force, Space Warfare Center, and Space Support Team have all attempted to fill the experience and doctrinal gap, but for a variety of reasons, have fallen short. Recognizing a problem is a beginning toward a solution, but the lack of a clear method for correcting the doctrinal shortfall is a weakness of the work. Stating that we need more doctrinal development falls short of stating who is to do it, on what sort of continuing cycle it is to be done, and in what forum it is to be developed—Air Force, joint, and/or combined. Furthermore, the contention that

> the inclination to be on the leading edge of technology often comes with a mutually strong penchant to disregard the teachings of the past

offers a false dilemma of either technological development or doctrinal development. The fact that space technological development leads its complementing doctrinal development does not mean that the former comes at the expense of the latter. Beyond these obvious limitations, Gallegos provides a useful summary of the major space lessons of the Gulf War. His articulation of the cold war space paradigm as a highly classified, strategic approach to space, which emphasizes technological research and development over doctrinal development and operational integration is accurate, and offers the next generation of space strategists an objective perspective. As emphasized in the *GWAPS*, space architectural development is one possibility such doctrinal development may support, a subject examined by the next author.

In "Blueprints for the Future: Comparing National Security Space Architectures," Christian C. Daehnick makes a credible argument that US posture toward developing a space architecture in support of national security is strongly biased by an historical inertia of organizational development, as opposed to a rational decision to produce the most efficient and effective architectures.[8] He defines the current approach to space architecture as a command-oriented approach and offers an alternative: demand-oriented space architecture. Command and demand architectures vary on three counts.

Physically, the current command-oriented architecture focuses on heavy lift for specialized cargos and requires big investments for a few large systems with extensive ground-based infrastructures. A demand-oriented architecture would involve lighter lift requirements not tailored to any specific cargo and would require dispersed investments in many systems with smaller ground-based infrastructures.

Temporally, the development cycle that supports the command-oriented architecture is restricted to incremental improvements in design, manufacture, and deployment, as the sunk costs in current systems compel future investments to support them. Once deployed, the paradigm is long-loiter, on-orbit capability with long-lasting mission-specific capability. The demand-oriented approach allows for radical change, as huge sunk costs in particular systems do not exist. Additionally, the paradigm can shift, allowing ground-to-space missions to meet situational requirements on demand, as opposed to maintaining predetermined capabilities on orbit.

The third difference between command-oriented and demand-oriented architectures is probably the most profound. Philosophically, the command-oriented approach grew out of a high-performance, 100-percent reliability aircraft manufacturing community. It was politically motivated by a controlled response to the USSR during the cold war. The demand-oriented architecture is a rational approach without zero-fault tolerance or cold war biases. It emphasizes responsiveness, flexibility, ease of operations, and cost attributes over high performance and reliability (most spacecraft, unlike most aircraft, are unmanned). While the command-orientation prescribes centralized command, control, and execution directed by specific group interests, demand-orientation allows for flexibility in command, control, and execution. Military use may require centralized command and control and decentralized execution analogous to the traditional method of allocating scarce air assets. Depending on the military situation, a demand-oriented architecture would allow for a more distributed network of space assets which would reduce each asset's vulnerability. Corporations, on the other hand, may see the low-cost communication space asset as a capability that is readily decentralized in command, control, and execution.

The strength of Daehnick's research rests in his presentation of a different approach, one that has not been previously considered and seems superior to the old way of doing business. By framing US current posture as a command-oriented paradigm, and offering an alternative, Daehnick sheds new light on long-held beliefs. For instance, duration is often seen as a characteristic advantage of space power. But on-orbit capability equates to spending limited monies on specific capabilities before the situation that generates the demand exists. By comparison, the demand-oriented alternative of an earth-to-space, tailored response diminishes the worth of durable, on-orbit capability. Daehnick discusses many strengths and weaknesses of space, and further recognizes that many of those weaknesses (life-cycle costs, inflexibility, timelines) are not a result inherent to the environment, but more a result of a prechosen architecture.

The weakness of Daehnick's work is that he presents the current command-oriented architecture in a negative light. He describes that architecture as a flawed approach to highlight the strengths of the demand-oriented approach rather than as a credible alternative. Ironically, had a strong case for command-oriented space architecture been made, the argument against it would have been more credible. To be fair, the author does not simply advocate a demand-only oriented space architecture. In his conclusion, he recognizes that a hybrid command/demand-oriented space architecture is possible and may be the optimum solution. The value of this work does not reside in the debate over command or demand orientation but lies in the recognition that alternative space architectures exist, which in turn frees future space planners from the command-orientation paradigm. This broad examination of space strategy, doctrine, and architecture provides an objective backdrop for the remaining papers.

Sanctuary/Survivability Perspectives

The SAAS is a professional military education facility. Not surprisingly, students interested in space-related research are apt to be space enthusiasts. Upon initially consolidating this volume, an overall weakness became apparent: No contributing author had made the case against pursuing space for military purposes beyond intelligence, surveillance, and reconnaissance (ISR). Although each research paper is balanced in its analysis, the balance is between command or demand architecture, or between one concept of operations for reusable launch vehicles or another. None of the papers questioned whether the US's pursuit of weaponizing space at this time in a sound military strategy. I challenged David W. Ziegler, a space enthusiast, to do just that.

In “Safe Heavens: Military Strategy and Space Sanctuary Thought,” Ziegler outlines the historical development of US space policy, and the lessons of that review reflect a tradition of American restraint. From that context, he makes the point that US interests in space are currently limited to surveillance, reconnaissance, intelligence (SRI), and signal relaying. Ziegler lays out the logic that currently and for the foreseeable future, we don’t live in space, there are no natural resources which can be cost effectively developed in space, nor is space a travel medium. Furthermore, the cost of accessing space is currently enormous—and that alone may be good reason for waiting until commercial exploitation of the medium drastically reduces the cost of getting there. The enormous-cost-now/cheaper-cost-later argument is further strengthened as the author takes a serious look at requirements and opportunity costs. Aside from competing social programs outside the DOD, the opportunity cost to other military programs, which could satisfy the same need or other significant need is staggering.

Ziegler then presents a line of reasoning that even the staunchest space enthusiast would agree to be novel. There is a lot of interest in emerging technologies that facilitate access to space. But what if equivalent investment was aimed at different, surface- or air-based solutions to meet the same requirements? In spite of unequal funding, advances in surface-based, fiber-linked telecommunications threatens high-cost/highly vulnerable space-based counterparts. Long-loiter unmanned aerial vehicles (UAV) are also beginning to fill ISR requirements in a more cost-effective, flexible, and responsive manner than equivalent space-based assets.

Beyond the lack of interest, huge opportunity costs, and substitute technologies, Ziegler has tapped the best available intelligence sources which estimate that the United States faces virtually no peer threat in space for at least 10 to 15 years. The author defines *peer threat* as a competitor that seeks to dominate space to the same level as the United States. Hence the author recognizes little utility in furthering the militarization. The author did find *challenging threats*, threats weaker than peer threats that seek to deny or destroy US capabilities but lack an ability to field similar capabilities.

Surface-based, directed-energy ASATs stand out as a potential weapon that a challenging threat could employ even if it lacks the technology to field space-based ASATs. This discussion serves to articulate the survivability viewpoint, and the author expounds upon significant limitations of space-based systems. Additionally, any attempt at this time to weaponize space threatens a renewed arms race in a realm that offers significant advantages over the air realm. There is no logic in escalating the armaments game.

Based on this analysis of historical precedents, US interests in space, the cost of access, the potential of substitute technologies, the lack of a peer threat, and the presence of challenging threats, Ziegler concludes by defining space as a credible *military sanctuary*, as a place where forces can be posited and trained, but an attack on that sanctuary changes the political nature of the conflict. Such a definition dominates US current posture in space. It distinguishes between the US current militarization of space and suggested weaponization of space. The author presents a credible argument that a sanctuary strategy in space has significant merits. The work also highlights the danger of blindly proceeding beyond the militarization threshold and plunging the United States into an era of space weaponization.

Ziegler effectively articulates the argument that favors a military sanctuary strategy regarding US use of space. The argument balances the remainder of the papers which, by-in-large, assumes a natural escalation to space weaponization.

Space Control Perspectives

James Lee, in "Counterspace Operations for Information Dominance," examines space strategy from the traditional perspective that space control is a military requirement, but he adds a nontraditional twist by emphasizing that control does not necessarily require the use of antisatellite weapons. The work shows space control in a new light that defines it in terms of information rather than the physical environment. Tracking the development of US space power, Lee highlights the fact that the US notion of space control grew out of the cold war paradigm, a path which led the United States to anticipate a peer competitor in space. Hence, space control developed as a notion of physically controlling the space medium. Making that notion stronger was its compatibility with previous experience. The development of sea power and airpower demonstrated that once access to those domains became common, it was necessary to physically dominate them during conflict.

A strength of Lee's work resides in his excellent summary of unclassified US and foreign satellite reconnaissance capability. He supports the argument that access to space surveillance and reconnaissance capabilities are essential to the employment of US military power and that those capabilities are spreading around the globe. Given these developments, Lee recognizes that the United States requires a space control strategy which can be tailored to particular threats and situations, and has the practical aim of controlling information traffic from space. He offers a three-dimensional model that considers the capability of the threat (extensive space access, limited space access, or purchased space information); the situation (peace, crisis, or war); and the space system to be manipulated or targeted (ground, up/down link, or orbital elements). While the paper makes sense in terms of giving the commander flexible options in the control of space information, the model seems to be over-simplified, particularly in its categorization of such human events as peace, crisis, or war. This is perhaps not so much a weakness of the work, as it is an opportunity for further research and thought. Clearly, the issue of space control in the information age is complex—a function of threat, capability, circumstance, domestic and international relations, and international law. With the advent of proliferating access, the space medium may be beyond the ability of any one nation to control, and perhaps Lee's notion of space control as a matter of controlling information is more practical. In any event, the United States will have to develop its space doctrine under the assumption that the adversary will have some space information access, or in the words of the next author, we will have to proceed under the assumption that "the enemy has our eyes."

"When the Enemy Has Our Eyes" by Cynthia A. S. McKinley is primarily intended for space operations personnel who are tasked with the challenge of becoming space strategists. It is also of value to individuals who seek unclassified information about reconnaissance satellites, an understanding of changes within the military space community, or an analysis of the space control mission. In reviewing the historical foundations of America's space-based strategic intelligence assets, McKinley identifies the visionaries who gave the United States its strategic eyes and the revolutionary technology that unnerved the US's closest competitor. Further, she discusses the use of strategic intelligence in theater warfare. The author offers a unique perspective for looking at the context in which national and international actors may prosecute warfare, which leads to illumination of the space control challenge facing the United States. To take positive steps toward meeting that challenge, McKinley offers an analytical approach for space control and applies the results to a commercial reconnaissance system. The author concludes that the space control mission is more challenging in today's multipolar world than it was during the cold war.

The strengths of McKinley's work include a practical analysis of space control and the military role in space for the next five to 10 years. The author compares a survey of the historical inertia which drives current space policy, capabilities, and force structure to the future context of warfare including a realistic estimate of future spaced-based capabilities. The merger leads the author to examine the significant role of imagery in future warfare and to recommend a space control strategy (access and denial). The most significant limitation of the study rests on the assumption that the enemy will have the same information as the United States. This is clearly pessimisstic.

Further, limitations of McKinley's effort are primarily a matter of scope. The thrust is limited to strategic intelligence and the role of space-based imagery with a primary focus on force enhancement. Additionally, the author's theory of warfare is well thought out, but may unnecessarily constrain the vision of the future role of space in military affairs. Finally, the potential of extensive space-based weapons with the primary function of force application is briefly mentioned, but not seriously considered.

High-Ground Perspectives

In "National Security Implications of Inexpensive Space Access," William W. Bruner III recognizes that the government of the United States is about to embark on an ambitious enterprise. As per Presidential Decision Directive/National Science and Technology Council (NSTC)-4, *National Space Transportation Policy,* 5 August 1994, the United States is planning to make a significant leap forward in repeatable and economical access to space. While routine access to orbit will give the United States a clear advantage in the ability to use near- earth space to serve national political, economic, and military interests, those responsible for making national space policy and writing military space doctrine are fallaciously doing so based upon the old assumption of infrequent and expensive space access. The author explains that the difficult and expensive access assumption is primarily a result of an expectations gap where early promises of space exploration, as well as recent promises of routine space access via the shuttle, have left the public somewhat disillusioned. He also cites (1) the erroneous notion that the United States will necessarily lead the way into space; (2) perceived treaty, policy, and legal limitations; (3) the *Challenger* accident; and (4) the lack of a coherent national space policy are reasons this country is dragging its feet in the space access effort. Bruner asserts that these impediments will wane due to new political, economic, and technological realities. His analysis is balanced, as it addresses the cases for and against standing down, the status-quo, pursuing expendable launch vehicles (ELV), and pursuing reusable launch vehicles (RLV). The cost-benefit analysis seems to favor the latter. The author emphasizes that life-cycle costs make the RLV more attractive than the ELV, while at the same time RLVs allow for the expansion of military capabilities.

The most significant strength of the paper lies in the author's ability to recognize military possibilities for an RLV concept beyond the limitations of expectations and policy, which are for the most part, self-imposed. His concept of using RLVs for on-orbit refueling shatters the old paradigm of orbital mechanics dictating inflexibility. The concept allows on-orbit upgrades, repairs, replacements, access to higher orbits, and capability for orbital maneuvers—traditionally assumed to be cost prohibitive.

Several inconsistencies appear. On the one hand, the author is optimistic regarding technology's ability to provide space access and assumes this access readily allows for military space-to-earth precision capabilities. On the other hand, the author is pessimistic regarding technology's ability to provide remote control to spacecraft, insisting that onboard human judgment is often a necessity. This is somewhat ironic in that progress in the technologies of remote control and virtual environments is to a large extent already proven, whereas the technological pursuit of ready access to space has been disappointing. Bruner's basic contention, that space offers an inherent energy advantage, is also optimistic from the spacelift perspective and, at the same time, ignores the possibility of other technologies. While his contention is true from a potential and kinetic energy standpoint, he does not address, for instance, the advent of directed energy technologies, which could very well turn the advantage of altitude/elevation into the disadvantage of exposure. Finally, toward the closing sections, the work takes somewhat of an Air Force parochial turn, degenerating into a discussion of which service should take the lead in space, the Navy or the Air Force. Although the discussion regarding the applicability of Navy and Air Force cultures to space is interesting, it is an aside from the main theme. Further, the analysis offers a false dilemma: Should the Navy take the lead from the environmental perspective of living and working in a stationary but hostile environment, or should the Air Force take the head from the functional perspective of employing military power from the third dimension? A separate space force is just one of many alternatives to the dilemma.

A primary limitation of the work is that while Bruner accurately recognizes what international laws and treaties do allow, he overlooks what domestic policy won't allow. Space as a sanctuary may not be part of international law, but that may be irrelevant, if domestic expectation demands it. Bruner reaches out 20 or 30 years and assumes the militarization of what he calls "decisive orbits" to be an accepted practice, without considering the broader context of domestic and international politics or nongovernmental commercial interests. Although this is a recognizable limitation of the work, it is also excusable. As part of his professional obligation as a military planner/strategist, Bruner is expected to plan contingencies that might warrant military action. In this regard, he has provided some of the best military vision of what space power could be in the future.

In "Concepts of Operations for a Reusable Launch Space Vehicle," Michael A. Rampino also pursues military concepts of operations (CONOPS) without answering fundamental questions regarding who is the threat and what are the requirements to negate that threat. As with Bruner's work, this is a justifiable planning approach from the military perspective. Militaries don't necessarily need to arm for contingencies, but they ought to plan to arm for contingencies. When that plan recognizes a need for long-term investment to arm appropriately, the issue of preparedness in the absence of a clear and present adversary has merit. Rampino's thesis emphasizes that the US military must be prepared to take advantage of reusable launch vehicles should the NASA-led effort to develop an RLV demonstrator prove successful.

The strengths of the work are many, the most obvious being the structured methodology. The author develops two different concepts of operations from a detailed investigation of military requirements and current paths to produce the capability to meet those requirements. The first concept attempts to make the fullest military use of a roughly half-scale notional RLV to accomplish not only traditional spacelift missions but also the additional missions of returning payloads from orbit, transspace operations, reconnaissance, and strike (in and from space). The second concept is based on the full-scale vehicles currently being proposed under the RLV program. It too attempts to make expanded use of RLVs, but military application is inhibited by design attributes and a focus on completely commercial operation. Both of these CONOPS are comprehensively described via their mission, the systems they require, the operational environment, the command and control links, the support they require, and the means by which they are employed in civil and military situations. Subsequent to the detailed descriptions, a comparative analysis of the two concepts proceeds with criteria which include capability, cost, operations efficiency and effectiveness, and political considerations.

Major conclusions are drawn from that analysis. RLVs are recognized to have military potential, yet the design choices for any operational RLV must be measured in terms of risk, cost, capability, and operations efficiency and effectiveness. Given this preliminary analysis, the choice of a larger vehicle is found to be accompanied by more risk. Beyond the RLV itself, supporting science and technology development is the crucial issue. Particularly, increased investment in propulsion technology is warranted. The final conclusion gives the entire space community a clear focus: The top priority for the RLV program, even from the DOD perspective, should remain cheap and responsive access to space.

Based on the conclusions, Rampino puts forth three recommendations. The US military should become a more active participant in the RLV program, the United States should not pursue development of operational RLVs before the technology is ready, and finally, it is not too early for the US military to think deeply about the implications of operational RLVs for war-fighting strategy, force structure planning, training, and doctrine.

As with any other research, this work has limitations of scope. While the author effectively extrapolates space capability to the 2012 time frame, he assumes a command and control structure dictated by current Air Force doctrine. This assumption places his 2012 space capabilities in a 1996 context. From a broader perspective, the requirements for a military RLV were garnished from the military environment. Asking the military to produce military requirements does not necessarily mean there is a genuine need. Of course, this ties back to the initial point of the military planner's role of developing courses of action in the event of military need.

The final paper, by Gregory Billman, also makes similar assumptions. "The Inherent Limitations of Spacepower: Fact or Fiction?" Billman squarely addresses the US approach to space. He finds it odd that many of the self-imposed limitations to exploiting space stand in light of twentieth century US airpower experience. The analogy seems strong: The first employment of airpower concerned a primary focus on observation and reconnaissance; it rapidly evolved into an offensive form of military power due to advantages of response, speed, and reach; and finally, doctrinal and organizational development followed the new capabilities. Billman compares space power with the forms of terrestrial powers by examining each across a set of military force characteristics that he generalizes into five distinct categories: strategic agility, commitment and credibility, economic considerations, military considerations, and political considerations.[9] While the latter three initially appear unclear and unfocused, Billman delineates them as a reasonable means of categorization. A weakness of the work is the lumping together of all terrestrial military powers (air, land, and sea), on the grounds that they all have gravitational limitations while space power uses gravity to its advantage. The grouping of terrestrial forces comes across more as a matter of analytical convenience rather than a technically justifiable assertion. It may have been beyond the scope of the work, but a similar analysis comparing space, air, land, sea, and perhaps even information power would be enlightening.

A strength of the analysis is Billman's recognition that as these five categories of characteristics apply to terrestrial and space forces, they must be measured at different phases of employment. Each military force characteristic will vary as the instruments of that force are home based, deployed, or engaged.

Billman's analysis strongly favors the advantages of space power under all five military force characteristics. Assuming space power to be predominantly in a deployed, or even engaged state, he supports the argument that it has strategic agility and commitment and credibility advantages without the economic, military, and political risks of terrestrial forces. This, coupled with the airpower/space power developmental analogy, leads the author to conclude that space power should develop as a separate capability which exploits the medium in all military roles, including the force application role. He asserts that space power must no longer be merely a supporting force.

While the air and space power analogy is useful on certain specific points, extrapolating the analogy into sweeping recommendations on the US's future approach to space is a fundamental breech of logic. On one count, the similarities between airpower and space power development were emphasized, without any serious effort to examine distinctions between the two. On a second count, numerous examples of using gross historical analogies in major policy decisions have been documented with a single resounding outcome: The decision they lead to is most often wrong.[10] The most significant weakness of the work is not a limitation of historical inference, though, but one of omission. The author establishes that the only limitations of US space power are self-imposed. He makes a strong case for the advantages afforded by a future space force unencumbered by those limitations. The shortcoming is that he never articulates why those self-imposed limitations exist. He loosely attributes their existence to policy, but policy is often made for good reasons. Those good reasons in this case include international law, domestic and international opinion, significant technical limitations, opportunity costs, and even military advantages of a sanctuary approach. While the author summarizes with three requirements to overcome the self-imposed limitations: a change of military perspective, space as a separate military area of operations (AOR), and military/civilian cooperative efforts, these recommendations are hollow in the absence of a detailed examination of why those self-imposed limitations exist.

Conclusion

There are perhaps two weaknesses that remain in spite of the synergy of this consolidated volume. First, although many of the works begin with a historical survey, the total leaves the impression of lacking context.[11] For example, some authors assume the space community to be distinct from the air community, yet to date those technical communities are one in the same, made up of such aerospace giants as Lockheed-Martin and McDonnell-Douglas. Exploring the contextual development of the space community reveals many current space trends, such as the preoccupation with zero-fault tolerance. Such trends may seem irrelevant for the space architect planning efficient unmanned operations, but it is a reality, as it is ingrained in an air community that for almost a century has had human cargo.

The second weakness, evident in several of the works, is the idea that advocating one position or another on space power must be done in the context of a zero-sum game. That is, it must be to the benefit or detriment of another form of military power. In some ways, the zero-sum game of economic funding forces this issue. This tends to overshadow the fact that new forms of military power have historically complemented one another, allowing missions that were unachievable from a single environment. Sea power did not supplant land power, airpower did not supplant land and sea power, nor will space power supplant air, land, and sea power.[12] The enlightened joint approach to the employment of military power recognizes that different environments require different forces, and all must work in harmony. It seems shortsighted to advocate a distinct military force for a new environment at the expense of other forces. It is the situation at hand, and not the physics or position of a particular environment, that dictates the dominance of one force over another. In advocating different aspects of the US role in space, it is not the intent of this editor or this learned group of air and space professionals for our material to be taken without an appreciation of the air, land, and sea roles in putting forth the most effective joint force in support of national security. The intent is a comprehensive examination of space power: the Ziegler and Billman works being extremes which illustrate the value of this collection of papers. While each may overlook the perspectives and assumptions of the other, collectively they comprehensively address the subject. What Bruner, Rampino, and Billman overlook or assume away is addressed in Ziegler, Mckinley, and Lee's work. The reverse is also true. Additionally, these sanctuary, survivability, control, and high-ground perspectives are balanced against a background of the most significant issues: space organization (Wright), doctrine (Gallegos), and architecture (Daehnick). As the collection of strengths addresses most of the weaknesses, this collection reflects a mature, documented consolidation of military thought on space power.

Notes

1. *Air Power & Space—Future Perspectives 1996*, Queen Elizabeth Conference Centre, Broad Sanctuary Westminster, United Kingdom, Airpower Conference, 12-13 September 1996.

2. Professor and Air Vice-Marshal R. A. Mason, "Characteristics of Aerospace Power," panel 1 presentation, *Air Power & Space*, 12 September 1996.

3. David E. Lupton, *On Space Warfare: A Space Power Doctrine* (Maxwell AFB, Ala.: Air University Press, 1988).

4. A broad definition of military *sanctuary* is a place where aggressive forces can be postured, but attacks in that sanctuary would change the nature of the conflict. Under this definition, China was a sanctuary for communist forces during the Korean War and more recently, Argentina was a sanctuary during the Falklands War. Space is considered to be such a sanctuary today as evidenced by the number of militarized space-borne systems and lack of weapons on these systems. To assume the sanctuary notion of space is now obsolete at least requires justification. What is lacking among the space enthusiast arguments is a discussion of opportunity costs—that is, not the fact that space access is affordable, but what other capabilities are not pursued at the expense of space access (such as airpower, research and development, or education). Given a study of opportunity costs, the space sanctuary concept may be a very valid military strategy. See Professor Lawrence Freedman, "Sanctuary or Combat Zone? Military Space in the 21st Century," *Air Power & Space*, 12 September 1996. The counter to Freedman's argument was given by US Air Force Vice Chief of Staff Gen Thomas S. Moorman: Violation of the space sanctuary will not be one of choice, but of a response to a projected threat.

5. AFM 1-1, *Basic Aerospace Doctrine of the United States Air Force*, 5 March 1992; AFDD-1, *Air Force Basic Doctrine*, draft, 1 September 1997; AFDD-4, *Space Operations Doctrine*, 22 May 1996; JDTTP 3-14, *Space Operations*, April 1992; FM 100-5, *Operations*, 14 June 1993, 2–16 through 2–18. It is worth noting that although capabilities-based planning has merit, threat-based and objective-based planning are other options worth consideration.

6. Paul B. Stares, *The Militarization of Space: US Policy 1945–1984* (Ithaca, N.Y.: Cornell University Press, 1985), 206–7. As early as 1978 miniature homing vehicles were successfully launched from F-15 platforms, using first stage boost via a modified Boeing short-range attack missile (SRAM) and second stage boost via a Vought Altair III.

7. Air Force Association, *"Special Report: Facing Up to the Space Problem,"* 1 November 1994.

8. The seminal work of Graham T. Allison, *Essence of Decision* (New York: HarperCollins Publishers, 1971), clearly distinguishes rational,

organizational, and bureaucratic decision making, and further discusses the benefits and shortcomings of each.

9. Gregory Billman takes a unique perspective on commitment and credibility. While standard deterrence theories assume commitment to be a function of an actor putting himself at risk, and credibility a function of demonstrated capability, Billman hypothesizes that the lack of risk and ease of use of space-based assets engender commitment and credibility. The lack of risk Billman focuses on is risk to the operator, and while that is reduced for an unmanned space asset, the risk in deterrence theory is risk of reprisal to the operators or any other form of political, economic, and military fall-out (pun intended). Much of that risk is retained when using a strategic weapon, manned or not.

10. Yuen Foong Khong, *Analogies at War: Korea, Munich, Dien Bien Phu, and the Vietnam Decision of 1965* (Princeton, N.J.: Princeton University Press, 1992).

11. Cynthia A. S. McKinley's article is the notable exception, being well grounded in historical development.

12. Maj Bruce M. DeBlois, "Ascendant Realms-Characteristics of Air and Space Power," in *Paths of Heaven: The Evolution of Airpower Theory* (Maxwell AFB, Ala.: Air University Press, 1997).

PART I

Space Organization, Doctrine, and Architecture

Chapter 1

An Aerospace Strategy for an Aerospace Nation

Stephen E. Wright

> *America is an aerospace Nation. Our aerospace technology and industry is a national treasure and a competitive edge, militarily and commercially. Assured access to air and space are as important to the Nation's economic well-being and security as access to the sea has always been. . . . Now, more than ever, we have the opportunity to mature the abilities of our air and space forces and make them even more useful tools for meeting our national security objectives.*
>
> *Global Reach—Global Power*

I agree. The purpose of this paper is to examine why former secretary Donald B. Rice is correct in his statement and to expand his focus of "air and space forces" to include the aerospace industry.[1] Together, the aerospace industry and its military counterpart combine to form United States (US) aerospace power. That capability requires a national aerospace strategy to exploit its potential in providing for the future economic and national security well-being of the United States. What factors then make a national aerospace strategy important for America's future?

To state that the world is changing its geopolitical course seems an understatement. Several world events occurred in 1991 that indicate global relations underwent changes on a scale not seen since the post–World War II years. The defeat of Saddam Hussein in Desert Storm infused Americans with

This work was accomplished in partial fulfillment of the master's degree requirements of the School of Advanced Airpower Studies, Air University, Maxwell AFB, Ala., 1996.
Advisor: Col Phillip Meilinger, PhD
Reader: Col Kenneth Feldman, PhD

confidence in their military forces. Never before had aerospace power so decisively dominated a conflict. The transformation of the Soviet Union ushered in a new political environment that alters the cold war paradigm of international relations. The changing geopolitical environment alone provides impetus for reconsidering US national security strategy; however, the need to review that strategy becomes essential in light of the economic imperatives facing the United States. Since the late 1980s, the US economy grew at a meager rate (one to three percent a year) while at the same time the national debt more than tripled. With yearly budget deficits exceeding $300–400 billion per year, domestic issues became the focal point for the 1992 presidential race that resulted in President Bill Clinton's election.

The newly elected Clinton administration quickly spotlighted the aerospace industry. The reductions in defense spending initiated by the Bush administration coupled with a poorly performing world economy resulted in a crisis situation in the aerospace industry. United States's airlines lost over $10 billion from 1990 to 1992 and layoffs in both the airlines and aerospace manufacturing were numbering in the thousands. In office just over a month, President Clinton traveled to Washington state to assure Boeing employees that he was concerned about the future of the vital aerospace industry.[2]

Today, both military and commercial aerospace struggle toward an uncertain future. What that future entails depends upon decisions made today. The United States must determine if and how it will remain the preeminent aerospace nation or falter and assume some lesser position. To begin this odyssey, one needs to ask some basic questions.

Is the United States the preeminent aerospace nation? American aircraft manufacturers control more than 80 percent of the worldwide, large commercial jet market. Further, with the political and economic downturn in the former Soviet Union, no nation provides the range of space services that the United States does. Desert Storm demonstrated America's military aerospace dominance—there are no competitors in the world today.

But, is the United States an aerospace nation? Navalists argue that the United States is a maritime nation. Their argument usually hinges on water and weight. First, water covers 70 percent of the globe and second, most of the cargo, by weight, is transported by ship. However, 100 percent of the globe is covered by air and by value for amount shipped, aerospace looms far ahead.[3] For example, less than one-third of one percent of goods (by weight) imported or exported to or from the United States do so by air. However, this tiny fraction of a percent in weight accounts for over 32 percent by value of those goods—a percentage value that doubled from 1970 to 1990. As a manufacturing industry, maritime concerns generate only one-eighth the product value of the aerospace industry. Perhaps we would be better served to say the United States is an aerospace nation with significant maritime interests.

If indeed the United States is an aerospace nation, how do its component parts, economic and military aerospace, relate to the future well-being of the United States; what problems exist that indicate the United States needs an aerospace strategy; and what ideas form the basis for such a strategy? These questions presage the rest of this paper.

The next section describes the importance of the aerospace industry to the US economy. This study then looks at the reasons that war remains a concern for national security considerations and discusses the political imperatives that will govern the application of military force in the future. The next section reviews the espoused strategies of the military services and examines them in light of the political imperatives and their reliance upon aerospace power for successful execution. The following section considers the problems facing the economic and military elements of aerospace power and offers ideas as to the nature of a national aerospace strategy.

The Economics of Aerospace

From the earliest theorists of airpower to current day aerospace strategists, many including economists and politicians have recognized the important relationship between the aerospace industry, the economy, and the government's aerospace forces. Giulio Douhet linked all three aspects in his seminal work, *The Command of the Air*.[4] In addition to forecasting a future for military aviation, he devoted considerable effort to explaining "aerial navigation" as a new form of transportation.[5] Gen William "Billy" Mitchell clearly understood the potential of airpower when he stated, "Those interested in the future of the country, not only from a national defense standpoint but from a civil, commercial and economic one as well, should study this matter carefully, because air power has not only come to stay but is, and will be, a dominating factor in the world's development."[6]

Another early airpower strategist, Alexander de Seversky, foresaw the necessity to couple the development of commercial and military aerospace. He stated that "their development must be scientifically meshed into the military-aeronautical structure" of the United States.[7] Then Secretary of the Air Force Rice noted the "great potential [for aerospace forces] to draw on advanced technologies" and the increasing importance of technology to national defense.[8] President Clinton and Ross Perot both acknowledge the importance of the aerospace industry to the well-being and competitiveness of the overall US economy. Finally, noted economists Robert Reich, Laura D'Andrea Tyson, and Lester Thurow point to aerospace as one of the key industries for the future.[9]

The linkage between commercial and military aerospace, the two components of aerospace power, differs fundamentally than those for land and sea power. No one connects tanks and the automobile industry by intimating that if the United States stopped building tanks it could no longer build automobiles. Likewise, this linkage is missing from the relationship between naval forces and the merchant marine. The United States has the premier navy in the world; yet, the US merchant marine ranks far from the top, and other than naval construction, commercial shipbuilding received only one order for a vessel larger than 1,000 gross tons in fiscal year 1991.[10] In contrast, Japan is the world's leading shipbuilder and has the largest merchant marine but a very limited navy.

Aerospace enjoys a unique position in the relationship between its industry and military components, the US government, and the economy. The relationship is synergistic in its effect within each of these elements. Three questions help us understand this unique relationship. First, what impact does the aerospace industry have on the US economy? Second, what links the aerospace industry and government aerospace components? Third, what explains the ties between these elements?

The Aerospace Industry and the US Economy

After World War II the aerospace industry experienced a growth streak that propelled it to the number one ranking export industry in the United States in 1991—exceeding even agriculture.[11] Over this time frame, the aerospace industry grew into an industrial sector of great importance to the overall US economy.

One key indicator of the industry's growth is sales. In 1948 the industry had sales of almost $1.5 billion; by 1991 this figure exceeded $134 billion.[12] Table 1 details this growth in sales and shows the almost 100-fold increase. Over the last 30

Table 1

Aerospace Industry Sales (millions of current dollars)

Year	Total Sales	DOD[a]	NASA & Other Government Agencies[b]	Other Customers[c]	Related Products
1948	1,493	1,182		117	134
1955	12,411	10,508		786	1,117
1965	20,867	11,396	4,490	2,816	2,165
1975	28,373	13,127	2,727	7,727	4,792
1985	96,571	53,178	6,262	21,036	16,095
1990	134,375	60,502	11,097	40,379	22,396

[a]Includes foreign military sales
[b]NASA formed in 1958
[c]Primarily nonmilitary aircraft sales

Source: Aerospace Industries Association (AIA) of America, Inc., *Facts and Figures.*

years, aerospace accounted for 2.5 to 3.5 percent of the US gross national product (GNP) and averaged nearly 4 percent of all manufacturing industries.[13]

Jobs are another measure of aerospace's impact on the economy. In 1990 aerospace provided 1.295 million jobs, about the same number of jobs as the automobile industry. Moreover, aerospace furnishes the kind of high-technology, high-skill, high-value jobs that economist Reich argues are critical to an improving standard of living.[14] During the post–World War II period, production workers in aerospace enjoyed on average a 10 percent advantage in hourly wages over the average worker in durable goods manufacture.[15]

Employment of scientists and engineers yields another indication of aerospace's economic power. Since the 1950s, one of every four scientists and engineers worked in aerospace. The fact that aerospace scientists and engineers received from 7.5 to 9.0 percent more pay than their contemporaries in other fields serves as another indicator of the importance of these workers to the national economy.[16]

Another key sign of aerospace's influence on the economy results from its position as the nation's top net exporter and its number six position in industry in terms of value of shipments in 1991.[17] The nearly $30 billion (net balance) in exports in 1991 surpassed even agriculture and accounted for nearly $1 in every $10 of US exports.[18] Table 2 contrasts

Table 2

Trade Balance of Selected Commodities (billions of dollars)

Commodity	Exports	Imports	Balance
Aerospace	39,083	11,801	27,282
Agriculture	40,003	22,099	17,904
Chemicals	36,485	20,752	15,733
Motor Vehicles	25,480	79,003	(53,523)

Source: AIA, *Facts and Figures 91–92* and *The Statistical Abstract of the United States.*

aerospace exports and imports with three other major product groups. Aerospace leads the nation in export balance.

A final indicator of the importance of the aerospace industry comes from its preeminent position in the world market for large jet aircraft. Figure 1 graphically portrays this trend.[19] Even today, the United States maintains a market share in excess of 80 percent of the world market despite Lockheed's withdrawal from the large jet manufacturer competition.

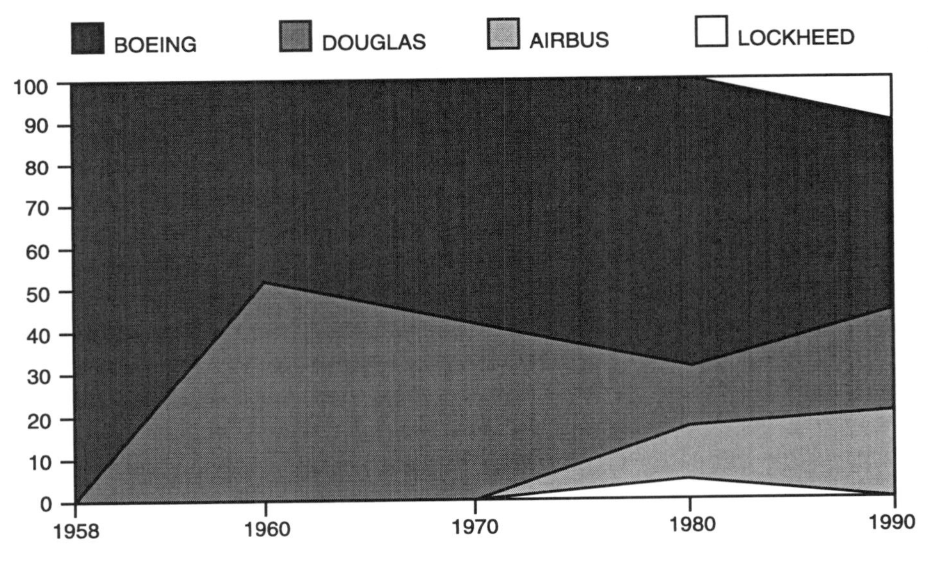

Source: James W. Chung, "Whither the U.S. Aerospace Industry?" *Breakthroughs,* Winter 1992–93.

Figure 1. World Market Share of Large Jet Airplane Deliveries

These indicators show the aerospace industry to be a crucial part of the overall health of the US economy. The president, economists, and of course the military all see aerospace as one of the key useful technologies for the future well-being of America. In the final decade of the twentieth century, aerospace can look forward to a projected total world air traffic growth of 5.4 percent.[20] Clearly, aerospace represents a crucial industrial field that is important to the future competitiveness of America's economy.

Linking the Aerospace Industry and Government Aerospace

A synergistic relationship exists between the aerospace industry and government aerospace. Tyson describes this effect stating, "The synergies between the military's emphasis on performance and flexibility and the commercial sector's emphasis on cost and reliability have been central to aircraft technology and innovation."[21] She goes on to note that "a competitive commercial aircraft industry thus contributes to a nation's military prowess."[22] The relationship Tyson describes is obviously driven by technology, and many examples abound to illustrate this connection.

A key area linking the two entities is engine technology. Engineers first designed jet engines for military aircraft in World War II, and their efforts continued in the postwar era. Boeing used its J-57 engine in its proposal for the B-52 and later coupled this same engine to the United States's first successful commercial jet aircraft, the Boeing 707.[23] The competition to develop jumbo jet technology to haul oversized military cargo resulted in the engine designs to power aircraft as large as the Lockheed C-5. Boeing put this technology to use on its Boeing 747. The 747 went on to become the greatest post–World War II success story in commercial aviation history.

Several other innovations mark this association between industry and government. Designers still use the swept-wing design of the B-47; the Boeing 707 being the first commercial jet aircraft to incorporate this innovation. Airbus incorporated fly-by-wire technology, originally pioneered in the F-16 fighter aircraft, into its A320 aircraft—the first commercial jet so equipped. Supersonic flight not only resulted in aircraft design introductions but also drove improvements in metallurgy and fuels. The composite materials found in the military's newest stealth aircraft have increasingly found their way into commercial aircraft. Composite structures not only add strength, but reduce weight resulting in more fuel-efficient aircraft.

The technology spin works in the other direction as well. The commercial sector improves and innovates many new systems that find their way into military use. The airline industry improved onboard radar capabilities originally developed by the military and produced specialized weather radar equipment. Many military aircraft, especially transport aircraft, incorporate this technology. The commercial industry enhanced the capabilities of cathode ray tube technology creating "glass cockpits" that enhance the presentation and type of information presented to pilots. Newer military aircraft, like the F/A-18 and F-117, incorporated this technology into their cockpits, increasing the performance of their flight crews. Although the highest risk technology still flows from government-to-industry, significant transfer occurs in both directions. Clearly a dedicated link exists between these two aspects of aerospace power. Thus far we have seen how important the aerospace industry is to the US economy and the linkage that exists between it and the government. The next section seeks to explain why this relationship exists.

Explaining the Linkage

The focal point in an explanation of the linkage between government and industrial aerospace is risk. In the United States the government reduced the risk accrued to aircraft manufacturers by underwriting their production costs via indirect and direct means. The primary indirect methods were research and development (R&D) funding and military aircraft purchases. Direct risk reduction resulted in the federal funding of the US space program; however, space accrued much higher political risks as a result of that arrangement.

After World War II the federal government continued to underwrite a large portion of aviation research and development. In the 1950s and 1960s, aerospace R&D exceeded 30 percent of all federally funded R&D dollars and approached almost 40 percent in the 1960s.[24] From the mid-1970s until the start of the Reagan military buildup, 50 percent of all federal R&D dollars went to aerospace and from 1984 to 1989 this percentage increased to over 60 percent.[25] Table 3 provides the details of the R&D dollars. The preponderance of aerospace R&D funding comes from the National Aeronautics and Space Administration (NASA) and the Department of Defense (DOD). From the early 1970s to the mid-1980s, NASA and DOD furnished approximately 97 percent of federal aerospace R&D funds.[26] Tyson refers to this national R&D effort as the "visible hand of government."[27]

Table 3 shows that three of every four aerospace research dollars comes from federal sources. If one breaks out aerospace funds from the rest of industry, one finds a federal-to-industry funding ratio of one-to-three, a virtual reversal from that of the aerospace industry.[28] Not only is the cost of R&D high in the aerospace industry; failure can be disastrous to the individual company. Of the $4–6 billion to produce a new aircraft product line, development expenses represent two-thirds of fixed costs.[29] These represent high entry barriers for any business, let alone one as volatile and risky as commercial aircraft manufacture. Tyson quotes the Office of Technology Assessment as estimating that, in 1991 dollars, it cost $3 million in 1936 to develop the McDonnell Douglas DC-3. Today, Boeing expects to pay over $10 billion to develop its Boeing 777.[30]

These facts serve to highlight the high cost of R&D in the aerospace industry and the risk that must accompany an

Table 3

US Government Research and Development Expenditures (millions of current dollars)

	All Industries	Aerospace Industry		
Year	Total	Total	Federal Funds	Company Funds
1950	1,143	*	1,080	*
1960	10,509	3,558	3,180	378
1970	18,062	5,245	4,032	1,213
1980	44,505	9,198	6,628	2,570
1990**	104,344	25,357	19,217	6,140

*Breakout of data not avilable
**Last year data available

Source: Facts and Figures.

investment of that magnitude. In effect, the risk of failure represents an all-or-nothing gamble that forces the builder to "bet the company" with each major aircraft venture.[31] Boeing sank every resource it had to launch the 747 program, nearly bankrupting the company. Lockheed's failure with the L1011 aircraft forced it out of the commercial aircraft manufacturing business altogether. The list is long for those companies that, like Republic, Wright, and Curtiss, great names in aviation, are no longer corporate entities.

The government takes direct action to support the aircraft industry by its purchase of military planes. Several companies like Lockheed, General Dynamics, and Northrop make their living primarily through government contracts. Many other firms rely upon the government for varying but significant portions of their revenues. At times government support has taken the form of loan guarantees like the $250-million loan guarantee to Lockheed in the 1970s.

A special risk results from government involvement in aerospace—political risk. Nowhere is this risk manifested so clearly as in the US space industry.[32] Through NASA, the government controls the price and schedule of the US space launch business. Further, NASA exerts additional oversight as the certification authority for flight payloads. By funding most of the US space program, the government virtually eliminates risk to space manufacturers. Risk enters in when political decisions result in severe handicaps for the industry. For example, prior to the *Challenger* accident, the United States made the decision to forego all other launch vehicles and rely solely on the space shuttle (this decision was made in an attempt to make the shuttle program more cost-effective). After the *Challenger* accident, the United States failed to launch another satellite for two years because it had no alternative launch capability. The resulting gap in American launch capabilities allowed European competitors (primarily France) to enter the space business as effective challengers.

The historical data shows that the federal government effectively reduced operating risk for the aerospace industry by funding R&D and purchasing military aircraft. In essence, this funding amounted to a subsidy of the industry and served to mitigate the risk involved in the development of high-technology, high-cost aircraft. This government support through R&D dollars underpinned the industry throughout its development and fostered the cross flow of technology from the commercial industry and the government (especially military) sector of aerospace. The government further supported its aerospace industry by purchasing large numbers of aircraft and funding the space program. With drastic cuts in defense procurement, industry risk will increase.

In the next section, the potential for future war and also some imperatives that will govern the application of military force are examined.

War and Political Imperatives

The second element of aerospace power is the military one. Prior to looking at how military aerospace capabilities influence the military strategies of the services, one must consider two questions. First, will war or conflict be a factor in the future conduct of nations? Second, if war and conflict persist in the future, what political imperatives might control a US response to a crisis? Understanding these two issues will prepare the reader to assess the role of aerospace power in the military strategies discussed later.

A Future of Armed Conflict

The nature of the international security environment is changing. In the former Soviet Union, Boris Yeltsin's supporters appear fewer in number, and he operates in a growing climate of unrest. Can Yeltsin hold onto the democratic reforms or will Russia return to communism? If the Russians do revert to communism, will it be with the same global ambitions seen during the cold war? How will the nations of the world deal with the violence in Bosnia-Herzegovina? What can these same nations do about growing ethnic unrest in the southern regions of the former Soviet Union? These questions, and the many more that could be asked serve to highlight the uncertainty the United States and the rest of the world face in building toward the future. There are, however, two questions that must be addressed before examining the military service strategies devised to meet the challenges of the future. First, will there be armed conflict in the future, and if so, why? Second, what political imperatives may drive the US response to potential conflicts?

The global unrest discussed above indicates that the occurrence of armed conflict is one of the few certainties the world faces in the future. Since the end of the cold war and Desert Storm, the United States, as part of ongoing United Nations (UN) efforts, sent over 20,000 troops into Somalia to feed people and restore law and order. The United States flew military aircraft in the Middle East to enforce the no-fly zone over Iraq. American forces conducted operations to impel UN economic sanctions on Iraq and Serbia. Also, the United States committed forces to implement the no-fly zone over Bosnia-Herzegovina. At the same time, the United States finds itself losing its "War on Drugs" and concerned about the "economic war" of the twenty-first century.[33] What then are the potential centers of conflict for the future?

To predict the future, sometimes a look to the past is beneficial. People/countries have fought wars for a variety of reasons. Historically, nations most commonly have gone to war for economic reasons. Agrarian societies sought the acquisition of more and better land. As trade became a more dominant feature of society, the issue became trade routes, resources, and colonies. Today, some argue that economic warfare involving the use of armed forces is a thing of the past. Is it? George Friedman and Meredith Lebard in their book, *The Coming War with Japan*, provide compelling arguments that a war between the United States and Japan is not just possible but "inevitable."[34] Their key tenet states that an immutable tension exists between Japan, needing to obtain resources and expand into markets for its products, and the United States, needing to protect its own economy from the ravages of trade deficits and declining economic power. According to Friedman and Lebard, the dynamics of each country, as it seeks to optimize its economic position, will propel the two countries toward conflict. The conflict described by Friedman and Lebard portends a shooting war of global proportions. Is this theory too far-fetched? One might ask: What happens if a country attempts to extort US financial markets by manipulating currencies or debt financing? In the summer of 1992, changes in German currency exchange rates greatly affected economies around the world (negatively for the most part). What would the United States response be if that kind of manipulation were purposefully directed at its economy to compel economic crisis? Would not the United States construe such action as an invasion of sovereignty and a possible threat to the "economic" survival of the nation? It appears plausible that a whole new world of economic coercion is possible in the global electronic marketplace of the future.

Ideological concerns represent a second rationale for conducting war. Several variations of this category exist. First, religious differences served as justification for bitter wars, the Crusades being an excellent example. A second variation, an offshoot of religion (and often enmeshed in religious differences), is ethnic friction. Cultural differences between people often result in conflict. In the Middle East, the Persian Iranians and the Arabs of Iraq fought one of the bitterest wars in history in the 1980s. In this case, the power of cultural differences exceeded the ties of religion. Iraqi Shiites fought with Iraqi Sunnis against their Shia brethren in Iran. Certainly the breakup of Yugoslavia illustrates both the religious and the cultural tensions that can produce war. A final source of ideological contention between countries results from differences in governmental processes. The cold war pitted communism and its totalitarian rule against the West's democracy. With the waning of communism, some strategists predict that this kind of conflict will subside. They pin their hopes on the tenuous assumption that democracies do not go to war against each other. Unfortunately, there are many "democratic" totalitarian governments in the world. In 1990, the United States invaded Panama to capture "elected" president Manuel Noriega and bring him to the United States to face drug-related charges. Richard Betts and Samuel Huntington argue convincingly that by the end of this century the world will face an increase in totalitarian regimes with potential instabilities resulting from expected power transition problems.[35] Thus, an assortment of ideological reasons may result in conflict for the United States.

A final category of rationales for war results from those leaders who seek some form of self-aggrandizement. These leaders seek to create their own personal legacy at the expense of their own people and the people of affected countries. Saddam Hussein provides a recent example of this kind of leader. Although no one knows his reasons for attacking Kuwait, a plausible hypothesis states that he sought to set himself up as the leader of the Arab world, much as Gamal Nasser attempted to do some 30 years before. Napoléon fits this mold, especially in the final years of his military career when the opposing coalition (British, Germans, Russians, and Austrians) sued for peace on generous terms, but he held out seeking one last great victory. The world political scene has rarely lacked some new Napoléon, Adolph Hitler, or Hussein.

While conflict still appears inevitable, not every disagreement will escalate to war; however, armed conflict seems more certain today now that the overwhelming fear of nuclear Armageddon has abated with the decline in tensions between the United States and the former Soviet Union. What political imperatives, then, will direct the responses, specifically the use of armed force, in crisis situations?

Political Imperatives for Future Conflicts

Carl von Clausewitz wrote that war was an extension of political intercourse; thus, it comes as no surprise that political imperatives (others may consider them to be restraints) govern the conduct of conflict. Whether conflict resolution involves an economic, diplomatic, or military solution, political imperatives will preside over the issue(s) in dispute. Nine dictums will govern the application of the military instrument in crisis situations in the future.[36] The first imperative results from the change in East-West relations. The monolithic threat of communism, reflected in the nuclear arsenals of the United States and Soviet Union, has lessened greatly with the breakup of the former Soviet Union and subsequent dissolution of the Warsaw Pact. The bipolarity indicative of the old international security paradigm has been altered to one reflecting greater multipolarity.

The second imperative is an extension of the first. In the future, the United States will focus on regional crises. The relaxation of tensions between East and West manifested itself in an explosion of third world ethnic violence. The southern border countries of Russia, the former Yugoslavia, and many African countries are experiencing great unrest and threaten international security. Burgeoning populations in Asia and Africa are increasing migratory pressures and increasing social tensions for improvements in the quality of life. The great disparity between the concentrations of wealth in the Northern Hemisphere versus the Southern Hemisphere exacerbate the cultural tensions that already exist. In the former Soviet Union, drastic changes must occur, otherwise the stabilizing effects of the nuclear standoff between the United States and the Soviet Union will be lost in a wave of regional upheaval. Thus, as the US national security and national military strategies state, the focus of future wars will be regionally based.

The third imperative flows from the two previous dictums. The global community will face more threats, although of lesser worldwide impact, in the future. As described above, the potential sources of conflict multiplied after the superpowers lifted the lid on East-West tension.

The next area of political direction is based upon the assumption that the United States desires to continue in its role as the leading power within the international community. With the many threats that exist in the world today and the interconnected relationships within the business community, the United States appears to have little choice but to remain engaged in the political process of nation-states.

The fifth imperative involves another assumption. It assumes that the desire to remain an economic power will serve to direct US policy. Americans will see this dictum reflected in further reductions of the defense budget, increased emphasis on job creation and training, and so forth. Economic concerns will indeed be a compelling force in political decision making.

The remaining four political imperatives deal exclusively with how the United States will employ force in the future. The sixth imperative assumes the United States will strive to wage short, decisive wars, and to avoid long, costly wars of attrition such as Vietnam. This dictum directly reflects the overarching concern for the economic welfare of the nation.

Another imperative that falls out from the concern for the economy is the employment pattern of US forces. In the past the United States forward deployed much of its active duty forces. The US Army had hundreds of thousands of troops in Europe, and the Air Force had hundreds of fighter aircraft and crews. The Navy has maintained a yeoman's schedule of fleet deployments in the Atlantic, Pacific, and Indian Oceans as well as in the Mediterranean Sea and the Arabian Gulf area. Now, however, the United States will continue to withdraw troops from overseas locations and reduce its naval commitments consistent with decreasing defense budgets and naval force structure. Clearly, America finds itself in a position that requires the use of forces that can project power from the United States to whatever geographical destination is required by circumstance. The United States simply will not be able to afford large, forward-deployed forces in the future.

The eighth political dictum issues from the previous imperative. Because fewer troops will be forward deployed, a capability to respond from the United States must be present to allow America to meet its treaty commitments with its allies. Historically, responses to the smaller, regional type crises envisioned for the future required a rapid response capability. Examples abound illustrating this demand, such as the Berlin airlift in the late 1940s, the Suez crisis in the 1950s, and on up to Grenada, Panama, and the Desert Shield portion of Gulf War II. These crises, and hundreds of other emergencies and disasters, demanded the rapid response of US forces to distant places to achieve the desired political outcomes of US policy.

The final imperative involves casualties and collateral damage. In the future, unless the war is one of survival for the United States, wars must minimize both casualties (United States and adversary) and collateral damage to the enemy's noncombatant structures. Lt Gen Buster C. Glosson, one of the key architects of Desert Storm's air campaign, recalled in an interview that President George W. Bush stated "in no uncertain terms" that Coalition forces needed to minimize the loss of life and damage to any of Iraq's cultural symbols or nonwar supporting facilities.[37] The requirements to minimize casualties and collateral damage will increase as a result of Desert Storm because of the accuracy exhibited by precision-guided munitions (PGM) and the precise bombing demonstrated by high-technology weapon systems like modern aircraft and cruise missiles. In tomorrow's conflict environment, the exigency for accuracy will be more demanding, requiring even more capable weapon platforms and munitions.

These imperatives underpin the military responses possible in future crises. Assuredly, as time goes by, some of these dictums will change. Certainly the president in office and the makeup of the Congress at the time of a given crisis will greatly influence which of these imperatives receives greater emphasis in a given situation. For the military services these imperatives serve to limit the strategies each service can employ and/or contribute to the kit bag of options for US political leaders.

Of Aerospace and Military Strategies

Each of the military services has sought to develop strategies that operate within the political imperatives discussed in the previous section. This section relates each strategy to the political imperatives discussed in the previous section and shows how dependent each strategy is upon aerospace power.

Naval Expeditionary Forces . . . *From the Sea*

On 28 September 1992, Secretary of the Navy Sean O'Keefe, Chief of Naval Operations Adm Frank B. Kelso II, and Commandant of the Marine Corps Gen Carl E. Mundy Jr. signed a white paper delineating the Navy-Marine Corps strategy of the future. They titled the strategy, *. . . From the Sea*.

This new construct refocuses the Navy away from a blue-water perspective towards regional, littoral operations. The Navy-Marine Corps team seeks, through forward deployment and presence, to provide on-call power projection and crisis response to littoral conflict.

In devising this strategy, the naval services assumed they had control of the seas; therefore, they could now concentrate on littoral warfare.[38] The concept calls for the "team" to seize and defend ports and naval bases, and/or to control coastal air bases to allow entry of US air and army forces as required.[39] Upon successful penetration, naval forces then turn the mission over to heavier Air Force and Army units. This reliance on Air Force and Army firepower coupled with planned reductions in Naval and Marine Corps capabilities indicates that the strategy envisions the team operating at the lower end of the low-intensity conflict spectrum.[40] Thus, . . . *From the Sea* is a limited focus strategy tightly linking the Navy and the Marine Corps in the projection of power upon littoral areas.

The new construct identifies four key operational capabilities necessary for success. First, the team recognized that command, control, and surveillance capabilities are essential to joint and combined operations.[41] The secretary of the Navy (SecNav) directed the Naval War College's Wargaming Center to evaluate the new strategy with respect to the Navy's Program Objective Memorandum (POM), the Navy's programmatic budget.[42] The Navy discovered that the entire architecture of command, control, communication, computers, information, and intelligence (C^4ISR) required increased attention. The war game identified key problem areas such as positive identification systems, real–time battlefield damage assessment, and multispectral surveillance. Further, the Navy found that it needed improved intelligence dissemination capabilities. These shortcomings reflect the increasing emphasis on the exploitation of space for the successful employment of naval strategy.[43]

The team identified battle space dominance as the second key operational capability. Naval forces consider this area the heart of naval warfare. The two components of the battle space are landward and seaward. Naval forces seek within the littoral area, to control the sea (on and below the sea), the air, and operations on the land. Space control receives emphasis, too. As the strategy states, "We must use the full range of US, coalition, and space-based assets to achieve dominance in space as well."[44]

Achieving battle space dominance makes possible the third key capability, power projection. The naval forces team expects to use its mobility, flexibility (tailorable forces), and technology to mass its strength against enemy weaknesses. Embedded in this aspect of the construct is the fourth capability, force sustainment. The . . . *From the Sea* strategy touts the Navy's ability to sustain deployed operations and its ability to remain on station for long periods.

The new naval forces expeditionary strategy does reflect most of the political imperatives discussed above. The strategy shifts its focus from a Soviet, blue-water threat to a regional, littoral one.[45] The complete refocus of the team to littoral warfare indicates implicitly that the naval services recognize the increase in lesser threats and that the United States will desire to maintain a leadership role in those areas. The new strategy recognizes the economic and threat imperatives resulting in downsizing of its force structure as it seeks to make its operational capabilities work in a more flexible manner. In the future, the team will increasingly operate surface action and amphibious readiness groups independent of carrier battle groups (CVBG). As stated in . . . *From the Sea*, the Navy Department "must structure a fundamentally different naval force to respond to strategic demands and these new forces must be sufficiently flexible and powerful enough to satisfy enduring national security requirements."[46] The new strategy recognizes the imperative for minimizing casualties as evidenced by its listing this goal as one of the seven key results in the SecNav Strategy-POM war game.[47]

At odds with the political imperatives is the strategy's reliance on forward deployment/presence to enhance response time to a crisis. As long as the Navy-Marine Corps can maintain forward basing in Japan, the Mediterranean, and the Indian Ocean (the Marines still have a significant amount of prepositioned equipment afloat there), the naval team can achieve power projection measured in days versus weeks. The move to lighten Marine forces will ease deployment and sustainment problems for the corps but, at the same time, reinforce a limited role at the lower end of low-intensity conflict. Thus, they will be used in short conflicts or as early on forces awaiting the arrival of heavier air and army units. Overall, within its stated focus, the . . . *From the Sea* strategy confirms and operates within the stated political imperatives.

The results from the Navy's Strategy-POM war game illustrate the areas the Navy-Marine team must focus on to "flesh out" its new strategy. The study also offers us a tool to show the dependence of this new strategy on aerospace power. Larry Bockman and Brad Hayes list seven major results from the game; six directly relate to aerospace power (the seventh emphasizes the importance of minimizing casualties in any future conflict).[48]

The first key result area recognizes the increasing importance of C^{4}ISR systems. Bockman and Hayes list requirements for command data links, position location gear, and super and extremely high-frequency communications. In the surveillance area, they note the need to exploit multispectral capabilities. All of these areas require extensive use of aerospace power. The global positioning system (GPS), used so successfully in Desert Storm, can provide immediate help to navigation capabilities. Improved capabilities in satellite systems like the Defense Satellite Communications System and Land Satellite System (LANDSAT) will enhance capabilities in global command, control, and communications (information handling) and multispectral imaging. Improving the links between operators and national intelligence satellites will facilitate the flow of intelligence information to the users most in need of their data.

The need for defensive capabilities against theater ballistic missiles (TBM) was the second key result area. This aerospace threat requires the ability to detect, target, and kill not only the missile but also the launcher. Such aerospace assets as the joint surveillance target attack radar system (JSTARS) and strategic surveillance, satellites will complement the Navy's effort to develop antiballistic missile defenses on its Aegis cruisers and provide the Navy with the initial tools to face this threat.

Third, the increased integration of PGMs for naval aircraft will provide the strike capability for attacking TBM launchers and other high-value, hard targets. Bockman and Hayes note the Navy seeks penetrating weapons in greater numbers than ever before.[49] Obviously, the Navy desires to increase the flexibility of its aircraft firepower.

To aid weapons delivery, the Navy-Marine team seeks to procure multimission, low-observable aircraft. This fourth key area coupled with the fifth area, the acquisition of unmanned aerial vehicles (UAV) illustrates the Navy's reliance on aerospace power to provide the penetration force of the naval forces team.

Finally, the Strategy-POM game reinforced the need to resolve the Marines' need for medium vertical lift; a problem exacerbated by the political haggling over the V-22. Once again, aerospace is at the forefront of naval power projection strategy.

Thus, reflected in this major evaluation of its new strategy, US naval forces recognized the absolute necessity of aerospace power for their ability to prosecute their strategy today and in the future. As the Germans learned at the Battle of Britain, and the Navy learned at Pearl Harbor, control of the air must be achieved before surface operations can be successfully conducted against an aerospace-capable adversary. The Navy and the Marine Corps clearly realize the need for space operations to enhance communications, navigation, and surveillance. Implicit in . . . *From the Sea* is the requirement for aerospace control and dominance. No one can imagine exposing amphibious or carrier forces to an environment where US or allied air control is lacking. The linkage of CVBGs to amphibious readiness groups to form the new naval expeditionary force team reflects the concern for gaining and maintaining air control in littoral warfare.

Army Operations

The Army's new doctrine, *Army Operations*, seeks to project strategically agile forces while providing the bulk of US forward presence on five continents.[50] Gen Gordon R. Sullivan, then Army chief of staff, notes several forces of change in the international environment: democracy, ethnic strife, ideological and religious tenets inimical to free markets and democracy, economic crises in many countries, proliferation of military technology, and threats from drug traffickers.[51] He goes on to note that these forces drive the Army toward a strategic power projection footing. Further, Sullivan sees two constants that result in the need for a capable Army. First, enduring American global interests of democratic and economic processes require access to critical resources and free economic and political interaction.[52] Second, there is the argument that 50 years of American world leadership cannot be abandoned. Field Manual (FM) 100-5, *Operations*, states the Army's role is to apply "force to fight and win quickly, with minimum casualties," and, as General Sullivan states, "With the Army, America signals that national interests are at stake."[53] To meet the challenges that General Sullivan poses in his world view, the Army developed a strategy geared to mobility and versatility. Based on a mobility study, the Army has set requirements to move one light and two heavy divisions from the United States to a conflict theater 7,500 miles away in 30 days. Further, the Army plans to transport the remainder of the corps and two more divisions to the theater within an additional 45 days. To accomplish this task, the Army wants to fund a $13-billion buy of 39 ships including medium roll-on, roll-off ships. To fight the war envisioned by Army strategists, the service developed a strategy to maximize the maneuverability of Army forces as seen during Desert Storm.

The Army's new strategy focuses on power projection as its central element.[54] To accomplish its mission, the Army plans to function within an eight-phase construct of force-projection operations. The phases may occur sequentially or run simultaneously depending on specific circumstances. The eight phases are predeployment activity, mobilization, deployment, entry, decisive operations, restoration, redeployment, and demobilization. The first three phases entail activities leading up to the embarkation of troops. These activities include training, requirements formulation, the assembling of troops and materiel, and deployment execution.

The entry phase may be opposed or unopposed. The Army wants a forced-entry ability capable of success under any conditions. "Speed is especially important" as the Army wants to seize the initiative.[55] The entry phase sets the stage for decisive operations by creating the environment within its area of influence to mass forces to destroy the enemy.

In the decisive operations phase, the Army brings speed, maneuver, shock action, and violent aggressive tactics to overwhelm the enemy with as little loss of US lives as possible. The strategists plan to attack only at critical times and emphasize offensive operations, using the defensive only as required. Key to accomplishing this phase is the use of massed fires to support maneuvering troops and massed combat service support to sustain operations. The supported land commander will require not only close air support (CAS) but interdiction fires short of, and beyond, the fire support coordination line.

The Army seeks to dominate the enemy through battlefield preparation and shaping. Preparation actions include establishing the detection area, using available detection sensors to define the battlefield, determining the location of high-value targets, and protecting the main battle force and logistics support elements. Army commanders seek to shape the battlefield to gain and maintain the initiative. To accomplish this task, they rely upon the heavy use of air assets and long-range fires to disrupt the enemy. By integrating tactical air support, battlefield air interdiction, and conventional weapons (and nuclear and chemical ones if required), the Army plans to mount a massive fire support effort to throw the enemy force off balance and keep them there. The planners also note the need to deliver logistics support to maintain the high tempo of operations.

The final three phases of restoration, redeployment, and demobilization occur after "the cessation of armed conflict."[56] In these phases, the Army plans to assist in the restoration of civil order including civil affairs activities and the clearing of military hazards (mines, ammunition, etc.). Prior to redeployment, the Army remains prepared to resume hostilities should the peace fail. Demobilization completes the transfer of Army units to a peacetime posture.

To employ this strategy in a war-winning manner, the Army adopted five key tenets that help establish conditions for victory.[57] Those tenets are initiative, agility, depth, synchronization, and versatility. To gain a greater understanding of the Army's strategy, we will briefly review each tenet.

In its first tenet, initiative, the Army imputes an offensive spirit in the conduct of all operations.[58] Using offensive strike, the Army seeks to never let the enemy recover from the shock of attack. If placed on the defensive, the Army seeks to quickly turn the tables on the attacker and reestablish offensive operations. For operations other than war (OOTW), Army forces seek to control the environment instead of allowing it to control operations.

The second tenet is agility.[59] Agility, the prerequisite for seizing and holding the initiative, is achieved by reacting faster than the enemy. The Army views agility as much a mental as a physical quality. The strategy plans to use greater quickness to rapidly concentrate strength versus enemy vulnerabilities.

Depth, the extension of operations in time, space, resources, and purpose, serves as the third tenet.[60] The Army envisions a three-dimensional maneuver battlefield extending up to 300 kilometers or beyond. This extension represents a vast projection in the depth of the battlefield from even the 150 kilometer moves in Desert Storm. For OOTW, the Army wants to extend area activities as above to affect and shape the environment to achieve the desired political resolution.

The fourth tenet, synchronization, seeks to achieve "the focus of resources and activities in time and space to mass at the decisive point."[61] The Army views synchronization as "both a process and a result." Synchronization incorporates such activities as intelligence, logistics, and fires with maneuver to achieve synchronized operations. In short, the Army wants to get the "maximum use of every resource where and when it will make the greatest contribution to success."

With versatility, the final tenet, the Army wants its units to have the capability "to meet diverse mission requirements."[62] Thus, Army forces could inherently adapt to different missions or tasks, even tasks that may not have been on the unit's original mission-essential task list. How, then, does the new "Operations" strategy reflect the new political imperatives, and how does it rely on aerospace power? General Sullivan provides clear insight into the development of this strategy. His view of global changes and the need to meet future challenges are reflected in the emphasis on deployability and maneuver.

In his acknowledgement of the constants requiring a highly capable Army, General Sullivan recognized the need to have forces capable of projecting US power to ensure that democratic and economic imperatives are met. Further, the deployability of the new Army appreciates the need to respond rapidly to regional crises. The focus of the Army's new operations manual, FM 100-5, to apply "decisive force to fight and win quickly, with minimum casualties" clearly recognizes the imperatives for short, minimum casualty wars. Thus, "Operations" clearly supports the new political imperatives facing the United States in the future.

The key new element in the Army's new construct clarifies just how reliant the strategy is upon aerospace power. Crucial to Army actions in the future is the replacing of close battle with deeper maneuvers employing joint operations, fighting at the maximum range of weapons. In short, the Army seeks to push out the engagement line to avoid casualties. To do this, the Army must employ aerospace power.

In entry- and decisive-operations phases of the new strategy, the Army needs the sophisticated "eyes and ears" of aerospace assets to conduct the intelligence preparation of the battlefield. Currently the Army uses Guardrail aircraft to conduct electronic and signal surveillance of the battle area. They also employ Mohawk aircraft to do close-in targeting of enemy forces out to some 50–70 kilometers. (JSTARS will provide the Army with the capability to do this mission virtually throughout the theater, as was evidenced in its performance in Desert Storm.) The Air Force aids this process by providing air and space systems to conduct intelligence gathering operations throughout a theater of operations, facilitating Army desires to function out to 300 kilometers. Conducting deeper operations, the Army will rely more heavily upon satellite communications systems as its units move beyond line-of-sight communications ranges. The Army discovered in Desert Storm that the GPS provides exceptionally accurate navigation data. This capability will expedite targeting, resupply, and battlefield management capabilities for ground forces.

As discussed previously, the conduct of decisive operations required significant amounts of aerospace power for interdiction and CAS. Of course, Army helicopters are a fundamental part of aerospace power on the battlefield. Recall that Army air assault brigades sealed off the roads out of Kuwait towards Iraq during Desert Storm. Improving helicopter technology is one of the four critical technology areas for the future Army, according to General Sullivan.[63] Furthermore, aerospace power provides the rapid airlift capability that allows the Army the logistics flexibility to mass for decisive operations. While Army attack helicopters will be involved increasingly with CAS, Army doctrine still views the principal function of its aviation brigades as a flexible maneuver force.[64]

Finally, to support Army deployment to and from the theater, aerospace power—through strategic and tactical airlift (to include helicopters)—provides the Army the ability to deliver high-value replacement equipment or parts (even repair units) exactly when and where needed. No other mechanism provides this combination of flexibility and response time.

Like the Navy and the Marine Corps, the Army of the future has set its sights on a strategy that demands the unique capabilities that aerospace power brings to the combat environment. Aerospace power inherently embodies each of the five key tenets for successful Army operations. Aerial power always seeks the initiative, uses its own agility and flexibility to deliver ordnance or beans throughout the combat theater, and offers the capability to choreograph the deep fires necessary to minimize casualties in future conflicts. Thus, throughout its new strategy, the Army weaves aerospace power into its operations to provide it with the decisive edge for war winning.

Global Reach—Global Power

The Air Force calls its strategy *Global Reach—Global Power*. As did the other services, the Air Force took notice of the end of the cold war and refocused its attention to regional issues. The Air Force adopted a strategy designed to provide "the quickest, longest range, leading edge force available to the President in a crisis."[65] The Air Force envisions itself as becoming the force of first choice and serving as the primary instrument of national military power.[66]

The Air Force foresees conflict based upon a regional threat. Complicating this focus are two factors. First, the declining force structure requires the Air Force to operate with fewer assets. Second, the proliferation of sophisticated weapons and technologies creates a dangerous threat environment for operations.[67] The heart of *Global Reach—Global Power* is encapsulated in the following quote from the 1992 white paper:

> The demands of our new military strategy play to the inherent strengths of air and space power. In an age of uncertainty, with the location and direction of future challenges almost impossible to predict, space forces allow us to monitor activities around the world and to know the battlefield even before our forces arrive. With smaller forces overall and fewer deployed overseas, airpower's ability to respond globally—within hours, with precision and effect—is an invaluable capability that is America's alone.[68]

Gen Merrill A. McPeak, then Air Force chief of staff, stated the mission of the Air Force in a speech at Maxwell Air Force Base (AFB), Alabama.[69] He said that "the job of the forces we bring to the fight is to defend the United States through control and exploitation of air and space." Five key objectives and five key tenets support this mission.[70] First, the objectives begin with the goal of sustaining deterrence, relying primarily upon nuclear forces. Next, the Air Force seeks to provide versatile combat capability through its ability to conduct and sustain theater power projection operations. Third, the Air Force wants to provide rapid global mobility via its airlift and air-refueling tanker aircraft. In fact, with the new regional focus, the Air Force envisions greater demands for both of these capabilities, especially for operations other than war.[71] Fourth, and perhaps most important, the Air Force wants to control the high ground of space and command, control, communications, and intelligence (C^3I). It seeks to do this by attaining and maintaining space dominance. In its last objective, the Air Force desires to enhance US influence abroad by strengthening security partners through deployments, exercises, and education and training programs.

To achieve these objectives, the Air Force relies on what it considered to be the "inherent" tenets of characteristics of aerospace power. These five tenets are composed of speed, range, flexibility, precision, and lethality.[72] As might be expected, the Air Force considers Desert Storm the validation of these tenets. The combination of stealth aircraft, crew training, PGM, air refueling (an "indispensable force multiplier"), and the introduction of space into combat operations affirm these characteristics.[73] For nearly 40 days, the world watched aerospace power dismantle Iraqi war-making capability with amazing deftness and finesse. General McPeak stated that the Air Force has become the "maneuver force par excellence."[74]

For the Air Force, space represents an area of increasing importance. The Air Force contributes 80 percent of the Department of Defense space budget and provides, as mentioned previously, some 98 percent of space manpower.[75] In *Global Reach—Global Power* the Air Force states that "space forces' superiority of speed and position over surface and air forces points to *control* of space as a prerequisite for victory. Space superiority has joined air superiority as a sine qua non of global reach and power."[76] Most important, control and exploitation of space provides the capability to achieve a level of battlefield situational awareness never before possible. Some of the fog of war has cleared from the battleground. As the strategy states, in the future the "control of the high ground will increasingly make space forces part of the versatile combat forces—decreasing the time required to respond to aggression and allowing us to strike anywhere with overwhelming but discriminate power."[77] Within the new Air Force strategy, *Global Reach—Global Power*, there is evidence of each of the future political imperatives. Up front in this strategy, the Air Force acknowledges the end of the cold war and the need to downsize its forces while changing to a regional focus. The extended quote presented above clearly reflects the imperatives of a new, regional focus with fewer forces (reflecting the economic imperatives at work in American politics). Another clear indicator of the Air Force's response to changing circumstances is its shift in viewpoint on strategic and tactical weapon systems. In the post–Desert Storm environment, the Air Force views its weapons platforms in terms of mission accomplishment, not by an arbitrary label. Fighters, previously labeled as tactical weapons, may accomplish strategic bombing while B-52s may conduct tactical strikes against troop concentrations.[78] In fact, the Air Force no longer refers to its units as fighter or bomber wings; it simply calls them wings (e.g., the 1st Wing, formerly the 1st Tactical Fighter Wing).

Global Reach—Global Power concentrates on the ability to project power from the continental United States (or a few forward bases) to any point on the globe. Clearly the Air Force recognizes the political emphasis on improving US economic competitiveness by decreasing defense costs. The Air Force's strategy supports that effort by seeking to provide forces that can do the job without the expense of forward basing and deployment. In time of crisis, however, the Air Force plans to take advantage of its airlift and air refueling capabilities to quickly project power when and where it is needed.

The Air Force is restructuring itself to provide forces that can "punch hard and terminate quickly."[79] A prime example of these efforts is the formation of composite wings providing ready force packages capable of delivering the hard punch. Key elements of the strategy serve to support US imperatives of short wars with minimal casualties. Former Air Force Secretary Rice targeted these aspects in one of his first writings on the new strategy.[80] He pointed out that the Air Force sought the ability to strike quickly with lethality and survivability. He credits stealth technology with providing this combined capability. The discriminate nature of PGMs provides the capacity to limit collateral damage.

Thus, the Air Force's new strategy clearly supports the new political imperatives driving national security policy. Naturally, the Air Force relies upon aerospace power to support national security objectives. But, as stealth and PGMs helped redefine the capabilities of aerospace power, space will redefine those capabilities in the future.

Space, then, will be the high frontier of military aerospace power, and the Air Force plans to "operationalize" space forces to benefit all war fighters.[81] Gen Charles A. Horner, one-time US Space Command commander, notes the stunning successes of Desert Storm in areas like navigation, weather, surveillance, missile warning, and communications.[82] He recognizes the need to improve upon these capabilities. The Air

Force leads the efforts to develop next-generation missile warning systems like the Follow-on Early Warning System (FEWS). The GPS not only provides superb navigational data but may help solve the friendly fire problem seen in Desert Storm. A major program, Talon Sword, seeks to take data from national reconnaissance assets and transmit that information directly to aircraft cockpit displays.

Space represents the future of the Air Force and, increasingly, aerospace power will be projected through space systems. Although the cost of operating from space is high, the force leverage gained is immense. Indeed, the Air Force is committed to providing the United States with the forces to control and exploit air and space.

Serious problems, however, face the aerospace nation. The next section examines the major problems confronting US aerospace power and offers the beginnings of a national aerospace strategy.

A National Strategy for the Aerospace Nation

In previous sections, economic and military aspects of aerospace were examined. These two components combine to produce aerospace power. The US aerospace industry is a business that provides a significant portion of the nation's high-value, high-technology manufacturing base. Militarily, the tremendous importance of aerospace to the future strategies of each of the military services was noted. If, as this thesis argues, aerospace power is crucial to the economic well-being and national security of the United States, then one would expect the United States to have a national strategy for aerospace power. No such strategy exists. Furthermore, current efforts aim only at either the economic or military components—no strategy exists to integrate these elements into a cohesive policy of national aerospace power.

Two questions, then, remain to be answered. First, what problems exist indicating the need for such a strategy? Second, what is entailed in a national aerospace strategy; what are its objectives and recommended processes?

Trouble in the Aerospace Nation

Earlier the importance of the aerospace industry to the US economy was discussed. However, serious problems abound for both the economic and military components of aerospace power. The aerospace industry concerns will be examined first and then the military ones. To discuss the industry problems, the discussion is limited to the aircraft manufacturing and airline subsets of the aerospace industry. Most of the problems facing these two concerns affect other aspects of the aerospace business. Together they account for over 50 percent of total aerospace sales US aircraft production supplies 80 percent of the world's large commercial jet aircraft. Thus, these two segments of the aerospace industry provide a good way to review the problems plaguing this vital industry.

In industry, the trouble starts with the bottom line. From 1990 to 1992, the world's airlines lost $10.8 billion; US carriers accounted for 73 percent of that total or some $7.85 billion.[83] Employment statistics further highlight the industry's woes. The aerospace business lost 87,000 jobs in 1991; production workers declined in number by more than 7 percent.[84] Boeing alone cut 10,000 employees in 1992 and plans to slash another 28,000 from its payroll by 1994.[85] Since mid-1990, Douglas Aircraft Company reduced its work force from approximately 43,000 to only 19,000. It expects to cut another four thousand jobs this year.[86] Worker reductions affect management, too. United Airlines recently announced it was trimming 20 percent of its senior officers in the face of continuing losses.[87] Further, United wants some $300 million in wage concessions from its employees in an effort to improve its financial picture (United alone lost almost $1.3 billion in 1991–92). Another factor is the declining market trend in military and commercial aircraft sales. Between a 1981 high point and 1991, the number of military aircraft delivered by industry fell by 30 percent.[88] Commercial aircraft sales turned downward in 1991. Both Boeing and Douglas scaled back production some 40 percent to meet the reduced demand.[89] Already this year aircraft manufacturers suffered $15 billion in cancelled orders.[90]

But these are just the symptoms; what are the roots of the problems? At the heart of industry's problems is the issue of competitiveness. The key to competitiveness in the aerospace industry is risk management. The American aerospace industry historically has used government military contracts and R&D funding (see table 3) to reduce its production costs, thereby reducing product risk. Table 4 illustrates the dramatic increase in development costs that federal contracts and R&D funding helped to offset.

Table 4

Changing Aircraft Production Costs

Aircraft Type	Year Entered Service	Development Costs (1991, $ millions)
McDonnell Douglas DC-3	1936	3
McDonnell Douglas DC-6	1947	90
McDonnell Douglas DC-8	1959	600
Boeing 747	1970	3,300
Boeing 777		10,000[a]

[a]Estimated

Source: Laura D'Andrea Tyson, *Who's Bashing Whom? Trade Conflict in High-Technology Industries* (Washington, D.C.: Institute for International Economics, 1992); and "Making Elephants Fly," *The Economist,* 23 January 1993, 77.

These traditional risk management supports are diminishing in the face of budget deficit pressures. As discussed earlier, military aircraft sales are in decline. Also, the Clinton administration proposes to realign the ratio of nondefense to defense R&D funding from the current 40:60 ratio to a 50:50 ratio.[91] How critical is federal research and development funding? Recall that federal funding comprises three of every four dollars expended on aerospace R&D (all other manufacturing industries receive only 1.4 in 10 dollars from federal R&D).[92] How will the US aerospace firms compete with foreign consortiums like Airbus, which has the financial backing of three powerful governments? What happens to the Far East market if Japan targets the aircraft building industry through the financial backing of its Ministry of International Trade and Industry? Eiju Toyoda, chief executive officer of Toyota Motor Corporation, told visiting Boeing executives that Toyota was "in the transportation business. It's our destiny to be in the airplane business."[93] The challenge to American leadership in aerospace is very real.

The US government exacerbates the competitiveness issue with inconsistent policies. For example, the Clinton administration's proposed energy tax will add approximately $1 billion in tax burden to the airline industry. Further, the cuts in federal R&D funds to aerospace described above can only worsen the very industry the president is committed to support. Additionally, the onset of Stage II noise restrictions may create a greater demand for quieter aircraft but will increase airline debt burden as companies are forced to buy new aircraft. Clearly, the industry requires a national strategy to integrate these facets of market and government policy.

Civilian and Department of Defense policy makers suffer from their own strategic dysfunctions. Each service has its own aerospace force dependencies; however, no DOD-level integration office exists to coordinate military aerospace power. In fact, as analysts for *The Economist* point out, the DOD remains the only Western military establishment with separate service acquisition systems.[94]

A more dramatic indication of military dysfunction is evident in the DOD response to Sen Sam Nunn's (D-Ga.) questioning of the efficacy of the military's having four air forces (meaning the four services).[95] The DOD response came in Gen Colin L. Powell's report on roles and missions.[96] The report argues that "the other services have aviation arms essential to their specific roles and functions but which also work jointly to project America's air power."[97] The debate argues that as it makes no sense to assign all radios or trucks to one service, so too it would not make sense to assign all aircraft to one service. Is this an aerospace rationale? Would we need aerospace forces to operate differently in the services' strategies if there were only one air service? Would we not be better served to describe what we want US forces (land, sea, and aerospace) to do and develop an integrated strategy to achieve some desired end state? For example, if the nation wants a highly mobile amphibious assault capability, it needs marines with airpower. If the nation wants sea control and power projection capabilities with minimal reliance on other nation support, it needs a navy with airpower in the form of carrier air wings. If the United States wants an army with the capability for sustained, heavy combat with low casualties, it will need aerospace power. If the nation wants to exploit air and space forces as it did in Desert Storm, it will need many air and space capabilities. Future service strategies depend on aerospace power. The political imperatives driving those strategies devolve upon aerospace capabilities. If the Defense Department is to answer Senator Nunn, it must answer within the context of a military aerospace strategy.

The ties linking the aerospace with its military counterpart were forged through two world wars, a cold war, Korea, Vietnam, and other lesser conflicts. Add to this crucible of the past the economic challenges of the future and one sees the desideratum of aerospace power. To achieve a position of predominance in aerospace, the United States requires a national aerospace strategy.

Whither the Aerospace Nation?

If this paper serves no other purpose, it must serve as a wake-up call, a call to action for the aerospace nation.[98] United States policy makers must view aerospace power as a national treasure. If such economists as Reich, Michael Porter, and Thurow are correct, the aerospace industry will be critical to America's future economic prosperity. Each argues that the future belongs to those nations with trained, skilled workers who add unique, high value to products. Each agrees that aerospace is one of those industries. Militarily we cannot operate without control of aerospace—all military strategies rely upon it. Aerospace dominance provides the capability for US forces to win within the political imperatives of the future, especially with reference to casualties. Aerospace power, both its economic and military elements, is under great pressure to succeed in the future. To do so requires a national aerospace strategy.

What, then, should be the goal of an aerospace strategy? The economic vision needs to be one that aspires to world leadership in aerospace technology. The military vision is clear—provide aerospace control and exploitation capabilities on demand, regardless of whether land, sea, or aerospace forces represent the predominant medium in any given circumstance. Together these two ideals combine to form the goals of the US aerospace strategy.

What are the broad objectives that work to achieve the goals stated above? To paraphrase Tyson, "Ultimately, the fate of the nation's [aerospace strategy] depends not on trade battles fought abroad but on the choices we make at home: in macroeconomic policy, education policy, technology policy, industrial policy (and national defense policy)."[99] Ms Tyson's framework is used herein to offer broad objectives and ideas for formulating a national aerospace strategy.

On a macroeconomic level, the national strategy should contribute to the economic well-being of the United States. Aerospace should help the United States improve the standard of living for its people. Further, improved economic well-being ensures the United States the capacity to support military capabilities to secure national security interests.[100]

The leading objective of US macroeconomic policy is to make the aerospace industry profitable and competitive in the world marketplace. Several policy options work to attain this goal. A key option task is to level the playing field of aerospace competition. As seen earlier, federal R&D funding and military aircraft purchases supported (subsidized) US commercial aerospace in an indirect manner. The European Community used direct subsidies (direct government financial support) to help Airbus break through the start-up barriers in the aircraft manufacturing field. Now other countries (like Japan) seem poised to take off. Bilateral/multilateral agreements need to account for these extra-market forces. The 1992 United States-European Community bilateral agreement on trade in civil aircraft provides a starting point. This agreement stipulates a set percentage (33 percent) for direct government funding of aircraft development. The agreement also states that "indirect (i.e., military) supports should neither confer unfair advantage . . . nor lead to distortions in international trade in such aircraft."[101] Trade agreement discussions with aspiring entrants to the aerospace industry (like Japan) would have to provide provisions for new players to overcome the high entry barriers to the aviation business.

Another key to macroeconomic policy is the question of foreign investment in US aerospace. The United States needs to develop consistent policies to accommodate foreign investment. In his book, *The Work of Nations*, Robert Reich lays out the argument that where investment dollars come from is irrelevant.[102] What matters is having the production and skilled workers in the United States. That way, if the foreign investors pull out, the United States still has the people and process. Naturally, one would have to consider security issues; however, the high cost of aerospace development is driving firms to seek joint ventures, consortium, and ad hoc arrangements to generate the skills and/or funds to produce new products. As Reich and others argue, globalization of the aerospace industry is a trend that is here to stay.

US tax structures provide another issue of concern for macroeconomic policy as it applies to aerospace. Obviously, in an industry that carries as much debt as aerospace, tax structure is very important. The aerospace strategy must produce a consistent tax plan that encourages civil research and development investment. At the same time, this new tax structure must recognize that commercial success from R&D expenditures is an inherently low-return proposition. Further, the strategy needs to avoid/resolve situations like the proposed energy tax that work at cross-purposes to other industry promoting efforts. Few industries can absorb a $1 billion tax mistake.

Education policy requirements are often overlooked in policy proposals. The aerospace industry needs highly skilled engineers, designers, and craftsmen to compete in the future. Likewise, the military requires highly qualified engineers, technicians, and flyers. The objective of US education policy must be to provide education and training to equip its workers with the skills to compete for and obtain the high-technology, high-wage jobs that result in an increased standard of living. This policy must not limit itself to college education but must be extended to include vocational training so that a supply of educated and trained technicians is available to the industry. Reich argues for "positive economic nationalism" focused on improving job skills through national education programs.[103] He argues that the educational (and financial) elites must accept the social responsibility to raise the educational and training standards of America's workers. Whatever mechanism the strategy adopts will impact not only aerospace but the nation as a whole.

The aerospace strategy should commit the United States to a technology policy seeking dominance in the aerospace field, commercial and military. As noted earlier, President Clinton directed US policy toward this objective by stating that certain technologies are more important than others if the United States is to compete in the future global economy. Aerospace is one of those "designated" technologies. Technology transfer between the commercial and military sectors lies at the heart of technology policy.[104] Currently, the United States is structured to deal only with the transfer of military technology to the commercial sector; the Defense Advanced Research Projects Agency (DARPA) leads this effort. This policy needs to be broadened to include transfers from the commercial sector to the military.

A concern exists, however, that the new DARPA focus degrades its primary job of developing new defense-related technologies.[105] Reports indicate DARPA suffers from undermanning and high personnel turnover, begging the question of whether or not DARPA is the best choice for this job. Several analysts recommend creation of a National Advanced Research Projects Agency (NARPA) to facilitate the transfer of defense and other technologies into the commercial sector freeing DARPA to continue to concentrate on its own projects. Separating the two agencies would minimize security concerns and allow NARPA to adopt a more visible role in sponsoring the commercial transfer of technology than DARPA. The two agencies could be linked by agreement or by formal structure to achieve the cross flow to make dual-use technology run both directions.

A fundamental industrial policy consideration concerns the legal framework within which industry and military aerospace operate. The industry needs a centralized methodology to guide industry and military programs. This methodology would help the administration and Congress develop and enact legal structures that provide a streamlined, consistent way for aerospace industries to move into and out of joint ventures, ad hoc partnerships, and so forth. Further, the legal construct should address investment, ownership, technology transfers, and government funding guidelines (this list is by no means all-inclusive). The development of these guidelines will require international agreement. International law and transparency regimes must be pursued to provide oversight capabilities. Militarily, these guidelines should serve a similar streamlining purpose to aid foreign military sales and foreign aid involving aerospace issues. Certainly, these legal concerns cut across most of the policy ideas offered in this paper.

The defense policy objective should seek to provide an integrated aerospace plan for congruous force application and programmatic support (development, acquisition, maintenance) of military aerospace. Instead of having four aviation and space programs, the Department of Defense needs to view its aerospace power as a single entity. As we have seen, aerospace power has a central role in each of the services' strategies. Further, the high cost of obtaining aerospace capabilities and continuing reductions in DOD budgets require the adoption of methods to eliminate needless redundancies without giving up needed capabilities. Programmatically, the Defense Department should consider combining its service acquisition systems, at least for aerospace.

The United States is not without an example in developing a broad construct under which to craft a national aerospace strategy. The president's National Space Council provided the space community the kind of oversight direction envisioned for an aerospace strategy.[106] The council, chaired by the vice president, sought to integrate all US space efforts for government, industry, and space customers (military and civilian).

The aerospace strategy requires a similar high-level process. That process must encompass both elements of aerospace power, industry and military, and include the governmental agents included on the space council. Thus, the space council construct provides an excellent methodology from which to initiate a national aerospace strategy.

The scope and effort required to develop and implement a national aerospace strategy will necessitate the realignment of many government organizations. A National Aerospace Council could provide the oversight/integration leadership to manage the many changes implicit in the development of a national aerospace strategy. The time to start this process is now. Aerospace power is too critical to the economic and national security well-being of the United States to be left to the chance direction of market forces and budgetary pressures.

Closing Remarks

The United States has undergone many starts and stops in both its economic and military elements in its development as an aerospace nation. This paper showed the absolutely essential contribution aerospace power makes to the security and well-being, economically and militarily, of the United States. There can be no doubt that America is an aerospace nation. However, many problems cloud US aerospace power and necessitate a national strategy that encompasses both elements of its power.

The aerospace industry provides the jobs, skills, and products that serve to increase the US standard of living. It serves as a visible symbol of the technological expertise and economic power of America. Militarily, the United States faces uncertainty about potential threats; however, as long as it can control and exploit aerospace at will, its future is secure from hostile intent.

Americans can be justifiably proud of what aerospace power has accomplished for the United States: the first man on the moon, worldwide dominance in aircraft and space manufacturing, and military aerospace forces capable of providing decisive results in combat. Now, the United States must go forward with a national aerospace strategy that secures the leadership role of the aerospace nation for the twenty-first century.

Notes

1. Secretary of the Air Force Donald B. Rice, *Global Reach—Global Power*, white paper (Washington, D.C.: Department of the Air Force, November 1992), 15.

2. Jim Impoco and David Hage, "White House Workout," *U.S. News & World Report*, 8 March 1993, 28.

3. Trend information presented here extracted from the US Bureau of the Census, *Statistical Abstract of the United States: 1992*, 111th edition (Washington, D.C.: Government Printing Office [GPO]), 1992.

4. Giulio Douhet, *The Command of the Air*, trans. Dino Ferrari (1942; new imprint, Washington, D.C.: Office of Air Force History, 1983).

5. Ibid. Douhet provides extensive analysis of the future of aerial transportation on pages 77–92. He details his views on the relationship between economics, industry, and national security within the context of aviation.

6. William "Billy" Mitchell, *Winged Defense* (New York: Dover Publications, Inc., 1988), 119.

7. Alexander de Seversky, *Victory Through Air Power* (New York: Simon and Schuster, 1942), 295.

8. Rice, 1.

9. Robert B. Reich, *The Work of Nations* (New York: Vintage Books, 1992); Laura Tyson, *Who's Bashing Whom? Trade Conflict in High-Technology Industries* (Washington, D.C.: Institute for International Economics, 1992); and Lester Thurow, *Head to Head, The Coming Economic Battle Among Japan, Europe, and America* (New York: William Morrow and Co., Inc., 1992).

10. United States Department of Transportation, *MARAD '91*, Maritime Administration 1991 Annual Report (Washington, D.C.: Maritime Administration, 1992), D, 2.

11. James W. Chung, "Whither the U.S. Aerospace Industry?" *Breakthroughs*, Winter 1992–1993, 12.

12. Aerospace Industries Associaton of America, Inc., *Aerospace Facts and Figures 92–93* (Washington, D.C.: Aerospace Industries of America, Inc.), 13. Hereafter referred to as *Facts and Figures*, *Facts and Figures 79–80*, 13; and *Facts and Figures 1960*, 10.

13. *Facts and Figures 92–93*, 48; *Facts and Figures 85–86*, 18; and *Facts and Figures 79–80*, 16.

14. Reich, 3.

15. *Facts and Figures 92–93*, 147; *Facts and Figures 79–80*, 131; and US Bureau of the Census, *Statistical Abstract of the United States: 1992*, 111th edition (Washington, D.C.: GPO, 1992), 410. Hereafter referred to as *Statistical Abstract of the United States: 19xx*. *Statistical Abstract of the United*

States: 1991, 413; *Statistical Abstract of the United States: 1975*, 366; and *Statistical Abstract of the United States: 1965*, 237.

16. *Facts and Figures 92–93*, 153; and *Facts and Figures 79–80*, 132.
17. Chung, 12.
18. Ibid.; and Impoco and Hage, 28.
19. Chung, 15.
20. Ibid., 16.
21. Tyson, 157.
22. Ibid., 160.
23. Robert J. Serling, *Legend and Legacy, The Story of Boeing and Its People* (New York: Saint Martin's Press, 1992), 107.
24. *Facts and Figures 92–93*, 105; and *Facts and Figures 79–80*, 101.
25. *Facts and Figures 92–93*, 105.
26. Ibid., 108.
27. Tyson, 157.
28. *Facts and Figures 92–93*, 105.
29. Tyson, 162–63.
30. Ibid., 163, table 5.3; and "Making Elephants Fly," *The Economist*, 23 January 1993, 77.
31. Tyson, 168.
32. John L. McLucas, *Space Commerce* (Cambridge, Mass.: Harvard University Press, 1991), 203–4.
33. Cable News Network reported on 15 April 1993 that a recent study showed that children, especially those in the eighth grade, are increasing their use of drugs, including a rising rate of LSD usage as the "new" drug of choice.
34. George Friedman and Meredith Lebard, *The Coming War with Japan* (New York: St. Martin's Press, 1991), first paperback edition, 1992, xiv.
35. Richard K. Betts and Samuel P. Huntington, "Dead Dictators and Rioting Mobs," *International Security*, Winter 1985–1986, 112–46.
36. Pieces of the imperatives that follow in the text are to be found in many articles. The key articles used to develop this section include, *Global Reach—Global Power*; FM 100-5, "Operations," final draft, January 1993; H. T. Hayden and G. I. Wilson, "Defining the Corps' 'Strategic Concept,'" *Marine Corps Gazette*, May 1992, 44–46; Carl E. Mundy Jr., "Expeditionary Forces: A Defining Concept for the Future," *Sea Power*, April 1992, 43–44, 48, 50, and 52; *. . . From the Sea—Preparing the Naval Service for the 21st Century* (Washington, D.C.: Department of the Navy, September 1992), hereafter referred to as *. . . From the Sea*; Stan Weeks, "Crafting a New Maritime Strategy," Naval Institute *Proceedings*, January 1992, 30–37; and Thomas C. Linn, "Naval Forces in the Post-Cold War Era," *Strategic Review*, Fall 1992, 18–23.
37. Edward O'Connell, "A Look into Air Campaign Planning" (unpublished research paper, US Defense Intelligence College, Washington, D.C., 1992), 17.
38. *. . . From the Sea*, 1–2.

39. Ibid., 4.

40. "Remarks of Gen Carl E. Mundy Jr., before Congress," *Marine Corps Gazette*, April 1992, 36. General Mundy stated that the Marines were cutting 50 percent of their tanks and 30 percent of their towed artillery to achieve a lighter, more agile force. This initiative results in a marked decrease in firepower. Further, the Navy has reretired its battleships thereby reducing the capabilities of naval gunfire.

41. . . . *From the Sea*, 7.

42. Larry J. Bockman and Brad C. Hayes, "Breathing Life into the Naval Service's New Direction," *Marine Corps Gazette*, February 1993, 48.

43. . . . *From the Sea*, 7.

44. Ibid., 8.

45. Ibid., 1-2.

46. Ibid., 2.

47. Bockman and Hayes, 49.

48. Ibid.

49. Ibid.

50. Gordon R. Sullivan, "'Vital, Capable and Engaged,'" *Army*, October 1992, 24 and 28.

51. Ibid., 24.

52. Ibid.

53. Sullivan, 28; and FM 100-5, 2-3.

54. The information for this chapter was taken primarily from chapter 3, FM 100-5.

55. FM 100-5, 3-10.

56. Ibid., 3-11.

57. Ibid., 2-11.

58. Ibid., 2-11 to 2-12.

59. Ibid., 2-12 to 2-13.

60. Ibid., 2-13 to 2-14.

61. Ibid., 2-14 to 2-15.

62. Ibid., 2-16.

63. Sullivan, 28.

64. Lt Col Mark P. Gay, USA, joint war-fighting instructor, Air War College, Maxwell Air Force Base (AFB), Ala., interviewed by author, 29 April 1993.

65. Rice, 4.

66. Ibid., 2–3.

67. Ibid., 2.

68. Ibid.

69. Merrill A. McPeak, "Does the Air Force Have a Mission?" lecture, Maxwell AFB, Ala., 19 June 1992, 5.

70. Rice, 1 and 3; and McPeak, 4.

71. Rice, 7–8.

72. Rice, 1; and McPeak, 4.

73. Rice, 7.

74. McPeak, 6.
75. Rice, 8.
76. Ibid., 8. Emphasis is in the original text.
77. Ibid.
78. Ibid., 3–4.
79. Donald B. Rice, "Punch Hard and Terminate Quickly," *Air Force Times*, 26 March 1990, 23.
80. Ibid.
81. James W. Canaan, "Space Support for the Shooting Wars," *Air Force Magazine*, April 1993, 34.
82. Ibid., 31.
83. Danna K. Henderson, "1993: Hesitant Optimism, Hopes for Joint Efforts," *Air Transport World*, January 1993, 24.
84. *Facts and Figures 92–93*, 138. *Facts and Figures* included employment data up to 1991; 1992 numbers were not available.
85. Henderson, 24.
86. Bruce A. Smith, "Commercial Strategy Developed for Douglas," *Aviation Week & Space Technology*, 22 February 1993, 26.
87. James T. McKenna, "United Cuts 20% of Senior Officers," *Aviation Week & Space Technology*, 22 February 1993, 37.
88. *Facts and Figures 92–93*, 40.
89. Michael A. Dornheim, "Cuts, Layoffs Affirm Transport Boom's End," *Aviation Week & Space Technology*, 1 February 1993, 22.
90. Henderson, 25.
91. Impoco and Hage, 28.
92. *Facts and Figures 92–93*, 105.
93. Dori Jones Yang and Andrea Rothman, "Reinventing Boeing Radical Changes Amid Crisis," *Business Week*, 1 March 1993, 61.
94. "Slimming the General," *Economist*, 16 January 1993, 64.
95. Rick Maze, *Air Force Times*, 13 July 1992, 3. Senator Nunn asked his question in a speech delivered to the Senate on 2 July 1992.
96. Colin L. Powell, *Chairman of the Joint Chiefs of Staff Report on Roles, Missions, and Functions of the Armed Forces of the United States* (Washington, D.C.: GPO, 1993), xv.
97. Ibid.
98. James W. Chung, "Whither the U.S. Aerospace Industry?" *Breakthroughs*, Winter 1992–93, 12.
99. Tyson, 296.
100. Paul Kennedy, *The Rise and Fall of the Great Powers Economic Change and Military Conflict from 1500 to 2000* (New York: Random House, 1887); and *The Rise and Fall of British Naval Mastery* (Malabar, Fla.: Robert E. Krueger Publishing Co., 1982).
101. Tyson, 208.
102. Reich, chap. 22.
103. Ibid., chaps. 23–25.

104. My reference here to technology transfer is somewhat different from the normal connotation. Normally one associates technology transfer as that occurring between states. Indeed I recognize that construct; however, I believe that issue needs to be discussed as part of industrial policy.

105. William B. Scott, "Caution Urges on DARPA Changes," *Aviation Week & Space Technology*, 1 February 1993, 28.

106. The charter of the National Space Council reads, "The National Space Council is responsible for advising the president on national space policy and strategy, and coordinating the implementation of the president's policies. It was authorized by an act of Congress in 1988 and was established as an agency of the government by President Bush on April 20, 1989." The council was chaired by the vice president and included the secretaries of State, Treasury, Defense, Transportation, Commerce, and Energy and the director of the Office of Management and Budget, the assistant to the president for National Security Affairs, the assistant to the president for science and technology, the chief of staff to the president, the director of Central Intelligence, and the administrator to the National Aeronautics and Space Administration.

Bibliography

Aerospace Facts and Figures. Washington, D.C.: Aerospace Industries of America, 1992.

"*Air Cargo World's* 1992 Listing of Airports." *Air Cargo World*, October 1992, 34–68.

Air Force Center for Studies and Analyses. *Methodology for Analyzing Global Reach—Global Power*. Washington, D.C., October 1990.

Air Force Manual 1-1, *Basic Aerospace Doctrine of the United States Air Force*. 2 vols., 1992.

"Airlines: Financial." *Air Transport World*, October 1992, 70–73.

"Airlines: Traffic." *Air Transport World*, October 1992, 74–78.

Air Transport Association of America. *Air Transport Facts and Figures*. Washington, D.C.: Air Transport Association of America, 1991–92.

"Aspin's FY94 Budget Lacks Bite." *Armed Forces Journal International*, May 1993, 17–18 and 21–22.

Baer, George W. "U.S. Naval Strategy 1890–1945." *Naval War College Review*, Winter 1991, 6–33.

Barreto, Dawn. "Airports Took Off in New Directions with Cargo's Arrival." *Air Cargo World*, October 1992, 20–22.

———. "Taming the Turbulent North Atlantic." *Air Cargo World*, July 1992, 32–39.

Baucom, Donald R. "Technological War: Reality and the American Myth." *Air University Review*, September–October 1981, 56–66.

Betts, Richard K., and Samuel P. Huntington. "Dead Dictators and Rioting Mobs." *International Security*, Winter 1985–86, 112–46.

Bhagwati, Jagdish. *The World Trading System at Risk*. Princeton, N.J.: Princeton University Press, 1991.

Bird, Julie. "About 350 U.S. Airmen Remain as Somalia Mission Winds Down." *Air Force Times*, 19 April 1993, 22.

Bingham, Price T. "The Air Force's New Doctrine." *Military Review*, November 1992, 12–19.

———. "Air Interdiction and the Need for Doctrinal Change." *Strategic Review*, Fall 1992, 24–33.

Blake, Dan. "Airline Ills Affecting Aerospace Industry." *The Montgomery (Ala.) Advertiser,* 27 February 1993.

Bockman, Larry J., and Brad C. Hayes, "Breathing Life into the Naval Service's New Direction." *Marine Corps Gazette,* February 1993, 47–49.

Boneu, Z. *The Interaction between Technology and Tactics—Can the Process Be Improved?* Potomac, Md.: C&L Associates, 5 August 1988.

Borowski, Harry R. "Leadership to Match our Technology." *Air University Review,* May–June 1984, 30–34.

———, ed. *The Harmon Memorial Lectures in Military History, 1959–1987.* Washington, D.C.: Government Printing Office (GPO), 1988.

Brown, Harold. "Technology, Military Equipment, and National Security." *Parameters,* March 1983, 15–27.

Brune, Lester H. "Foreign Policy and the Air Power Dispute, 1919–1932." *The Historian: A Journal of History.* Vol. 23. August 1961, 449–64.

Burns, Robert. "Clinton Plans High-Tech Economic Bridge." *Montgomery (Ala.) Advertiser,* 11 May 1993.

Bush, Vannevar. *Modern Arms and Free Men.* New York: Simon and Schuster, 1947.

Cahn, Robert. "Dr. Theodore von Karman Gemutlicher Genius of Aeronautics." *Air Force Magazine,* October 1957, 41–48.

Canaan, James W. "A Watershed in Space." *Air Force Magazine,* August 1991, 32–37.

———. "Space Support for the Shooting Wars." *Air Force Magazine,* April 1993, 30–34.

———. "The Instrument of Airpower." *Air Force Magazine,* April 1993, 12–13 and 15.

Cannon, Michael W. "Just Meeting a Requirement?" *Military Review,* August 1992, 63–69.

Cate, James L. "Development of Air Doctrine 1917–41." *Air University Review Quarterly,* Winter 1947, 11–22.

Cheney, Dick. *Annual Report to the President and the Congress.* Washington, D.C.: GPO, January 1993.

———. *Defense Strategy for the 1990s: The Regional Defense Strategy.* Washington, D.C.: Department of Defense, 1993.

Chung, James A. "Whither the U.S. Aerospace Industry?" *Breakthroughs*, Winter 1992–93, 12–18.

Clay, Lucius D., Jr. "Shaping the Air Force Contribution to National Strategy." *Air University Review*, March–April 1970, 3–9.

Clayton, Anthony. *The British Empire as a Superpower, 1919–1939*. Athens, Ga.: University of Georgia Press, 1986.

Cogar, William B., ed. *New Interpretations in Naval History, Selected Papers from the Eighth History Symposium*. Annapolis, Md.: Naval Institute Press, 1989.

Collins, John M. *Grand Strategy Principles and Practices*. Annapolis, Md.: Naval Institute Press, 1973.

———. *U.S. and Allied Options Early in the Persian Gulf Crisis*. Washington, D.C.: Congressional Research Service, 1990.

Congressional Research Service. *World-wide Space Activities*. Prepared for the Subcommittee on Space Science and Applications of the Committee on Science and Technology, US House of Representatives, 97th Congress. Washington, D.C.: GPO, 1977.

Conver, Stephen K. "Defense Industrial Base: Shaping the Downsizing." *Armed Forces Journal International*, March 1993, 48–52.

Corn, Joseph J. *The Winged Gospel: America's Romance with Aviation*. New York: Oxford University Press, 1983.

Correll, John T. "One Air Force." *Air Force Magazine*, April 1993, 2.

Crutsinger, Martin. "U. S. Trade Deficit Improves Dramatically as Exports Rise." *Montgomery (Ala.) Advertiser*, 18 December 1992.

"DASA and Fokker Conclude Buyout Pact." *Aviation Week & Space Technology*, 22 February 1993, 27.

Davis, Bob, and Bruce Ingersoll. "Aiding Aviation Firms, As Clinton Vows to Do, Will Be Very Tricky." *Wall Street Journal*, 8 March 1993.

Dentzer, Susan. "Clinton's High-Tech High-Wire Act." *U.S. News & World Report*, 29 March 1993, 44.

de Seversky, Alexander P. "Air Power in the Modern World." *New York Times*, 27 December 1942.

———. *Victory Through Airpower*. New York: Simon and Schuster, 1942.

Diehl, Alan E. "Create a U.S. Aerospace Service." *Aviation Week & Space Technology*, 25 January 1993, 71.

Dornheim, Michael A. "Cuts, Layoffs Affirm Transport Boom's End." *Aviation Week & Space Technology*, 1 February 1943, 22–24.

Douhet, Giulio. *The Command of the Air* (translated by Dino Ferrari). Reprint ed. Washington, D. C.: Office of Air Force History, 1983.

Drew, Dennis M. "Desert Storm as a Symbol: Implications of the Air War in the Desert." *Airpower Journal* 6, no. 3, Fall 1992, 4–13.

———. "The Airpower Imperative: Hard Truths for an Uncertain World." *Strategic Review*, Spring 1991, 24–31.

Dugan, Michael J. "Operational Experience and Future Applications of Air Power." *RUSI Journal*, August 1992, 35–38.

Earle, Edward M. "The Influence of Air Power Upon History." *Yale Review*, June 1946, 577–93.

———, ed. *Makers of Modern Strategy*. 2d ed. Princeton, N.J.: Princeton University Press, 1973.

Emme, Eugene M., ed. *The Impact of Air Power*. Princeton, N.J.: Van Nostrand, 1975.

———. "The NASA Space Program." Lecture. Air War College, Maxwell Air Force Base, Ala., 1 February 1963.

———. "The Relationship of Air Power to National Policy." Lecture. Air War College, Maxwell AFB, Ala.: 11 September 1958.

———. "Technical Change and Western Military Thought." *Military Affairs*, Vol. 24, 6–19.

Fawcett, John M. "Which Way to the FEBA?" *Airpower Journal*, Fall 1992, 14–24.

Federal Maritime Commission. *30th Annual Report Fiscal Year 1991*. Washington, D.C.: Federal Maritime Commission, 1992.

Fiedler, David M. "AirLand Operations, The Army's Answer to a Changing World." *National Guard*, February 1992, 26–29.

Flint, Perry. "Partners Preferred." *Air Transport World,* February 1993, 59, 60, and 62.

Freeman, Waldo D., Randall J. Hess, and Manual Faria. "The Challenges of Combined Operation." *Military Review,* November 1992, 2–11.

Friedman, George, and Meredith Lebard. *The Coming War with Japan.* New York: Saint Martin's Press, 1991.

Friedman, Norman. "The Future Shape of the U.S. Navy." *Annals of the American Academy of Political and Social Science,* September 1991, 106–19.

Fulghum, David A. "ALCMs Given Nonlethal Role." *Aviation Week & Space Technology,* 22 February 1993, 22–23.

"Full Naval Review." *Aviation Week & Space Technology,* 27 February 1989, 9.

Futrell, Robert F. *Ideas, Concepts, Doctrine: Basic Thinking in the United States Air Force 1907–1960.* Vol. 1. Maxwell AFB, Ala.: Air University Press, 1989.

———. *Ideas, Concepts, Doctrine: Basic Thinking in the United States Air Force 1961–1984.* Vol. 2. Maxwell AFB, Ala.: Air University Press, 1989.

Gilpin, Robert. *War and Change in World Politics.* Cambridge: Cambridge University Press, 1981.

Glosson, Buster C. "Aerospace Power in Modern Warfare: After the Storm." Unpublished paper. Washington, D.C.: US Air Force, December 1992.

———. "Military Capabilities for the New Covenant the Aerospace Contribution." Unpublished paper. Washington, D.C.: US Air Force, December 1992.

Goodman, Glenn W., Jr. "Powell's Roles and Missions Report Retains Services' Major Redundancies." *Armed Forces Journal International,* March 1993, 10.

Graydon Michael. "Roles and Changing Priorities." *The RUSI Journal,* August 1992, 28–34.

Greeley, Brendan M., Jr. "Carrier Air Wings Trained for Coordinated Strikes." *Aviation Week & Space Technology,* 27 February 1989, 46–47.

Green, Brian. "Down Another $126.9 Billion." *Air Force Magazine,* April 1993, 9.

Guinn, Gilbert S. "A Different Frontier: Aviation, the Army Air Forces, and the Evolution of the Sunshine Belt." *Aerospace Historian,* March 1982, 34–45.

Gunzinger, Mark A. *Power Projection: Making the Tough Choices.* Maxwell AFB, Ala.: Air University Press, June 1992.

Hampton, Ephraim M. "Air Power, Global Force in a Global Struggle." *Air University Quarterly Review,* Spring 1955, 68–77.

Hayden, H. T., and G. I. Wilson. "Defining the Corps' 'Strategic Concept.'" *Marine Corps Gazette,* May 1992, 44–46.

Healy, John P. "Air Power and Foreign Policy." *Air University Quarterly Review,* Fall 1948, 15–26.

Henderson, Danna K. "1993: Hesitant Optimism, Hopes for Joint Efforts." *Air Transport World,* January 1993, 22–26, 30, 32–34, 36–38, and 41.

Hittle, James D. "Roles and Missions Fight Is Endless." *Marine Corps Gazette,* December 1992, 21–22.

Howard, Michael. "The Forgotten Dimensions of Strategy." *Foreign Affairs,* Summer 1979, 975–86.

Hughes, David. "Videoconferencing May Cut Air Travel." *Aviation Week & Space Technology,* 8 February 1993, 31 and 33.

Hurley, Alfred F., and Robert C. Ehrhart., eds. *Air Power and Warfare.* Washington, D.C.: GPO, 1979.

Impoco, Jim, and David Hage. "White House Workout." *U.S. News & World Report,* 8 March 1993, 28–30.

Jackson, Brendan. "Air Power." *RUSI Journal,* August 1992, 27.

Jacquemin, Alexis. *The New Industrial Organization: Market Forces and Strategic Behavior* (Translated by Fatemah Mehta). Cambridge, Mass.: MIT Press, 1987.

Jones, John F., Jr. "Giulio Douhet Vindicated Desert Storm 1991." *Naval War College Review,* Autumn 1992, 97–101.

Jones, Neville. *The Beginnings of Strategic Air Power, A History of the British Bomber Force 1923–39.* London: Frank Cass & Co., 1987.

Kaufman, Robert G. "A Paradigm for a Post–Postwar Order." *Naval War College Review,* Winter 1991, 83–97.

Kaufmann, William W., and John D. Steinbruner. *Decisions for Defense: Prospects for a New World Order*, Washington, D.C.: Brookings Institution, 1991.

Kelso, Frank B., II. "Looking Ahead—Where We Go From Here." *Sea Power*, April 1992, 31–39.

Kendall, Frank. "Exploiting the Military Technical Revolution: A Concept for Joint Warfare." *Strategic Review*, Spring 1992, 23–30.

Kennedy, Paul M. *The Rise and Fall of British Naval Mastery*. Malabar, Fla.: Robert E. Krueger Publishing Co., 1982.

———. *The Rise and Fall of the Great Powers: Economic Change and Military Conflict from 1500 to 2000*. New York: Random House, 1887.

Kutyna, Donald J. "The State of Space." Prepared statement of General Kutyna, USAF, commander in chief, US Space Command, to the Senate Armed Services Committee, 23 April 1991. In *Defense Times* 6, no. 14, 1–8.

Lancaster, John. "Military Reshaping Plan Is Short of Clinton Goals." *Washington Post*, 13 February 1993, A4.

Layton, P. B. *The Strategic Application of Air Power in the New World Order*. Air Power Studies Center Paper no. 9. Fairbairn Base, Australia: Royal Australian Air Force, January 1993.

Lee, Joseph W. "Boost U.S. Lab's Role in Technology Transfer." *Aviation Week & Space Technology*, 7 December 1992, 59.

Lenorovitz, Jeffrey M. "USAIR Opposes Delays in BA Investment Plan." *Aviation Week & Space Technology*, 15 February 1993, 32.

Linn, Thomas C. "Naval Forces in the Post–Cold War Era." *Strategic Review*, Fall 1992, 18–23.

Loh, John M. "Advocating Mission Needs in Tomorrow's World." *Airpower Journal*, Spring 1992, 4–13.

Lopez, Virginia C., and David H. Vadas. *The U.S. Aerospace Industry in the 1990s: A Global Perspective*. Washington, D.C.: Aerospace Industries Association of America, 1991.

Luttwak, Edward N. *Strategy: The Logic of War and Peace*. Cambridge, Mass.: Harvard University Press, 1987.

Mahan, Alfred T. *Retrospect and Prospect: Studies in International Relations, Naval and Political, 1902*. London: Sampson Low, Marston and Co., 1902.

———. *The Influence of Sea Power upon History 1660–1783*. New York: Dover Publications, 1987.

———. *The Problem of Asia and the Effect upon International Policies*. Boston: Little, Brown & Co., 1905.

Mahnken, Thomas G. "Planning U.S. Forces for the Twenty-First Century." *Strategic Review*, Fall 1992, 9–17.

"Making Elephants Fly." *Economist*, 23 January 1993, 77–78.

"Marine Training Prepares Crews for Combined Operation." *Aviation Week & Space Technology*, 27 February 1989, 48–49.

"Maritime Priorities Shifting." *Officer*, January 1993, 10–13.

Markusen, Ann, and Joel Yudken. "Building a New Economic Order." *Technology Review*, April 1992, 23–28, and 30.

———. "The Birthing of Aerospace." *Technology Review*, April 1992, 29.

Mattson, Roy M. *Projecting American Air Power: Should We Buy Bombers, Carriers, or Fighters?* Maxwell AFB, Ala.: Air University Press, June 1992.

Maze, Rick. "Aspin Warns of Dangers Ahead for U.S. Military." *Air Force Times*, 13 July 1992, 10.

———. "Nunn Urges Overhauling of Military Roles, Missions." *Air Force Times*, 2 July 1992, 3.

McCarthy, James P. "New Directions in U.S. Military Strategy." *Parameters*, Spring 1992, 2–10.

McKenna, James T. "United Cuts 20% of Senior Officers." *Aviation Week & Space Technology*, 22 February 1993, 20–22.

McLucas, John L. *Space Commerce*. Cambridge, Mass.: Harvard University Press, 1991.

McNeill, William H. *The Pursuit of Power: Technology, Armed Force, and Society*. Chicago: University of Chicago, 1982.

McPeak, Merrill A. "Does the Air Force Have a Mission." Lecture. Maxwell AFB, Ala. 19 June 1992.

Mehuron, Tamar A. "Starting Points for the New Defense Budget." *Air Force Magazine*, April 1993, 10–11.

Meilinger, Phillip S. "The Problem with Our Air Power Doctrine." *Airpower Journal*, Spring 1992, 24–31.

Mitchell, William. *Winged Defense*. New York: Dover Publications, 1988.

Moore, R. Scott. "The Army Plans Its Future." *Marine Corps Gazette*, January 1992, 48–49.

Morocco, John D. "Defense Cuts Made in Policy Void." *Aviation Week & Space Technology*, 15 February 1993, 20–21.

———. "JCS to Backtrack Decisions in Roles, Missions Study." *Aviation Week & Space Technology*, 18 January 1993, 58–59.

———. "Pentagon Faces $88-Billion Cut." *Aviation Week & Space Technology*, 22 February 1993, 22–23.

———. "Somalia to Impact Debate on Reshaping U.S. Forces." *Aviation Week & Space Technology*, 14–21 December 1992, 23–24.

———. "Shifting Economic, Political Tides Force Reevaluation of Navy's Strategic Role." *Aviation Week & Space Technology*, 27 February 1989, 38–39 and 44–45.

Mundy, Carl E., Jr. "Expeditionary Forces: A Defining Concept for the Future." *Sea Power*, April 1992, 43–44, 48, 50, and 52.

———. "Marines Continue to be America's 'Force-in-Readiness.'" *The Officer*, August 1992, 40–42.

———. "Naval Expeditionary Forces and Power Projection: Into the 21st Century." *Marine Corps Gazette*, January 1992, 14–17.

———. "Naval Expeditionary Forces: Stepping Lightly." *Marine Corps Gazette*, February 1993, 14–15.

———. "Redefining the Marine Corps' Strategic Concept." Naval Institute *Proceedings*, May 1992, 66–70.

National Space Council. *National Space Council Report to the President*. Washington, D.C., May 1992.

"Now for the Really Big One." *Economist*, 9 January 1993, 57–58.

Nunn, Sam. "Roles, Missions Under Scrutiny." *The Officer*, February 1993, 20–24.

Ochmanek, David, and John Bordeaux. "Comparing Airpower Assets for Power Projection," RAND Draft Report DRR-166-AF. Santa Monica, Calif.: RAND, January 1993.

O'Connell, Edward. "A Look into Air Campaign Planning." Unpublished research paper, Washington, D.C.: US Defense Intelligence College, 1992.

O'Connell, Robert L. *Sacred Vessels, the Cult of the Battleship and the Rise of the U.S. Navy.* Boulder, Colo.: Westview Press, 1991.

Ott, James. "USAIR/BA Alliance Poses Global Issues." *Aviation Week & Space Technology,* 8 February 1993, 28–29.

———. "U.S. Airlines Await Clinton Directions." *Aviation Week & Space Technology,* 25 January 1993, 34–35.

Paret, Peter, ed. *Makers of Modern Strategy from Machiavelli to the Nuclear Age.* Princeton, N.J.: Princeton University Press, 1986.

Perin, David A. *A Comparison of Long-Range Bombers and Naval Forces.* Alexandria, Va.: Center for Naval Analyses, December 1991.

Perry, William. "Fallow's Fallacies." *International Security,* Spring 1982, 174–82.

Porter, Michael E., ed. *Competition in Global Industries.* Boston: Harvard Business School Press, 1986.

Powell, Colin L. *Chairman of the Joint Chiefs of Staff Report on Roles, Missions, and Functions of the Armed Forces of the United States.* Washington, D.C.: GPO, 1993.

———. "Information-Age Warriors." *Byte,* July 1992, 370.

———. "U.S. Forces: Challenges Ahead." *Foreign Affairs,* Winter 1992–93, 32–45.

Racz, Gregory N. "Airline Strength Is Held 'Critical' for Economy." *Wall Street Journal,* 25 May 1993.

Reich, Robert B. *The Work of Nations.* New York: Vintage Books, 1992.

Reinhardt, George C. "Air Power Needs Its Mahan." US Naval Institute *Proceedings,* April 1952, 362–67.

"Reinventing Boeing: Radical Changes Amid Crisis." *Business Week,* 1 March 1993, 60–62 and 67.

"Remarks of Gen Carl E. Mundy, Jr., before Congress." *Marine Corps Gazette,* April 1992, 34–37.

Rice, Donald B. "AF's Future Strategy: 'Punch Hard and Terminate Quickly.'" *Air Force Times*, 26 March 1990, 23.
———. "Airpower in the New Security Environment." Lecture, Washington Strategy Seminar, US Congress, Washington, D.C., 7 May 1991.
———. "The Air Force in Transition." Lecture, Air Force Association National Convention, Washington, D.C., 15 September 1992.
Roucek, Joseph S. "Geopolitics and Air Power." *Air University Quarterly Review*, Spring 1955, 52–63.
Scott, William B. "Caution Urged On DARPA Changes." *Aviation Week & Space Technology*, 1 February 1993, 28.
"Selective Realism." *Economist*, 9 January 1993, 59.
Serling, Robert J. *Legend and Legacy, The Story of Boeing and Its People*. New York: Saint Martin's Press, 1992.
Shifrin, Carole A. "US Air/BA Pact Faces New Fight." *Aviation Week & Space Technology*, 1 February 1993, 29.
Siegel, Adam B. *U.S. Navy Crisis Response Activity, 1946–1989: Preliminary Report*. Alexandria, Va.: Center for Naval Analyses, November 1989.
"Slimming the General." *Economist*, 16 January 1993, 63–64.
Smith, Bruce A. "Commercial Strategy Developed for Douglas." *Aviation Week & Space Technology*, 22 February 1993, 25–27.
"Spacecraft Played Vital Role in Gulf War Victory." *Aviation Week & Space Technology*, 22 April 1991, 91.
Stanton, Shelby L. *Anatomy of a Division: The First Cav in Vietnam*. New York: Warner Books, 1987.
Sullivan, Gordon R. "Doctrine: A Guide to the Future." *Military Review*, February 1992, 3–9.
———. "'Vital, Capable and Engaged.'" *Army*, October 1992, 24–26, 28, 30, 32, and 34.
Sweetman, Bill. "Why Not Just Satellites?" *International Defense Review*, February 1993, 107.
Thurow, Lester. *Head to Head, The Coming Economic Battle Among Japan, Europe, and America*. New York: William Morrow and Co., 1992.
———. "New Rules for Playing the Game." *Phi Kappa Phi Journal*, Fall 1992, 10–13.

Toguchi, Robert M., and James Hogue. "The Battle of Convergence in Four Dimensions." *Military Review*, October 1992, 11–20.

Toti, William J. "Sea-Air-Land Battle Doctrine." *Proceedings*, September 1992, 70–74.

Trenchard, Hugh M. *Airpower: Three Papers*. London: Directorate of Staff Duties, Air Ministry, December 1943.

"Trends." *Air Transport World*, October 1992, 1.

———. *Air Transport World*, January 1993, 1.

Tyson, Laura D'Andrea. *Who's Bashing Whom? Trade Conflict in High-Technology Industries*. Washington, D.C.: Institute for International Economics, 1992.

US Air Force. *Global Reach—Global Power*. White Paper. Washington, D.C.: US Air Force, November 1992.

———. *Reaching Globally, Reaching Powerfully: The United States Air Force in the Gulf War*. Washington, D.C.: Department of the Air Force, September 1991.

———. *The Air Force and U.S. National Security: Global Reach—Global Power*. Washington, D.C.: Department of the Air Force, June 1990.

———. *The Evolving Air Force Contribution to National Security: Global Reach—Global Power*. White Paper. Washington, D.C.: Department of the Air Force, November 1992.

———. *The United States Air Force and U.S. National Security: A Historical Perspective 1947–1990*. Washington, D.C.: Secretary of the Air Force, 1991.

US Air Force Headquarters, Strategy Division. *Aerospace Power in Modern Warfare*. Briefing. Washington, D.C.: US Air Force Headquarters, Strategy Division, 1992.

US Atlantic Command and Air Combat Command. *Joint Force Air Component Commander (JFACC) Concept of Operations for the U.S. Atlantic Command and Air Combat Command*. Norfolk, Va., and Langley Air Force Base, Va.: US Atlantic Command and Air Combat Command, 18 September 1992.

US Atlantic Command and US Pacific Command. *Joint Force Air Component Commander (JFACC) Concept of Operations*. Norfolk, Va., and Honolulu, Hawaii: US Atlantic Command and US Pacific Command, 15 January 1993.

US Bureau of the Census. *Statistical Abstract of the United States: 1992,* 111th edition. Washington, D.C.: GPO, 1992.

US Department of the Navy. . . . *From the Sea - Preparing the Naval Service for the 21st Century.* Washington, D.C.: Department of the Navy, September 1992.

———. *Conduct of the Persian Gulf Conflict An Interim Report to Congress.* Washington, D.C.: Department of Defense, 1991.

———. *Retaining Competitive Advantage.* Briefing slides and notes. Washington, D.C.: Department of the Navy, n.d.

———. *The United States Navy in "Desert Shield" "Desert Storm."* Washington, D.C.: Department of the Navy, 1991.

US Department of Transportation. *MARAD '91* (Maritime Administration Report 1991 Annual Report). Washington, D.C.: Maritime Administration, 1992.

US General Accounting Office. *1993 Air Force Budget—Potential Reductions to Research, Development, Test, and Evaluation Programs.* GAO/NSIAD-92-319BR. Washington, D.C.: US General Accounting Office, September 1992.

US Space Command. *United States Space Command Operations Desert Shield and Desert Storm Assessment.* Peterson AFB, Colo., January 1992.

van Creveld, Martin. *Technology and War, From 2000 B.C. to the Present.* New York: Free Press, 1989.

Velocci, Anthony L., Jr. "French Firms In 'Crisis'; U.S. Policies Assailed." *Aviation Week & Space Technology,* 15 February 1993, 40–45.

———. "TWA Set to File New Business Plan." *Aviation Week & Space Technology,* 15 February 1993, 28.

Warden, John A., III. "Exploiting Air Power into the Twenty-first Century." Draft paper for Air Command and Staff College, Air University, 3 January 1993.

Weeks, Stan. "Crafting a New Maritime Strategy." *Naval Institute Proceedings,* January 1992, 30–37.

The White House. *National Security Strategy of the United States.* Washington, D.C.: GPO, 1993.

Winton, Harold R. "Reflections on the Air Force's New Manual." *Military Review,* November 1992, 20–31.

Woolsey, James P., Anthony Vandyk, Arthur Reed, and Danna K. Henderson. "CEOs: Cooperate and Profit." *Air Transport World,* January 1993, 43–45, 48 and 49.

Wylie, J. C. *Military Strategy: A General Theory of Power Control.* Annapolis, Md.: Naval Institute Press, 1989 (reprint).

Yang, Dori Jones, and Andrea Rothman. "Reinventing Boeing Radical Changes Amid Crisis." *Business Week,* 1 March 1993, 60–62 and 67.

Chapter 2

After the Gulf War: Balancing Space Power's Development

Frank Gallegos

> *It is a military axiom to "take the high ground"—and space is the ultimate high ground. In the Gulf War, US space forces were virtually unopposed, but in the future that may not be the case. . . . Without question, it was fortunate that there were six months to get ready. The next time, that luxury may not exist, and we must be prepared. . . . The first need is a key element—development of space doctrine to provide guidance and direction at all levels of war, across the full spectrum of conflict.*
>
> —Lt Col Steven J. Bruger

Early military applications of space-based assets bore little resemblance to their successful use in "the first information war."[1] The United States developed most of its early space systems to serve the cold war nuclear deterrence strategy. The need to protect space sources and methods resulted in a high degree of secrecy and organizational compartmentalization. As a result, when Operation Desert Shield began, the highly fragmented leadership of the space community lacked coherent doctrine, operated with an inherited top-down "technology push" for system requirements, and had little space power experience.[2]

Space power was simply unprepared to support the theater commander in chief (CINC) in other than the cold war strategic role.[3]

The experiences of the Gulf War confirmed these characteristics—the majority of the documented lessons concerned

This work was accomplished in partial fulfillment of the master's degree requirements of the School of Advanced Airpower Studies, Air University, Maxwell AFB, Ala., 1996.
Advisor: Col Dennis Drew (retired)
Reader: Dr James Corum

a lack of doctrine or a lack of space literacy or experience. In the development of space power, doctrine and experience have evolved much more slowly than the pace of technology. In the interim, have the US participants redressed the imbalance that existed in the development of space power as witnessed in Operation Desert Shield/Storm? At issue for space policy makers is the question of whether or not reforms in technology, experience, or doctrine will move the US military space program toward a more robust war-fighting capability.

From its meager beginnings in the Vietnam conflict, space power evolved dramatically. In Vietnam the military used space-based platforms primarily for weather forecasting, navigation assistance, and communications support. During Operation Urgent Fury in Grenada, US forces used the Fleet Satellite Communications (FLTSAT) and Leased Satellite Communications (LEASAT) Systems in a command and control role for the first time in a joint operation. Operation El Dorado Canyon in Libya and Operation Just Cause in Panama were the first major operations in which US forces used information from space-based national intelligence systems.[4] In addition, Operation El Dorado Canyon was the first operation in which a space system developed as a Tactical Exploitation of National Capabilities Program (TENCAP) project was used.[5]

United States war fighters were not able to use the full array of civil, military, commercial, and national intelligence satellites until the Gulf War. Space-based assets carried over 80 percent of all messages to and from the US Central Command's (USCENTCOM) area of responsibility (AOR). Satellite intelligence data was essential for planning the air campaign, critical for early warning of surface-to-surface missile system (Scud) ballistic missile attacks, and aided in determining enemy positions and activities.[6] For the first time in any military campaign, Global Positioning System (GPS) satellites provided precise position information essential for navigation over an almost featureless desert terrain. Arguably, space "came of age" for war fighters in the Gulf War, but the situation was far from perfect.

US Space Command (USSPACECOM) traced some of the most significant problems from the Gulf War to a core issue—normalizing space operations for theater operators.[7] For example, since very little basic and operational doctrine existed, space preplanning for wartime situations lagged well behind space technology. Because USCENTCOM had not articulated how space power ought to be used in its AOR and USSPACECOM was not fully prepared to provide "normalized" support, US military forces were largely uninformed and unprepared for using space power when Operation Desert Shield began. The normalization of space operations for theater operations was still not complete as of 1995. Space power doctrine and experience are still significantly lagging behind space technology. All three of these threads of development—technology, doctrine, and literacy/experience—are crucial, but the lack of balance is particularly important because it points to the focus of what should be the next phase of development in military space policy.

A definitive guide to the future focus of space power development requires sophisticated cost-effectiveness and operational analysis. However, it is possible to make a useful, qualitative analysis based on recent experience and general assumptions about the relative costs and leverage of reforms. Are funds better spent on acquiring technology, improving experience, or developing doctrine? Which solution offers more leverage for the future?

After the Gulf War, the Air Force, Army, and Navy moved quickly to provide better space power support to the war fighters. Senior Air Force leadership founded the space numbered Air Force (Fourteenth Air Force), activated the AF Space Warfare Center (SWC), and established space support teams (SST). Following the Air Force lead, the Army and Navy established their own space support teams. In general, USSPACECOM, all service components, and the national intelligence agencies attempted to provide better support to the combatant commands and more efficient preplanning of existing space forces.[8]

Fourteenth Air Force is now responsible for war planning, readiness, and execution while serving as the Air Force warfighting component to USSPACECOM.[9] The Air Force activated the SWC to refine doctrine, develop tactics, formulate concepts, and demonstrate systems and technologies that improve military operations and the employment of space forces in warfare. Finally, all service components, USSPACECOM, and intelligence organizations currently deploy space support teams to help conduct integrated space operations for the theater CINC.

In contrast to the significant reorganization of space forces, doctrinal changes were less dramatic. At the time of this writing, Air Force Doctrine Document (AFDD) 4, "Space Operations Doctrine" is still in coordination and may be approved in 1995. Arguably the most important doctrinal manual, Joint Doctrine, Tactics, Techniques, and Procedures (JDTTP) 3-14, *Space Operations*, was in coordination prior to the Gulf War and is still at least a year away from closure.[10] The space support teams mentioned above are available to deploy and support the war fighter; however, joint doctrine is still not available to guide their actions four years after the end of "the first information war."[11] Indeed doctrine lags, suggesting important near-term focus for policy. The thesis of this study is that a lack of space power doctrine and experience caused the majority of the space-related problems in the Gulf War. Further, while the space community has made efforts to normalize space operations since the war, the lack of doctrine and experience is still the major impediment to effective war fighting today and for future conflicts.

Focus

This study focuses on basic and operational Air Force and joint space doctrine which was available to the principal space participants (USCENTCOM and USSPACECOM) before, to, and during the Gulf War, including operation plans (OPLAN). Equally important, this study relies largely on the unclassified portions of the after action reports from these two unified commands, the Joint Chiefs of Staff Joint Universal Lessons Learned System (JULLS), the *Gulf War Airpower Survey (GWAPS)*, and the *Conduct of the Persian Gulf War: Final Report to Congress (CPGW)*. When possible, these documents were verified with primary sources.

Assumptions

The Gulf War validated the operational worth of space systems. Space-based communications, weather, navigation, reconnaissance, and intelligence offered the war fighter capabilities unparalleled in earlier conflicts. The Gulf War provided a glimpse of how space control in the next century could be as crucial as air and sea control have been in this century.

In the next century, space will contribute significantly to national economic, political, and security objectives. National, civil, and commercial space agencies have a need to develop space systems in a complementary, not competitive process. Within the Department of Defense (DOD), cooperation is essential so that the information received from space assets continues to benefit war fighters. Outside the DOD, trust, space power literacy, and cooperation are critical to ensure efficient use of all space systems. The impact of space power for the future makes the thesis of this study all the more important.

Methodology

This study uses an inductive examination of evidence to support the author's thesis. The following section illustrates the USCENTCOM and USSPACECOM space lessons from the Gulf War and generalizes these experiences into three threads of development: technology, experience, and doctrine. From that perspective, a description of the efforts to solve the problems from the war is offered. Subsequent to that, observations from this study lead naturally to future implications.

Establishing the Framework: Lessons from the Gulf War

History, whatever its value in educating judgment, teaches no 'lessons'. . . . Alternatively one might argue that a given conflict teaches many lessons: unfortunately, most of them are wrong.

—Sir Michael Howard

This section establishes a framework for analysis by organizing the lessons from after action reports, the *GWAPS*, the *CPGW: Final Report to Congress*, and other nonofficial works into three broad categories of space power development: technology, experience, and doctrine.[12] A lesson requiring the acquisition of new technology to resolve the issue is included in the technology thread. A lesson leading to or requiring the accumulation of new knowledge, literacy, skill, or reorientation is organized in the experience thread. For example, airpower strategists learned from World War II experience that the first requirement for nearly all military operations was air superiority. Finally, a problem indicating a lack of a codified, sanctioned body of propositions to guide how space power ought to be used is attributed to a lack of doctrine. For the purposes of this study, doctrine includes not only formal, published doctrine, but also directives, manuals, and other official published guidance.

These common threads of the development paradigm are not foolproof; they offer a simple framework for analysis and a point of departure for future investigations. Using this three-part framework, it quickly becomes obvious that the majority of the space power problems encountered during the Gulf War can be attributed to a lack of doctrine and experience. Unfortunately, the development of US space technology continues to outpace both doctrine and experience.

US Space Command After-Action Report

"Normalizing space support for the war fighters" is the common theme echoed by the authors of USSPACECOM's after-action report.[13] The writers of this report made an obvious effort to address the importance of establishing and updating detailed space annexes (annex N) in the war-fighting CINC's operation plans. Table 5 illustrates the lessons from the viewpoint of USSPACECOM and the corresponding category in the space power development process.

More preplanning is required; the supported CINC's OPLANs need work; and communication requirements should be included in OPLANs. Space annexes to OPLANs either did not exist or were underdeveloped before the Gulf War. Prior to

Table 5

USSPACECOM Lessons

Lesson	Category
More preplanning required—May not have six months of buildup for the next war.	Doctrine
Supported CINC OPLANs need work.	Doctrine
Include communication requirements in OPLANs.	Doctrine
Normalize all space support.	Doctrine and Experience
Normalize tactical warning support.	Experience and Technology
Operational control of military satellite communication systems remains fragmented.	Doctrine and Experience
Maintain the US multispectral imagery capability.	Experience

Source: USSPACECOM After Action Report, 31 January 1992.

Operation Desert Shield, US Central Command's OPLAN did not address how space power would be used in the AOR.[14] In remarks to the Eighth National Space Symposium in April 1992, Lt Gen Thomas S. Moorman Jr., the vice commander of Air Force Space Command during the Gulf War, confirmed this fact. He commented that if the US military learned anything from the Desert Storm example it was that preplanning is essential. "The best example of the lack of planning that we had is that General Horner went to war without a space annex—he did not have in his US Air Forces, Central Command (CENTAF) operations plan a space annex."[15] As a result of the lack of preplanning, weather vans, ground antennas, intelligence terminals, and other space-related ground equipment were omitted from the time-phased force and deployment list (TPFDL).[16] Inadequate preplanning is a theme common to all the reports analyzed for this study.

Forces should normalize all space support and tactical warning support. USSPACECOM did not fully realize or plan for the important role space power would play in missions other than strategic ones. By normalizing space support at the theater level,

USSPACECOM envisions operating its space systems as the Air Force operates its aircraft on a day-to-day basis. Through the documentation of these lessons, the authors not only highlighted the value of normalizing space support to the theater war fighter, they also ensured readers would understand the significance of theater ballistic missile warning for the future. Gen Charles A. Horner, who had the unique experience of being the joint forces air component commander during the Gulf War and CINC USSPACECOM after the war, declared that the number one lesson of the Gulf War was that the US must develop a ballistic missile defense system capable of directly supporting the requirements of deployed forces as well as North America.[17] Normalizing space operations mandates the development of doctrine so that forces may organize, train, and equip to prepare for future wars.

Operational control of military satellite communication systems remains fragmented. Participants experienced the frustrations caused by a lack of centralized control of space communication systems. While USCINCSPACE is given combatant command (COCOM) by the chairman of the Joint Chiefs of Staff, no formal relationship exists between USSPACECOM and the managers of the several military satellite communication systems.[18] The operational control of these satellite systems remains fragmented among the various space agencies, services, and commands. This experience highlights the need for a centralized satellite communication structure in peacetime and war.[19]

The United States must decide whether to maintain its only multispectral imagery (MSI) capability, the aging LANDSAT, or to continue to rely on other nations for MSI support.[20] MSI proved to be beneficial by providing US and Coalition forces the opportunity to better understand and react to changes in the battlefield terrain. It will also offer future war fighters the ability to rehearse their missions, determine optimum tactics, and identify major threat lanes or attack axes to more effectively exploit training and technology in combat.[21] Finally, if the US Commerce Department continues to control LANDSAT on a day-to-day basis, agreements must be maintained to allow for peacetime military training and wartime control.

While this lesson covers all three threads of the development process, experience is the core issue.

USCENTCOM After Action Report

The war fighter's perspective was somewhat different than USSPACECOM's perspective. US Central Command developed five hundred JULLS after the war.[22] While USSPACECOM emphasized normalizing space operations, the supported command accented the need for better doctrine, training, and support from the experts. Table 6 is a compilation of the USCENTCOM lessons and the corresponding thread of space power's development process. The lessons highlighted are not the only USCENTCOM lessons related to space operations; however, at the unclassified level they represent the vast majority of the space power problems discovered by USCENTCOM during the Gulf War.[23]

US forces need better preplanning for space support doctrine on the use of ground mobile force (GMF) terminals. After the war, USCENTCOM planners were acutely aware of how little useful space power doctrine existed. Space power doctrine was either nonexistent or inadequate for the Gulf War. Through innovation and ingenuity during the six-month buildup of Operation Desert Shield, many forces made space power work. However, a six-month buffer is a luxury the United States may not have in future conflicts.[24] In addition, as the Gulf War developed and grew, military forces needed more GMF satellite communication terminals than doctrine prescribed and the TPFDL provided. The VII and XVIII Corps experienced shortages as a result.[25]

USSPACECOM needs a liaison to CINCs. The Space Demonstration Program and National Military Intelligence Support Team (NMIST) are critical for timely battle damage assessment (BDA). These lessons provided the impetus for the postwar SST concept.[26] Based on the Gulf War, USCENTCOM planners realized they did not have the expertise to effectively use space power. Their solution was to import the knowledge from the different space sectors for peacetime exercises and to continue having experts provide operational demonstrations of the capabilities provided by space power.[27]

Table 6

USCENTCOM Lessons

JULL	Category
Better preplanning required for effective space support.	Doctrine
Doctrine required on the use of ground mobile force terminals.	Doctrine
USSPACECOM liaison to CINCs required.	Experience
Space Demonstration Program.	Experience
NMIST critical for timely battle damage assessment.	Experience
Centralized control of theater communications must be exercised.	Experience
Space launch responsiveness.	Technology

Source: USCENTCOM After Action Report, 15 July 1991.

Forces need centralized control of communications. Because of the many sectors involved with satellite communications, initial control was, at best, fragmented.[28] Early in Operation Desert Shield, US Central Command assumed control of the validation process for all long-haul strategic communications. Without centralized control, early deploying units might have used all available resources before hostilities began.[29] Unity of command in allocating the limited resources, satellite capacity, and frequency spectrum, in particular, was vital to subsequent unit deployments.[30] The Gulf War validated the importance of exercising centralized control of theater communications.

USSPACECOM did not have a booster to meet a CENTAF request to accelerate the launch of the next Defense Satellite Communications System (DSCS) satellite.[31] The DSCS satellite would have improved USCENTCOM's overly taxed communications capability significantly. The inability of the United States to launch satellites in a short period of time is a serious weakness.

Gulf War Airpower Survey

The *GWAPS* authors focused on describing the "space product" and its operational impact. Even though the classified space power research by the GWAPS personnel is much more detailed, the unclassified report used here tells a story consistent with that of the classified reports. This unclassified report addressed five central themes.

Planning and Training for the Use of Space Systems. In the areas where space capabilities were not fully integrated with doctrine and tactics (e.g., BDA and other intelligence functions), the importance of the five and one-half months of Desert Shield preparation cannot be overemphasized.[32] While some annexes to USCENTCOM's Operation Plan 1002 were ample, weaknesses or omissions in other areas were inadequated for training or real-world events.

In the cases where adequate doctrine existed, space power was used effectively. In cases where doctrine did not exist or was inadequate, the results of space operations reflected the absence of in-depth preplanning.[33]

Space Mobilization. The time to mobilize space power varied across the board. In some cases, the equipment was immediately available due to peacetime requirements (e.g., F-16s equipped with GPS receivers). In other cases, the time to mobilize depended on preplanning, launch variables, and the availability of trained personnel.[34] If any one of these variables was deficient, there was a corresponding deficiency in mobilization.

Military Utility Space Systems. The contribution of space power was evident in terms of concrete war-fighting results. In some cases, however, desired results could only be achieved by crossing functional boundaries. For example, the detection of Scuds by the Defense Support Program (DSP) constellation required action from several of the Coalition forces to destroy these mobile targets. The lesson here is that doctrine must provide the flexibility to cross functional boundaries.

Command and Control of Space Systems. The highly classified, strategic focus of the US military space community was not suitable for the tactical environment of the Gulf War. The cold war mentality of the space community oriented its support to strategic customers prior to the war (e.g., National Command Authorities [NCA] and various intelligence agencies). Complicating this predicament, many of the key intelligence-related assets were not controlled by the war-fighting commander.[35]

After Operation Desert Storm, the space community realized wars in the future will likely require theater-level support from space forces. This lesson also implies that centralized control of space systems by the war-fighting commander is preferred over other arrangements.

The Role of Commercial Space Systems and Receiver Equipment. Commercial space systems played a significant role augmenting the military Coalition forces. In addition, the Coalition members cooperated to deny Iraq access to satellite imagery from France's commercial *Systeme Probataire pour l'observation de la Terre* (SPOT).[36] Military forces not only experienced the value of using commercial satellite systems, they now better understand the value of denying the enemy's use of commercial satellite systems.

Conduct of the Persian Gulf War: Final Report to Congress

As expected, the writers of the *CPGW* described the lessons and observations from the war in a much broader context than the sources previously cited.[37] They were also more interested in describing weapons and technology than operational concepts. Table 7 illustrates the space-related shortcomings and issues from volume II, appendix K, of the report.

The United States does not have a reactive space-launch capability. This observation is a common theme addressed by the majority of the studies referenced for this chapter. US space launch, responsive or otherwise, continues to be a national problem.

Tactical warning capabilities must be improved. While USSPACECOM emphasized the lack of experience and the need for doctrine in this area, the writers of the CPGW illustrated the need for improved technology to solve the tactical ballistic missile warning problem. Specifically, they believe that in the future, an improved sensor to replace the DSP is appropriate.[38]

Table 7

Persian Gulf War Space Power Shortcomings and Issues

Shortcoming/Issue	Category
The United States does not have a reactive space-launch capability.	Technology
Tactical warning capabilities must be improved.	Technology
GPS and most satellite communication (SATCOM) are vulnerable to exploitation.	Experience
The aging LANDSAT system under Commerce Department control must be replaced.	Experience and Technology
DSCS connectivity remained fragile due to age and condition of satellites and ground stations.	Experience and Technology
For future operations, planners must consider the challenges of operating within another nation's command, control, communications (C^3) infrastructure.	Doctrine and Technology
Military doctrine and training must institutionalize space-based support to operational and tactical commanders and incorporate it into operational plans.	Doctrine

Source: CPGW Final Report to Congress, Vol. 2, April 1992.

GPS and most satellite communications are vulnerable to exploitation. The Gulf War confirmed the need for the production, distribution, and integration of GPS receivers incorporating selective availability decryption. The Gulf War experience also proved the value of fielding the Military Strategic and Tactical Relay (MILSTAR) satellite system and installing anti-jam modems for super high frequency (SHF) fixed-base satellite terminals and tactical ground mobile terminals.[39]

The aging LANDSAT system under Commerce Department control must be replaced. The writers of the *CPGW* and USSPACECOM's after action report agree on this issue. The Gulf War experience validated the importance of maintaining an MSI capability available for military use.

DSCS connectivity remained fragile due to age and condition of satellites and ground stations. In the opinion of these authors, the older DSCS satellites and ground terminals require modernization. The experience from the war warrants an increase in the number of military satellites providing worldwide command and control coverage. In addition, procurement of smaller more mobile ground terminals, similar to a prototype used by the XVIII Airborne Corps, is needed to aid in transport to and within the theater.[40]

For future operations, planners must consider the challenges of operating within another nation's C^3 infrastructure, and military doctrine and training must represent institutionalized space-based support to operational and tactical commanders and be incorporated into operational plans. The last two issues from the *CPGW* are similar to previous lessons from USSPACECOM and the *GWAPS*.

Status of the Lessons

USSPACECOM and US Central Command are the only two sources discussed with any type of formal approach to tracking the lessons of the Gulf War. However, either through omission or by design, none of the space power lessons from the Gulf War are actively monitored by either of the unified commands today.[41]

After the Gulf War, USSPACECOM initiated action on many issues attributed to the Gulf War, even though they did not actively monitor the status of any of their lessons through a formal process. While issues such as space support teams and better OPLANs received considerable attention and each lesson was assigned a point of contact (POC), no agency was assigned the responsibility for resolving the fate of those lessons. Because of this, it is difficult to determine with confidence which Gulf War experiences USSPACECOM considered lessons for the future and which experiences were discarded after some scrutiny. Without question the USSPACECOM lessons did receive some level of hearing immediately after the war. USSPACECOM initially disseminated 97 copies of its report to 13 agencies including all war-fighting CINCs.[42] While there was wide distribution of the lessons, the point is that no mechanism existed to either discard a lesson as an anomaly, develop a solution, or elevate the problem to the Joint Chiefs of Staff for resolution.

In contrast, US Central Command inserted its lessons from the war into the JULLS. This process required the command to evaluate the five hundred lessons from the war and recommend what action should be taken for each. The recommendations ranged from designation as a noted item to flagging a lesson as a remedial action project (RAP) requiring periodic monitoring until resolved.[43] However, after the space power lessons were routed through the JULLS process, none were designated remedial action projects.[44] This does not mean the space-related lessons were not considered important, only that other processes or programs may already incorporate a solution to those problems. The lessons from USCENTCOM received much wider dissemination due to their inclusion in the JULLS database. While neither of the principal unified commands during the Gulf War currently monitors its respective lessons for resolution, USCENTCOM's lessons were adjudicated through a formal process.

Synthesis of the Lessons

In the development of space power, it is apparent from the studies examined that technology continues to surpass the progress of doctrine and experience. Arguably, the majority of lessons examined here were related to a lack of doctrine or a lack of experience (80 percent). The imbalance between space technology, doctrine, and experience is not a new phenomenon, but it is commonly overlooked.

Gen Charles A. Horner synthesized the most important space power problems from his unique perspective as the joint force air component commander during the war and as commander in chief of USSPACECOM after the war. The first major problem he noted was the lack of experience US forces had in using space assets, especially with respect to intelligence systems.[45] US forces simply were not familiar with using information obtained from satellite constellations like the DSP and GPS. The second significant problem General Horner noted was the overclassification of space information.[46] The classification of satellite products initially undermined the relationship between the United States and the Coalition forces and was a major impediment in getting information to the war fighters. In General Horner's opinion, the way to resolve these problems is to shed the cold war strategic heritage of space and to tear down the walls of classification the space intelligence community has built around itself.[47]

In a separate work, Mackubin Thomas Owens reviewed a number of Gulf War studies and distilled all of the lessons to three principles. "On first examination, these principles might seem so broad as to be trivial. Yet our lack of success in Vietnam demonstrates that we have not always paid as much attention to these principles as we should have. These lessons can be summarized as follows: people and organization matter; technology matters; and ideas (doctrine) matter."[48]

Technology, experience, and doctrine do matter. To maximize the potential of space power for future conflicts, it is evident from the material presented here that the United States needs to reassess the level of effort placed in developing space power doctrine and experience. Unfortunately, the inclination to be on the leading edge of technology often comes with a mutually strong penchant to disregard the teachings of the past.[49] The next section describes the efforts made since the war to improve these three developmental threads.

After the Gulf War—Uneven Improvement

The Air Force has a well understood, war-tested military doctrine for air power. The crux of the problem is Air Force insistence that the same doctrine applies to space.

—Kenneth A. Myers

It seems that the majority of the space power problems encountered during the Gulf War resulted from a lack of space power doctrine and experience. Since the Gulf War, the development of space power remains uneven—doctrine and experience continue to trail behind technology. While the search for superior systems is required, until space doctrine is on an even plane with the emerging technology, the employment of space power will not be optimized.

Space operation plans have improved; however, joint space doctrine remains unpublished. For example, while various SSTs are training regularly with war fighters, no joint doctrine exists to guide them on command relationships or how the space portion of the next war ought to be waged. Finally, new organizations designed to educate, train, and support the war fighters are making headway to normalize space operations. The US military is making progress in all three threads of space power development, but at uneven rates of advance, with technology clearly in the lead—a circumstance due in part to the legacy of space power.

Space Power's Legacy

The genesis of the American military space community's focus on research and development (R&D), vice operational support, began in response to the Soviet launch of sputnik in 1957. Following this event, the United States quickly became the world's leader in space power. However, the United States linked most military space development to support cold war nuclear deterrent strategies. High strategic stakes caused tight security and aggressive technological development. Space became a highly classified technology-oriented operation, characterized by restricted access to information about satellite capabilities that created impediments to supporting political and economic leadership in the United States.[50] This approach may have been appropriate for the cold war; however, Operation Desert Storm and a different world environment indicated a change was in order. Changing this mentality has not come easily, nor is the process close to completion. In a major study after the Gulf War, commonly referred to as "The Wilkening Report," distinguished authors advised Dan Quayle, then the vice president, of this reality.[51] They warned that the cold war security requirements continued to contribute to the inefficiencies in the conduct of the nation's space program.[52] The origin of space power in the

United States established a pattern of development that has proven difficult to overcome.

The experience of space operators has also varied. In the early years, many aviators with extensive flying experience in World War II and Korea were the core space operators. This changed in the mid-1960s when the requirements of the Vietnam War stripped the space community of its flyers and hence its operational focus.[53] Since then, the highly classified space program developed the reputation for breeding a R&D vice operational mentality that has been difficult to overcome.

The Gulf War was a turning point in revitalizing the operational focus for space power. In addition, to infuse more operational thinking into the space community, the Air Force merged intercontinental ballistic missile (ICBM) operators into Air Force Space Command.[54] Although considerable effort has gone into overcoming the R&D heritage of the United States space community, the transformation is incomplete.

What Lessons Apply to the Future?

Before examining where senior military space leadership focused development efforts after the Gulf War, it is important to determine if the pursuit of a resolution is worthwhile. Pertinent to this question is the well-known analysis of World War I airpower "lessons" developed by I. B. Holley Jr. "These lessons are much the same as those which might have been derived equally well from the Civil War or, for that matter, from any other war. As was true of former conflicts, World War I emphasized the necessity for a conscious recognition of the need for both superior weapons and doctrines to ensure maximum exploitation of their full potential."[55] In other words, wherever military leaders fail to emphasize the need for better weapons in lieu of more weapons, they usually suffer serious disadvantage. When military leaders fail to formulate doctrine to exploit innovative weapons, they suffer further disadvantages.[56] In terms of technological development, the analysis thus far highlights the need for space power leadership to develop a responsive launch capability for the United States, ensure war fighters retain the ability to acquire MSI, and develop a new system to provide theater ballistic missile warning. But equally important, this analysis suggests senior leadership should develop forward-looking space power doctrine to guide and educate war fighters.

In an era when space power is envisioned to perform many new missions with very limited resources, Dr. Holley's advice rings true. If the majority of the problems related to space power in the Gulf War fall into the categories of experience and doctrine, military leaders should be making every effort to formulate military doctrine to match the innovative space weapons. New doctrine will not only provide a direction for waging the next war, it can be used to train and educate war fighters on the applications space power can provide. Failing this, the nation may repeat the regretful pattern of the air weapon after World War I, recklessly groping forward with each technological innovation.[57] The salient question is, have US military leaders apportioned space power development efforts appropriately among technology, experience, and doctrine since the Gulf War?

Technology

Space power leadership is aggressively seeking resolution to the technological problems encountered in the Gulf War. In general, the senior leadership continues to expand R&D of new space technologies. For example, funding for TENCAP, which contains the major classified and unclassified Air Force technology projects, has increased by an order of magnitude. At the unclassified level, the budget for TENCAP is now $35 million per year versus $3 to 4 million prior to the Gulf War.[58] While resolution of the technological problems is far from complete, technology continues to receive an unbalanced portion of attention in the development of space power.

After the Gulf War, Air Force Space Command established the SWC to support combat operations through a variety of functions. One of its charters was to take the lessons learned in the Gulf War and apply them to day-to-day operations and wartime support. [59] Of note here is that TENCAP, well established prior to the Gulf War, dominates the SWC's functions and finances. After the war, TENCAP expanded its operation to leverage the billions of dollars spent on "national technical means."[60] The TENCAP system is organized using the previously classified code word *Talon* in six separate programs. The four principal technology divisions are command, control, communications, computers, and intelligence (C^4I) (Talon Command); mission support (Talon Ready); force application (Talon Shooter); and special operations (Talon Night). Talon Touch and Talon Vision provide communications connectivity and processing power support to all the programs.[61] These technology programs dominate the SWC's day-to-day activities.

To normalize tactical warning support, the 11th Space Warning Squadron recently reached a milestone in theater missile warning. Its Attack and Launch Early Reporting to Theater (ALERT) system reached initial operating capability (IOC) on 10 March 1995.[62] The ALERT program was developed following the Gulf War to find better ways of using the DSP satellites for theater ballistic missile defense.[63] The technology acquired to secure this capability under the Talon Shield program responds to some of the lessons illustrated earlier. The ALERT program is a technological attempt to normalize and improve tactical warning support to the war-fighting CINCs.

The lack of a responsive space launch capability is the subject of many studies and debates, but a decision addressing a long-term resolution to the problem is at least a year away.[64] This decision could result in an operational vehicle by 2005.[65] As described previously, the need for a responsive space launch capability in the United States was a significant lesson from the Gulf War. As a result, the fiscal year 1994 defense bill tasked the secretary of defense to provide a plan to improve the US launch capability. The result was Gen Thomas S. Moorman's Space Launch Modernization Plan which, in turn, led to Presidential Decision Directive/NSTC 4, "National Space Transportation Policy," issued on 5 August 1994.[66] The policy calls for a two-track effort. First, the short-term solution requires continued access to space by supporting and improving existing space launch capabilities—namely the space shuttle and current expendable launch vehicles (ELV). Second, the long-term goal is to pursue reliable and affordable access to space through focused investments in, and orderly decisions on, technology development and demonstration for next-generation reusable transportation systems.[67] President Clinton assigned responsibility for the next-generation reusable technology development/demonstration program to the National Aeronautics and Space Administration (NASA).[68]

To solve the problem of the United States's aging MSI and other national intelligence, surveillance, and reconnaissance (ISR) capabilities, USSPACECOM is working with the Office of the Secretary of Defense. MSI was extremely beneficial during Operation Desert Shield/Desert Storm providing US and Coalition forces the opportunity to better understand and react to changes in the terrain. It also offers future war fighters the ability to rehearse their missions, determine optimum tactics, and identify major threat lanes or attack axes to more effectively exploit training and technology in combat.[69] However, the failure of LANDSAT 6 coupled with the DOD decision to stop funding for LANDSAT 7 leaves the military dependent on the aging LANDSAT 5 and foreign sources, such as the French SPOT system, to satisfy MSI requirements.[70] In fact, during the Gulf War, we relied exclusively on the French for MSI requirements.[71] The MSI working group has not resolved this issue but is committed to resolve the problem by the turn of the century.[72]

Experience

After the Gulf War, several significant organizational fixes were geared to improve space power experience and to normalize space support to the theater commanders. To solve some of the major problems witnessed in the Gulf War, senior Air Force leaders created the Fourteenth Air Force, the SWC, the National Test Facility within the SWC, and the SST concept.

On 1 July 1993, the Air Force established Fourteenth Air Force as its operational space component to USSPACECOM to integrate space support for theater warfare, organize space support to theater operators, and to train/exercise with space systems.[73] For the first time, airpower leaders organized space power in a familiar manner to mirror the way the rest of the Air Force operated. Fourteenth Air Force is now responsible for war planning, readiness, and execution. It serves as the war-fighting component to USSPACECOM for satellite control, missile warning, communications, navigation, space surveillance, and space launch opertions.[74]

Establishing Fourteenth Air Force was one piece of the organizational solution enacted to resolve the problems identified during the Gulf War. In December 1993, the Air Force conceived the Space Warfare Center. The SWC's charter is to refine doctrine, develop tactics, and formulate concepts and capabilities to better apply space for all war fighters. Integral to the SWC are the war-gaming and analytical capabilities embodied in the National Test Facility, also located at Shriever Air Force Base (AFB), Colorado. The National Test Facility is responsible for helping educate, train, and prepare war fighters for joint warfare by providing space scenarios for military exercises worldwide.[75] General Horner, then the CINC AFSPACECOM, originally envisioned the SWC to be Air Force Space Command's version of Red Flag and the Air Corps Tactical School all under one roof. He saw a need for an organization to develop the "space tactics and doctrines" while developing prototype programs under the TENCAP program.[76] In reality, SWC personnel are developing many new space technology ideas but very little space power tactics and doctrine.

Air Force Space Command implemented the final organizational change by developing Air Force Space Support Teams (AFSST).[77] USSPACECOM service components and intelligence agencies followed with their version of this concept.[78] The AFSSTs will normally work with the joint force air component commander to provide space support.[79] At a minimum, SSTs from each of the three service components, USSPACECOM, and the National Reconnaissance Office (NRO) deploy to support all of the theater CINCs. War-fighting CINCs requested support from the SSTs in 20 exercises during 1994.[80] In a more recent exercise in South Korea, more than 15 separate SSTs deployed.[81] Many agencies are now spring-loaded to support the war fighter, but without the aid of joint space doctrine to describe the relationship between the SSTs.[82]

The Space Warfare Center is also conducting space courses for different levels of training. First, the Space Tactics School

(STS) completed its inaugural class in July of 1994.[83] This school (formerly the Space Tactics Instructor Course) was conceived by General Horner to give the career space and missile officers an avenue to improve their professional knowledge. In another attempt by General Horner to pattern space power after airpower, the STS was designed after the USAF Weapons School.[84] Its mission is to foster interagency "cross-pollination" so the best techniques and experiences can be transferred among the different elements of the space community.[85] The Air Force developed another training course for the Air Force Space Support Teams. This course is chartered to increase space power awareness and instruct personnel who assist the theater air component commanders and their staffs. Finally, a third space power training opportunity offers a three-to-four-day orientation course designed for audiences with broad backgrounds, including senior leadership.[86] All of these courses are attempts to increase space power experience and literacy.

Doctrine

War-fighting commanders and service components are developing doctrines to guide the use of space power in the next war. In spite of these steps forward, doctrine remains well behind the gait of space power's technological development. With the help of USSPACECOM, Fourteenth Air Force, the SWC, and the service components, war-fighting CINCs have made progress in developing their individual OPLANs.[87] "Space Operations Doctrine" (AFDD-4) is nearing completion after years of coordination.[88] *Air Force Basic Doctrine* (AFDD 1) is in the early stages of a major revision and is probably several years away from completion. Finally, "Joint Space Doctrine" (Joint Pub 3-14) has been in the coordination process since before the Gulf War.[89]

US Central Command OPLAN 1002-95. Prior to the Gulf War no doctrine was available to guide or educate USCENTCOM war fighters on space power. Since the war, USCENTCOM planners have incorporated a space power annex (annex N) in their OPLAN describing specific space assets available for future planning.[90] While not a replacement for basic or operational space doctrine, annex N to this OPLAN is a small step in the right direction. Nevertheless it does not provide the guidance needed to maximize space power's robust capabilities.

Air Force Manual 1-1. The current version of AFM 1-1, March 1992, assumes the same basic doctrine that applies to airpower applies to space—"aerospace power."[91] The next version of AFM 1-1, is expected to overturn this decision.[92] The drafters of the new version expect to separate airpower and space power into distinct roles and missions. This separation is a complete reversal of policy provided to the authors of the 1992 version. Based on the recommendations of the "Blue Ribbon" Todd Commission on Space, the writers of the 1992 version of AFM 1-l were instructed to totally integrate air and space.[93] The Air Force's indecision on integration of air and space is yet another reason why space doctrine continues to flounder. As outlined, the new version will take the position that space capabilities cannot be derived by simply applying the term *aerospace* to what is an otherwise comprehensive airpower doctrine.[94]

Major Air Force commands will have an opportunity to include applicable space power experiences from the Gulf War into AFDD 1. It is difficult to predict when AFDD1 will appear, but if it follows the same pattern as its predecessor it may be years away from completion.[95] It is too soon for the authors of AFDD1 to predict how the space power experiences from the Gulf War will affect the new document.[96]

AFDD 4. If approved as currently written, AFDD 4 offers a small doctrinal step for space command, but a huge leap for the military space community. This document has been in coordination since the Gulf War.[97] If AFDD 4 is approved as currently written, it will address many of the space power experiences from the Gulf War. For example, AFDD 4 describes command of space forces, roles and missions of space forces, space employment concepts, space power for the theater campaign, and education and training. All of these topics are directly related to the experiences of the Gulf War.[98]

In fact, of the space power doctrinal documents examined in this study, the draft of AFDD 4 is the only reference with a general description of the relationship between the warfighting CINCs and the space support teams.[99] Although the current draft of AFDD 4 is a less robust version of previous drafts, it offers some relief in the doctrinal stalemate.

Joint Doctrine, Tactics, Techniques, and Procedures (JDTTP) 3-14, *Space Operations*. Arguably the most important doctrinal document, Joint Pub 3-14, is no closer to completion than it was four years ago. The Joint Chiefs of Staff issued the program directive for Joint Pub 3-14 on 30 March 1990. USSPACECOM initiated plans to distribute the first, fully coordinated version of Joint Pub 3-14 by May 1991.[100] Unfortunately, the publication is mired in the coordination process and will be rewritten prior to another coordination cycle.[101]

Joint Pub 3-14 is the most important doctrinal reference, not only because future operations are likely to be joint efforts but also because the chairman of the Joint Chiefs of Staff recently included a statement in all joint publications stipulating they will be followed except when in the judgment of the commander, exceptional circumstances warrant otherwise.[102] This is especially important for joint space operations because of service, unified, and national space support teams augmenting the joint force commander's staff during war.

Space Power's Development after the Cold War

Efforts to address the problems encountered during the Gulf War are evident in all phases of the development of space power, but it is apparent that technological innovations still receive an unbalanced share of space power attention. The development of Air Force basic doctrine, Air Force operational space doctrine, and joint space doctrine is embarrassingly far behind innovative space technologies.

The disdain of space doctrine is a well-documented fact. In January 1988, Colin S. Gray made the following comment about space doctrine: "It has been 43 years since the first spacecraft was launched (Germany's V-2 rocket) and 30 years since Sputnik, yet today there is no doctrinal literature worth reading on the subject of battle field space."[103] Gray's statement is as accurate today as it was in 1988. Later, Lt Col Alan J. Parrington made similar comments in the *Airpower Journal*: "The United States has not decided what it wants to do in space, how it can achieve its aims, or what equipment it needs for future space exploration. If the US government is to eliminate confusion and give direction to the space program, it must first develop a cohesive military spacedoctrine."[104]

Col Edward C. Mann III supports Parrington's declaration by summarizing the short shrift many Air Force officers give Air Force basic doctrine in a recent publication, *Thunder and Lightning:* "Boring or not, when the popes (chief of staff), cardinals (four-star generals), and archbishops (three-star generals) disdain doctrine, the faithful will follow suit."[105] Finally, Lt Col Steven J. Bruger describes the actions needed to prepare US space forces for the next space war. Bruger states, "The first need is a key element—development of space doctrine to provide guidance and direction at all levels of war, across the full spectrum of conflict."[106] The development of space doctrine at all levels has been and continues to be the largest impediment facing the military space community today.

Conclusion

We need joint doctrine that clearly defines control and force application to support the evolution of space systems from a pure supporting role into a menu of joint space force options whose stated purpose is to ensure overall US space superiority.

—George Moore, Vic Budura, and Joan Johnson-Freese

Summary of Findings

The overwhelming majority of the documented lessons in the Gulf War concerned either a lack of doctrine or a lack of space literacy/experience. The military space community is years away from internalizing these experiences. While the space community pursues ideas to normalize space power operations, doctrine is an afterthought—"dull, boring, and useless," or "important but not read by warriors."[107] Specifically, the lack of doctrine continues to impede efforts to maximize effective war fighting with space power assets. Less costly reforms in doctrine could offer more leverage for the future US military space program when combined with the existing space power technology. The synergy of improvements to AFDD 1, approval of AFDD 4, and the creation of joint space doctrine offers a cost-effective boost to the advancement of space power for the future. Gen Thomas S. Moorman Jr., vice chief of staff of the Air Force, feels that the complete internalization of space power lessons from the Gulf War is at least a generation of war fighters away.[108] More focus on doctrine can accelerate the internalization of recent space power experiences. The impact of redressing the imbalance existing in the development of space power makes the thesis of this study a prime consideration for the next logical step in future space power policy.

Primary Conclusions

1. The majority of space power lessons from the Gulf War resulted from a lack of doctrine and experience.
2. Technology remains the military space community's primary focus—doctrine and experience continue to lag well behind technology in the development of space power.
3. Space doctrine development is long overdue.
4. USSPACECOM did not have a formal process of monitoring the space power lessons after the Gulf War.[109]
5. Space power advancement is still impeded by the cold war mentality and the extreme security requirements associated with this era.

Recommendations

The US space community should focus on redressing the imbalance among doctrine, experience, and technology in space power's development. Among the Gulf War lessons, the USSPACECOM exercise database, and the JULLS, sufficient historical information is available to help write useful space power doctrine. In particular, Joint Pub 3-14 is urgently needed to help guide the influx of space support teams in theater exercises. After approval, "Space Operations Doctrine" (AFDD 4) can potentially serve as an accurate guide for the rewrite of the space power portion of Air Force Basic Doctrine (AFDD 1). Finally, the US military space community is dangerously close to completely discarding forward thinking in space doctrine. We must reverse this mind-set to ensure that doctrine guides the development and employment of future space systems.

The development of space doctrine and the liberation of the space community from the security restrictions of the cold war paradigm will spur education concerning the attributes of space power. All services will benefit from the development of space doctrine because it can serve as the basis for space power professional military education (PME). An aggressive space power PME program, from basic training to the senior service schools, is the only way to fully internalize space power lessons. In addition, a major step forward in educating the force and establishing core competency would tear down the walls of classification the military space intelligence community has built around itself. The United States will be better served by establishing a single military space sector with representation from all the services. The current ultra-secret intelligence space sector is very resilient but inefficient.[110] In short, the United States should "give the warfighting CINCs more control over intelligence support."[111]

The integration of all military and intelligence space activities will not only increase the war-fighting CINC's influence on space power support, it will help centralize the acquisition, control, and tasking of satellites. The military space community must continue to search for superior weapons and force multipliers—this is an essential requirement. However, current acquisition and management of national satellites are fragmented. The recent Report of the Commission on Roles and Missions of the Armed Forces supports this finding. The commission recommends that the secretary of defense integrate the management of military and intelligence space activities, assign the development of the integrated architecture of military space systems to a joint service office, and designate the Air Force as the primary (not sole) agency for acquisition and operation of multiuser space-based systems.[112] These changes will make the already aggressive development of space power technology much more efficient.

Notes

1. Many authors reference the Gulf War as the "first space war"; however, since we have used space assets in warfare since Vietnam, it seems more appropriate to call Operation Desert Storm the "first information war." This is the first time a war revealed just what impact information management can have. James A. Winnefeld, Preston Niblack, and Dana J. Johnson, *A League of Airmen: US Airpower in the Gulf War* (Santa Monica, Calif.: RAND, Project Air Force, 1994), 4, 181–84.

2. In AFM 1-1, space power is defined as "that portion of aerospace power that exploits the space environment for the enhancement of terrestrial forces and for the projection of combat power to, in, and from space to influence terrestrial conflict." This definition originated in a draft to AFM 2-25 which no longer exists. Another definition is found in the current draft of AFDD 4: "Spacepower is the capability to exploit civil, commercial, intelligence, and national security space systems and associated infrastructure to support national security strategy and national objectives from peacetime through combat operations." This study uses the AFDD 4 definition. Air Force Manual (AFM) 1-1, *Basic Aerospace Doctrine of the United States Air Force,* vol. 2, March 1992, 300. AFDD 4, "Space Operations Doctrine," draft, 1 May 1995, 3.

3. Many of the reports analyzed for this thesis use the words *strategic* and *tactical* to differentiate between missions to support the nuclear deterrence strategy of the United States and other than nuclear missions respectively. *Strategic* and *tactical* are more appropriately used in terms of levels of war or effects during war. For a useful definition, see Col John Warden, *The Air Campaign* (New York: Pergamon-Brassey's, 1989), 2–3.

4. Lt Col Mike Wolfert, address to the Space Issues Team on Roles and Missions, Washington, D.C., 14 November 1994, slide S2-OVER 3.

5. *Gulf War Air Power Survey*, vol. 4, "Weapons, Tactics, and Training and Space Operations" (Washington, D.C.: Department of the Air Force, 1993), 169. (Hereafter cited as *GWAPS*.)

6. Gen Merrill A. McPeak, address during the SPACE TALK '94 Briefing, 16 September 1994.

7. USSPACECOM, *Operation Desert Shield and Desert Storm Assessment* (U) (Peterson AFB, Colo.: USSPACECOM, 31 January 1992), 65–67 (Secret/NoForn); and *US Central Command After Action Report, Operation Desert Shield/Storm* (U) (MacDill AFB, Fla.: USCENTCOM, 15 July 1991), 37–47 (Secret/NoForn). Information extracted from both reports is unclassified. (Hereafter cited as USSPACECOM after action report [AAR] and USCENTCOM AAR.)

8. Joint Doctrine, Tactics, Techniques, and Procedures (JDTTP) 3-14, *Space Operations*, 15 April 1992, V1-5.

9. Lt Gen Thomas S. Moorman Jr., "Space Acquisition Conference Remarks," 27 May 1994, 2.

10. At the most recent Joint Space Doctrine working group meeting, USSPACECOM/J5, Maj William Doyle stated that Joint Pub 3-14 will be

rewritten. He projected the document to be in final coordination 12 to 14 months from this meeting. Maj William Doyle, "Joint Space Doctrine Working Group," Peterson AFB, Colo., 31 May–1 June 1995.

11. Moorman, 7.

12. This concept was adopted from Col Dennis M. Drew, USAF, Retired. Colonel Drew presented this framework on 2 May 1995 during the School of Advanced Airpower Studies Course 680—Airpower Theory II. For a similar framework see Mackubin Thomas Owens, "Lessons of the Gulf War," *Strategic Review,* Winter 1992, 51.

13. A useful definition comes from James W. Canan in "'Normalizing' Space," *Air Force Magazine,* August 1990, 12. Canan defines *normalizing* as follows: "This means launching and operating its space systems as matter-of-factly and purposefully as it does its aircraft and treating those systems as workaday and warfighting tools, not as showpieces in the sky." I would add "acclimating the cold war space culture into everyday operational life." USSPACECOM AAR, 65.

14. USCENTCOM OPLAN 1002-90, United States Air Force Historical Research Agency, Maxwell AFB, Ala.

15. Lt Gen Thomas S. Moorman, "Remarks to the Eighth National Space Symposium," Colorado Springs, Colo., 2 April 1992, 1.

16. Ibid.

17. Gen Charles A. Horner, "Space Seen as Challenge, Military's Final Frontier," *Defense Issues* 8, no. 34 (22 April 1993): 1.

18. USSPACECOM AAR, 67.

19. Ibid.

20. "Joint Universal Lessons Learned System (JULLS) Database" (U) on CD-ROM, Navy Tactical Information Compendium (NTIC) (Washington, D.C.: Department of the Navy, December 1994), disk 2, JULLS 92659-18177 (Secret/NoForn). Information extracted is unclassified. Also, *Conduct of the Persian Gulf War: Final Report to Congress,* vols. 1 and 2 (Washington, D.C.: Department of Defense, April 1992), K 50-51. (Hereafter cited as the *CPGW.*)

21. Moorman, "Space Acquisition Conference Remarks," 4.

22. USCENTCOM AAR, 37.

23. Ibid.

24. Michael M. Garrell, *There Are No Space Wars, How Do CINC's Fight Using Space Forces?* (Newport, R.I.: Naval War College, 17 June 1994), 17. Garrell argues that it is clear from the postwar analysis that the successful use of space power was due largely to innovation, creativity, and ad hoc procedures, not operational thinking.

25. JULLS 31538-21500.

26. USSPACECOM Joint Space Support Team (JSST) Briefing, USSPACECOM/J33S, undated, slide J3-1-30-10.

27. JULLS 50352-59445 and 91747-98856.

28. Lt Col Steven J. Bruger, "Not Ready for the First Space War, What about the Second?" *Naval War College Review* 18, no. 1 (Winter 1995): 76. Also, *CPGW*, K 31.

29. JULLS 15242-11100.

30. Ibid.

31. Ibid., 50612-8818.

32. *GWAPS*, v.

33. Ibid.

34. Ibid., v–vi.

35. Ibid., vi.

36. Ibid.

37. *CPGW*, ix and K 50.

38. Ibid., K 49.

39. Ibid., K 48, 49.

40. Ibid., K 48.

41. Lt Col Robert E. Miller, chief of USSPACECOM's Joint Training and Simulation Section, interview by author, 1 June 1995.

42. USSPACECOM AAR, Attachment 2.

43. Mark G. Cooney, JCS Evaluation and Analysis Division, interview by author, 1 May 1995. For a detailed description of the JULLS process, see CJCS Instruction 5716.01, 1 October 1994, B-l.

44. Ibid.

45. Gen Charles A. Horner, interview by author, 28 April 1995.

46. Ibid.

47. Ibid.

48. Owens, 51.

49. For an enlightening view of doctrine, see Col Dennis M. Drew, "Of Trees and Leaves: A New View of Doctrine," *Air University Review*, January–February 1982, 40–48.

50. Lt Gen Thomas S. Moorman, "Creating Tomorrow's Space Forces," speech to the San Francisco Commonwealth Club, San Francisco, Calif., 1 December 1993.

51. The writers of the Wilkening Report include Laurel L. Wilkening, appointed by President Ronald Reagan in 1985 as vice chairman of the National Commission on Space; Lt Gen James A. Abrahamson, USAF, Retired, first director of the Strategic Defense Initiative; Edward C. "Pete" Aldridge, former secretary of the Air Force; Joseph P. Allen, former astronaut with NASA; Daniel J. Fink, with over 40 years aerospace engineering experience; John S. Foster Jr., former director of Lawrence Livermore National Laboratory; Edward Frieman, former director of Energy Research with the Department of Energy (DOE); Don Fuqua, served 12 terms as a US congressman; Gen Donald J. Kutyna, USAF, Retired, former commander of NORAD and AF Space Command; John M. Logsdon, author of *The Decision to Go to the Moon: Project Apollo and the National Interest;* and Bruce C. Murray, former director of NASA/California Institute of Technology Jet Propulsion Laboratory. *A Post Cold War Assessment of US Space Policy; A Task*

Group Report (Washington, D.C.: Department of Defense, 17 December 1992), appendix 11. (Hereafter cited as the Wilkening Report.)

52. Wilkening Report, 23.

53. Lt Col Mike Wolfert, chief of Air Force Space Command Strategy, Policy, and Doctrine, interview by author, 1 June 1995.

54. Lt Gen Thomas S. Moorman, "Public Affairs Background for Query Response," 30 June 1993. Effective date of the transfer was 1 July 1993.

55. I. B. Holley Jr., *Ideas and Weapons* (Washington, D.C.: Office of Air Force History, 1983), 175.

56. Ibid., 175–76.

57. Ibid., 178.

58. Maj Joe Squatrino, Space Issues Committee, Air Force Roles and Missions, interview by author, 15 May 1995.

59. Lt Col Kip Hunter, "TALON Programs Overview," *Space Tactics Bulletin* 1, no. 1 (June 1994): 5.

60. Col Mike Francisco, "SWC Support to Warfighters," *Space Tactics Bulletin* 1, no. 1 (June 1994): 3.

61. Hunter, 5.

62. Capt John Kennedy, "Theater Missile Warning Unit Reaches Operations Milestone," *The Guardian*, April 1995, 13. Also see Sean D. McClung, "TALON SHIELD Declares Victory!" *Space Tactics Bulletin* 2, no. 1 (November 1994): 3.

63. Kennedy, 13.

64. *Space Launch Modernization Study* (Washington, D.C.: Department of Defense, 18 April 1994), 17–19.

65. *NASA Implementation Plan for the National Space Transportation Policy*, 7 November 1994, 13.

66. Ibid., 3.

67. Ibid.

68. Ibid.

69. Moorman, "Space Acquisition Conference Remarks," 4.

70. Horner, "Testimony before the Senate Armed Services Committee," March 1994, 25.

71. Wolfert interview.

72. Ibid.

73. Moorman, "Space Acquisition Conference Remarks," 2.

74. Ibid.

75. Lt Gen Thomas S. Moorman, "Presentation to the Committee on Appropriations, Subcommittee on Defense, US House of Representatives," Washington, D.C., March 1994, 5.

76. Moorman, presentation to the House, 5–6.

77. AFDD 4, 10.

78. Gen Joseph W. Ashy, commander in chief, USSPACECOM, "Orlando Air Force Association," speech, Orlando, Fla., undated, 6.

79. Ibid.

80. USSPACECOM JSST briefing, slide J3-1-30-12.

81. Wolfert interview.

82. Maj William Doyle, Joint Space Doctrine Working Group, Joint Pub 3-14 Working Group, USSPACECOM/J5X, Peterson AFB, Colo., 31 May–1 June 1995.

83. Lt Col Dan Chapman, "Space Tactics School (STS) Completes Inaugural Class," *Space Tactics Bulletin* 2, no. 1 (November 1994): 4–5; Capt David Koster, "Space Training—Coming Soon to a Theater Near You!" *Space Tactics Bulletin* 2, no. 1 (November 1994): 5–6.

84. Chapman, 4.

85. Ibid.

86. Brig Gen David L. Vesely, "Commanders Corner," *Space Tactics Bulletin* 2, no. 1 (November 1994): 1.

87. Horner interview.

88. Wolfert interview.

89. Joint Space Doctrine Working Group.

90. Moorman, "Space Acquisition Conference Remarks," 4. USCENTCOM OPLAN 1002-95, undated, submitted on 1 June 1993. This OPLAN was never approved. Annex N is a general list of capabilities but does not provide guidance on how to use space power to fight the next war.

91. AFM 1-1, vol.1, 5.

92. AFDD 1, outline chap. 3.

93. Col Dennis M. Drew, USAF, Retired, project team chief and principal author of the 1992 version of AFM 1-1, interview by author, June 1995.

94. Col Kenneth A. Myers and Lt Col John G. Tockston, "Real Tenets of Military Space Doctrine," *Airpower Journal,* Winter 1988, 55. Also, see Maj Grover E. Myers, "Aerospace Doctrine We're Not There Yet," *Air University Review,* September–October 1986, 91–99.

95. AFM 1-1, vol. 1, i. The time period between the latest two versions of AFM 1-1 was eight years and two months. If the next version follows the same pattern, AFDD 1 will be available at the turn of the century.

96. AFDD 1, outline. Also, Mr. Wayne R. Williamson, principal author of AFDD 1, Air Force Doctrine Center, interview by author, 15 June 1995.

97. Wolfert interview.

98. Ibid.

99. AFDD 4, 8–10. Also, Joint Space Doctrine Working Group.

100. Joint Chiefs of Staff Program Directive for Joint Pub 3-14, Message date time group, 301638Z March 1990, Washington, D.C., 1.

101. Joint Space Doctrine Working Group.

102. It is also important to note that "if conflicts arise between the contents of this publication and the contents of Service publications, this publication (Joint Pub 3-14) will take precedence for the activities of joint forces." Joint Pub 3-14, ii. It is the opinion of this author that the doctrinal community as a whole has taken these statements as license to avoid development of forward-looking doctrine. During the most recent Joint Space Doctrine Working Group, all service representatives indicated that their senior leadership has interpreted these "directive" statements to mean

that doctrine cannot include futuristic guidance. If a space mission (force application) is not possible today because of politics or funding, it should not be described in any doctrinal publication. This thinking is a step backward in doctrinal development, especially for space-based assets. Space power's operational potential will be maximized in the future. The Gulf War is only a glimpse of what the United States will benefit from robust space power capabilities.

103. Colin S. Gray, "Space Warfare: Part I, the Need for Doctrine," *National Defense,* January 1988, 25.

104. Lt Col Alan J. Parrington, "US Space Doctrine: Time for a Change?" *Airpower Journal,* Fall 1989, 51.

105. Col Edward C. Mann III, *Thunder and Lightning: Desert Storm and the Airpower Debates* (Maxwell AFB, Ala.: Air University Press, 1995),164–65. Also see Colonel Mann's description on page 181: "As such, war is subject to all the vagaries of the human mind, spirit, and will. So long as this is true, ideas, concepts, philosophies, and doctrines will always matter."

106. Bruger, 79.

107. Mann, 164.

108. Gen Thomas S. Moorman Jr., vice chief of staff of the Air Force, interview by author, Maxwell AFB, Ala., 5 June 1995. I agree with General Moorman that until Air Force personnel can communicate the importance of space power from the ranking general officer to the basic airman, and until we have space power advocates or heroes (as we did with airpower), only then will we fully internalize the lessons from the war.

109. Recently, USSPACECOM instituted a program to monitor lessons from any exercise to which the joint space support teams deploy. USSPACECOM will monitor all lessons from these exercises and will submit significant findings to the Joint Chiefs of Staff for inclusion in the Joint Universal Lessons Learned System. Miller interview.

110. Reports indicate that the ultrasecret space programs are likely to remain tightly veiled, especially in the National Reconnaissance Office. James M. Gifford, "New Clinton Policy Aims to Reduce Government Secrecy," *Space News,* April 24–30 1995, 14.

111. "Directions for Defense; Report of the Commission on Roles and Missions of the Armed Forces" (Arlington, Va.: Commission on Roles and Missions of the Armed Forces, 1995), 2-6, 2-7.

112. Ibid., 2-7.

Selected Bibliography

Books and Reports

Baird, Henry D. *Is It Time for a Joint Space Component Commander?* Newport, R.I.: Naval War College, 19 June 1992.

Blackwell, James, Michael J. Mazarr, and Don M. Snider. *The Gulf War: Military Lessons Learned.* Washington, D.C.: Center for Strategic and International Studies, 1991.

Butterworth, Robert L. *Space Systems and the Military Geography of Future Regional Conflicts.* Los Alamos, N.Mex.: Los Alamos National Laboratory, January 1992.

Campen, Alan D., contributing editor. *The First Information War.* Fairfax, Va.: AFCEA International Press, 1992.

Clapp, William Q. *Space Fundamentals for the War Fighter.* Newport, R.I.: Naval War College, 22 June 1994.

Collins, John M. *Military Space Forces: The Next 50 Years.* Washington, D.C.: Pergamon-Brassey's, 1989.

Drew, Dennis M., and Donald M. Snow. *Making Strategy.* Maxwell AFB, Ala.: Air University Press, 1988.

Garrell, Michael M. *There Are No Space Wars: How Do CINC's Fight Using Space Forces?* Newport, R.I.: Naval War College, 17 June 1994.

Hays, Peter L. "Struggling towards Space Doctrine: US Military Space Plans, Programs, and Perspectives during the Cold War." PhD diss., Fletcher School of Law and Diplomacy, 1994.

Holley, I. B., Jr. *Ideas and Weapons.* Washington, D.C.: Office of Air Force History, 1983.

Library of Congress, Congressional Research Service. *Military Space Programs in a Changing Environment: Issues for the 103rd Congress.* Washington, D.C., 1 December 1992.

Lowery, Craig Z. *Joint Space Operations in the 21st Century.* Maxwell AFB, Ala.: Air Command and Staff College, 1993.

Lupton, David E. *On Space Warfare: A Space Power Doctrine.* Maxwell AFB, Ala.: Air University Press, 1988.

Mann, Edward C., III. *Thunder and Lightning: Desert Storm and the Airpower Debates.* Maxwell AFB, Ala.: Air University Press, April 1995.

Mantz, Michael R. *The New Sword: A Theory of Space Combat Power.* Maxwell AFB, Ala.: Air University Press, 1995.

Middleton, Gordon R. *Space Is a Different Place.* Maxwell AFB, Ala.: Air War College, 5 May 1992.

Toma, Joseph S. "Desert Storm Communications." In *The First Information War.* Edited by Alan D. Campen. Fairfax, Va.: AFCEA International Press, 1992.

Wentz, Larry K. "Communications Support for the High Technology Battlefield," In *The First Information War.* Edited by Alan D. Campen. Fairfax, Va.: AFCEA International Press, 1992.

Wilkening, Laurel L., et al. *A Post Cold War Assessment of US Space Policy: A Task Group Report.* Washington, D.C., December 1992.

Winnefeld, James A., Preston Niblack, and Dana J. Johnson. *A League of Airmen: US Airpower in the Gulf War.* Santa Monica, Calif.: RAND Project Air Force, 1994.

Articles

Air Force Association. "Space Almanac." *Air Force Magazine* 77, no. 8 (August 1994): 44–59.

Alison, John R., et al. "Facing Up to Space." *Air Force Magazine* 78, no. 1 (January 1995): 50–54.

Anson, Peter. "First Space War: The Contribution of Satellites to the Gulf War." *RUSI Journal*, Winter 1991, 45–53.

Bruger, Steven J. "Not Ready for the First Space War, What about the Second?" *Naval War College Review* 18, no. 1 (Winter 1995): 73–83.

Campen, Alan D. "Gulf War's Silent Warriors Bind US Units via Space." *Signal* 45, no. 12 (August 1991): 81–84.

Canan, James W. "A Watershed in Space." *Air Force Magazine* 74, no. 8 (August 1991): 32–37.

———. "Normalizing Space." *Air Force Magazine* 73, no. 8 (August 1990): 12–14.

———. "Space Gets Down to Earth." *Air Force Magazine* 73, no. 8 (August 1990): 30–34.

———. "Space Support for the Shooting Wars." *Air Force Magazine* 76, no. 4 (April 1993): 3–34.

Chapman, Dan. "Space Tactics School (STS) Completes Inaugural Class." *Space Tactics Bulletin 2,* no. 1 (November 1994): 4–5.

Covault, Craig. "Desert Storm Reinforces Military Space Directions." *Aviation Week & Space Technology,* 8 April 1991, 42–47.

Francisco, Mike. "SWC Support to Warfighters." *Space Tactics Bulletin* 1, no. 1 (June 1994): 3–4.

Gifford, James M. "New Clinton Policy Aims to Reduce Government Secrecy." *Space News,* 24–30 April 1995, 14.

Herres, Robert T. "The Military in Space: A Historical Relationship." *Space Policy,* May 1987, 92–95.

Horner, Charles A. "Offensive Air Operations: Lessons for the Future." *RUSI Journal* 138, no. 6 (December 1993): 19–24.

———. "Unpredictable World Makes US Space Capabilities Critical." *Defense Issues* 9, no. 43 (20 April 1994): 1–7.

Howard, Michael. "Military Science in an Age of Peace." *RUSI Journal,* March 1974, 3–11.

Howard, William E. "Space Importance Grows as a Future Battleground." *Signal,* August 1993, 57–60.

Hunter, Kip. "TALON Programs Overview." *Space Tactics Bulletin* 1, no. 1 (June 1994): 5–7.

Jennings, Frank W. "Doctrinal Conflict over the Word *Aerospace.*" *Airpower Journal* 4, no. 3 (Fall 1990): 46–58.

Kennedy, John. "Theater Missile Warning Unit Reaches Operations Milestone." *The Guardian,* April 1995, 13.

Koster, David. "Space Training—Coming Soon to a Theater Near You!" *Space Tactics Bulletin* 2, no. 1 (November 1994): 5–6.

Lynch, David J. "Spacepower Comes to the Squadron." *Air Force Magazine* 77, no. 9 (September 1994): 66–68.

Martin, Robert John. "Space—Building the Foundation for our Future." *Military Engineer* 82, no. 538 (September–October 1990): 42–46.

McClung, Sean D. "TALON SHIELD Declares Victory!" *Space Tactics Bulletin 2,* no. 1 (November 1994): 3–4.

Moore, George M., Vic Budura, and Joan Johnson-Freese. "Joint Space Doctrine: Catapulting into the Future." *JFQ: Joint Force Quarterly,* Summer 1994, 71–76.

Moorman, Thomas S., Jr. "The Space Component of Aerospace." *Comparative Strategy*, July–September 1993, 251–55.

———. "Space: A New Strategic Frontier." *Airpower Journal* 6, no. 1 (Spring 1992): 14–23.

Morrocco, John D. "From Vietnam to Desert Storm." *Air Force Magazine* 75, no. 1 (January 1992): 68–73.

Myers, Grover E. "Aerospace Doctrine: We're Not There Yet." *Air University Review*, September–October 1986, 91–93.

Myers, Kenneth A. "Military Space Control Reality Check." *Space News*, 7–13 November 1994, 15.

Myers, Kenneth A., and John G. Tockston. "Real Tenets of Military Space Doctrine." *Airpower Journal* 2, no. 4 (Winter 1988): 54–67.

Owens, Mackubin Thomas. "Lessons of the Gulf War." *Strategic Review*, Winter 1992, 50–54.

Parrington, Alan J. "US Space Doctrine: A Time for Change?" *Airpower Journal* 3, no. 3 (Fall 1989): 51–61.

"Space Comes into Its Own." *Leading Edge*, June 1991, 14–15.

Sweetman, Bill. "Making Sense out of Military Space." *International Defense Review*, September 1993, 705–10.

Vesely, David L. "Commanders Corner." *Space Tactics Bulletin* 2, no. 1 (November 1994): 1.

Waghelstein, John D. "Some Thoughts on Operation Desert Storm and Future Wars." *Military Review* 7, no. 22 (February 1992): 80–83.

Department of Defense Publications

Air Force Manual 1-1. *Basic Aerospace Doctrine of the United States Air Force.* 2 vols., March 1992.

Air Force Doctrine Document (AFDD) 1. "Air Force Basic Doctrine." Draft outline, 21 February 1995.

AFDD 4. "Space Operations Doctrine." Draft, 1 May 1995.

Conduct of the Persian Gulf War: Final Report to Congress Pursuant to Title V of the Persian Gulf Conflict Supplemental Authorization and Personal Benefits Act of 1991 (Public Law 102-25). Washington, D.C.: Department of Defense, April 1992.

Directions for Defense: Report of the Commission on Roles and Missions of the Armed Forces. Washington, D.C.: Commission on Roles and Missions of the Armed Forces, 1995.

Gulf War Air Power Survey. Vol. 4, *Weapons, Tactics and Training and Space Operations.* Directed by Eliot A. Cohen. Washington, D.C.: Department of Defense, 1993.

Joint Chiefs of Staff Program Directive for Joint Publication 3-14. Message date/time group, 301638Z MAR 90, Washington, D.C., 30 March 1990.

Joint Doctrine, Tactics, Techniques, and Procedures (JDTTP) 3-14. *Space Operations,* 15 April 1992.

Joint Universal Lessons Learned System (JULLS) Database. CD-ROM, Disks 1–3, Navy Tactical Information Compendium (NTIC). Washington, D.C.: Department of the Navy, December 1994.

NASA Implementation Plan for the National Space Transportation Policy. Washington, D.C.: NASA, 7 November 1994.

Space Launch Modernization Study. Washington, D.C.: Department of Defense, 18 April 1994.

US Central Command After Action Report, Operation Desert Shield/Storm. MacDill AFB, Fla.: USCENTCOM, 15 July 1991.

US Central Command Operation Plan 1002-95. MacDill AFB, Fla.: USCENTCOM, undated.

USSPACECOM. *Operation Desert Shield and Desert Storm Assessment.* Peterson AFB, Colo.: USSPACECOM, 31 January 1992.

Briefings/Speeches/Lectures

Ashy, Joseph W. "Orlando Air Force Association Speech," undated.

Doyle, William. "Joint Space Doctrine Working Group." Joint Publication 3-14 Working Group, Peterson AFB, Colo., 31 May–1 June 1995.

Drew, Dennis M. School of Advanced Airpower Studies Course 680—Airpower Theory II. Lecture. Maxwell AFB, Ala., 2 May 1995.

Horner, Charles A. "Space Seen as Challenge, Military's Final Frontier." *Defense Issues* 8, no. 34 (22 April 1993): 1–10.

McPeak, Merrill A. "Desert Shield/Storm." Briefing to National War College. Washington, D.C., 6 March 1991.
———. "SPACE TALK '94," 16 September 1994.
Moorman, Thomas S., Jr. "Creating Tomorrow's Space Forces." Speech to the San Francisco Commonwealth Club, 1 December 1993.
———. "The Future of United States Air Force Space Operations." *Vital Speeches,* 15 March 1994, 325–29.
———. "Presentation to the Committee on Appropriations, Subcommittee on Defense, US House of Representatives," March 1994.
———. "Public Affairs Background for Query Response," 30 June 1993.
———. "Remarks to the Eighth National Space Symposium." Colorado Springs, Colo., 2 April 1992.
———. "Space Acquisition Conference Remarks," 27 May 1994.
"US Space Command Joint Support Team (JSST) Briefing." USSPACECOM/J33S, undated.
Wolfert, Michael L. "Presentation to the Space Issues Team on Roles and Missions." Briefing before the Roles and Missions Committee, 14 November 1994.

Chapter 3

Blueprints for the Future: Comparing National Security Space Architectures

Christian C. Daehnick

In recent years it has become a cliché to speak of the growing importance of space systems and their capabilities to US national security in general and to military operations in particular. At the very least, the changing national security environment and our experiences in the Gulf War have caused a more open discussion of what those space-based capabilities are and what they should be. Along with a greater awareness of space has come realization that the systems often seem unresponsive to the needs of some users and that gaps exist in our capabilities. Many see the current US space architecture as fragmented and inflexible. At the same time, decreasing budgets mean that the solution to any problems cannot simply be the purchase of additional capability; the times demand more efficient answers.[1]

Complacency about our space capabilities at this point would be dangerous. Although the United States presently has the best space systems in the world and military peer competitors or threats to our national survival are beyond the horizon, there is a danger that efforts over the coming years will not adequately address the shortfalls of the current space architecture. Space systems that remain unresponsive, fail to live up to expectations, or fail to evolve toward new capabilities will disillusion national and military policy and strategy makers, who might then either ignore space capabilities en-

This work was accomplished in partial fulfillment of the master's degree requirements of the School of Advanced Airpower Studies, Air University, Maxwell AFB, Ala., 1996.

Advisor: Lt Col Robert Owen, PhD

Reader: Dr James Corum

tirely or back other, possibly less effective solutions. Ultimately, such a situation will hurt the United States.

Broadly speaking, there are two approaches to making the national security space architecture more effective. The first is incremental, working to eliminate inefficiencies and expand access to space systems and capabilities in a gradual fashion. It would by and large retain the command, control, and tasking arrangements, communications channels, organizational structures, and space system design and operating procedures of the current architecture. A less conservative approach would involve a shift to a fundamentally different architecture based on decentralization and improved responsiveness. Which approach will produce the best capabilities for the United States, given limited resources?

Answering this question begins with a clearer understanding of the alternatives. The current space architecture is primarily command-oriented: centralized, driven by specific performance requirements and employing a push approach to providing services. Numerous initiatives are under way to modify current space systems and make them more responsive, but fundamental changes would be needed to make the architecture demand-oriented. Demand orientation implies a more decentralized organization, a user-pull approach to providing services, and a focus on responsiveness.[2]

The basic question of this study is whether command- or demand-oriented architectures can make better use of space for national security purposes, and better respond to a changing security, technological, and budgetary environment. The question is complicated by real tensions between the characteristics of command- and demand-oriented systems. They do not perform all functions equally well, and each approach requires some compromises. For example, a command-oriented system requires investment in large, complex systems and only permits incremental changes in the architecture.

The incremental approach may not be a satisfactory long-term solution. Although attractive, and to some extent necessary, because it makes best use of what the United States has already invested, it begs several questions. Does such an approach attempt to defy fundamental trends in technology, operational requirements, and budgets? Because of the basic philosophy underlying current space systems, will we remain tied to small numbers of large, complex, expensive, and vulnerable systems spread ever more thinly trying to satisfy multiple users? Will these users then grow more dissatisfied with the responsiveness of space systems to their needs and seek other solutions? Will space systems take so long to design, build, and deploy that they are technologically out of date as soon as they are deployed? If the answers to these questions are "yes," we may do less, not more in space in the future, to the detriment of our national security.

The radical alternative is to shift to a demand-oriented architecture; one that more directly responds to the needs of today's primary users and can adapt more readily to changes in requirements or technological opportunity. The primary elements would be smaller, more distributed, and autonomous space systems that could be tasked directly by the users and more closely integrated with other military operations. Such tailored, distributed constellations of space systems would both be enabled by advances in microelectronics, miniaturization, automation, and modularity, and offer a better way to keep our space systems modern and effective. This approach also appears to fit better with a world of global commitments and pop-up crises than our current systems. Unfortunately, such a shift in architectures does not come without cost, nor will it satisfy all requirements. A demand-oriented architecture will require a more responsive space launch capability than we currently have. It will also require a change in satellite design philosophy to emphasize rapid production and deployment, perhaps at the expense of spacecraft lifetime. These trade-offs may reduce performance in some areas, which might be acceptable to some customers but unacceptable to others.

Problems will arise if recognized issues of coverage, responsiveness, timeliness, and so forth are not or cannot be addressed by the space architecture. If our space system design and operational philosophy remains closely linked to a cold war environment, our space architecture will likely be inadequate for the world of the next century.[3] Demands on space systems are rising as budgets decrease. Unfortunately, the acquisition, deployment, and to some extent operation of our space systems may remain caught in a vicious cycle of upwardly spiraling cost, complexity, and time, making it difficult to accommodate the changed circumstances. The technical problems will be compounded if institutional inertia and organizational turf battles are allowed to impede constructive change. What is needed is an objective method for deciding if the challenges can be better met by a command- or demand-oriented approach, or if elements of both are required.

This work is an effort to develop a methodology for comparing different space architectures. Since an overriding issue is how and why the question of space architecture matters to future national security, the work begins by describing the capabilities and limitations of space systems. This begs the question, though, of whether those limitations are absolute and intrinsic—unavoidable consequences of some characteristic of the space environment—or actually the result of the design choices made in creating the existing cold war based space architecture. Building on these basic issues, this effort paper next describes command- and demand-oriented space architectures in terms that allow objective comparison. Next, the work describes the fundamental factors—requirements, technology, and budgets—that determine future space architectures, and how these determinants affect different types of architectural approaches. The two approaches (command- and demand-oriented) are compared against a test case involving theater reconnaissance, surveillance, and target acquisition (RSTA). Though not comprehensive, this test case provides broadly useful insights into future options for national security space doctrine and policy.

Describing Space Architectures

Architecture: n. Construction or structure generally; any ordered arrangement of the parts of a system.

—Webster's Illustrated Contemporary Dictionary

A space system architecture, shaped by the determinants of requirements, technology, and cost at the time of its design, has inherent capabilities and limitations. Comparing architectural alternatives is the best way to highlight strengths and weaknesses of different approaches to developing a system of systems, but this requires a common framework. This section describes the advantages and limitations of space systems, asserts that not all the drawbacks traditionally associated with space systems are intrinsic, and closes by presenting a way of categorizing and comparing space architectures that are used in the rest of the work.

Types of Advantages and Limitations

A proper evaluation of alternative approaches to an issue begins with an objective discussion of the advantages and limitations of each approach.[4] Advantages and limitations of a class of environment-based systems (air, sea, land, or space) are either fundamental or derived.[5]

The first type (fundamental), which is based on the physics and phenomenology of the environment or medium, could also be called enabling or constraining. In other words, fundamental advantages (or limitations) cannot be altered, only overcome or exploited.

The second type (derived) is based on our ability to exploit the environment, which in turn depends on technology, doctrine, and cost.[6] Derived advantages and limitations, though related to fundamental characteristics, are subject to change as military forces for example, acquire new physical abilities and knowledge.

Distinguishing between fundamental and derived advantages or limitations can be difficult, especially when a way of operating has become so entrenched that its genesis and rationale are obscured. Failure to do so, however, may mean that the most effective solutions to a problem are not considered.[7] Thus, the ability to compare begins with an understanding of the recognized advantages and limitations of space systems and a realization that these are produced from an interaction of fundamental or environmental qualities with design choices.

The Advantages of Space Systems

Perhaps because the use of space for military operations, and particularly unclassified discussion of it, is a relatively recent phenomenon, and because applications of space power continue to evolve, there are nearly as many lists of the advantages of space systems as there are authors. For example, Joint Doctrine, Tactics, Training, and Procedures (JDTTP) 3-14, *Space Operations*, refers to the various missions space systems can perform (communications, navigation, surveillance, etc.) as space system capabilities.[8] More to the point, it describes space characteristics (extent, vantage, gravity, composition, radiation, temperature, and propagation) and operational considerations (difficult access, placement, long-duration flight, maneuver, global coverage, decisive orbits, weapons range, and organization).[9] While recognizing in the text both that the environment affects the characteristics of the systems and that this environment offers both opportunities and constraints, the JCS pub does not explain the concept completely. For example, it does not make clear what the net effect of the characteristics of extent and composition with weapons range and platform speed (an unmentioned feature) might be.[10] It also, probably necessarily, oversimplifies such concepts as orbit predictability. Except for some rather optimistic and unsupported statements, time and timeliness are hardly dealt with at the unclassified level as factors in space operations. Finally, the operational considerations are clearly based on existing systems; a valid approach, but one that may inhibit thinking about alternatives.

Evolving doctrinal discussions at US Air Force Space Command focus on the unique attributes of space systems: concentration, timeliness, continuity, and perspectives.[11] This list appears to be a step in the right direction, but it still contains some troublesome embedded assumptions about the architecture. For example, the attribute of continuity "relates to the long operational duration of spacecraft" implying "there is no need to generate forces during a period of increased tension or readiness."[12] This of course assumes we have (and can afford) all the capability we will ever need on orbit at all times, and also that we won't lose some of that capability (to mishap or hostile action) at unfortunate times.

The SPACECAST 2020 study conducted at Air University cited two "paramount advantages of space—unparalleled perspective and very rapid access to [distant points on] the Earth's surface."[13] These seem close to being fundamental. Perhaps significantly, the advantages were not asserted a priori, but culled from the ideas presented in the study.

Each of the authors or organizations impose particular biases on the use of space in describing space attributes and doctrine. These biases affect their interpretation of the advantages (and limitations) of space, so each list is somewhat incomplete. A reasonable synthesis of the fundamental advantages of space is shown in table 8.

Table 8

Advantages of Space Systems

Space Advantage	Reason
Nonterritorial operations	No worries about overflight rights or provocations in prehostility phases of a crisis.
Vantage point:	The ultimate high ground providing the following three features:
- Viewing angle	- Ability to avoid any obstructions as necessary
- Wide area perspective	- Ability to see an entire area of interst at once, potential for synoptic coverage
- High energy states	- High speed, useful for rapid transit or potentially to enhance weapons effects
Global access	Ability to get to any region on earth, support operations in separated regions.

These advantages are based on two characteristics of space. The first is that space operating restrictions are determined by the function of the spacecraft, not its location (unlike national airspace or territorial waters).[14] The second is that the physics of space systems place them higher than other systems and give them access to large areas of the earth in a relatively short period of time. These two features, manifested in table 8, seem both generic enough to allow further refinement and broad enough to capture the truly distinctive characteristics of space. The list is undoubtedly open to debate, but at this point only one difference from other lists will be highlighted: longevity (or continuity) is deliberately excluded. This is a design choice based on orbit selection and spacecraft characteristics, not an inherent quality of all space systems. Also, this "advantage" does not come without costs, as discussed later.

Of course, none of the advantages are unqualified, nor are they necessarily unique to space. Combinations of features (global access and nonterritoriality, for example) point out the unique contribution space can make, and provide the rationale for pursuing space solutions, even in the face of significant disadvantages and limitations.

The Limitations of Space Systems

Few authors, particularly in the space community, discuss the disadvantages or limitations of space systems in any detail. Such points are usually left to the advocates of alternate approaches (e.g., airborne or surface-based) as they compete for funding. As a result, several features of space systems that are more closely tied to design choices or even specific system concepts than to the environment itself have become accepted as generic disadvantages of space.

Space systems have perceived shortcomings in their ability to conduct routine, sustained, and effective military operations (table 9).[15] Efforts to overcome these limitations can take several forms: upgrades, mission diversion, or architectural change. The first, focusing on process and procedures, does not seek to address any fundamental limitations, but to improve space system performance at the margins, or in kinder terms, to take full advantage of existing capabilities. The second, mission diversion, involves replacing, augmenting, or avoiding the use of space-based systems through the use of such alternative means as airborne platforms for surveillance and reconnaissance, and terrestrial fiber-optic links for communications.[16] The architectural change response is the most radical and has arguably not been tried in the national security arena.[17] To explain how deliberate architectural choices affect space system characteristics, the study needs a framework for comparison.

Table 9

Perceived Disadvantages of Space Systems

Perceived disadvantage	Meaning
Distance	Space systems must operate remotely.
Predictability	Enemy knows when satellites will be overhead.
Poor continuity	Lack of dwell time and gaps in revisit time.
Poor responsiveness	Ability to respond to crises they weren't designed for (strategic) and to theater requirements (operational).
Inflexibility	Long planning lead times, difficulty of making changes.
Unsatisfactory timeliness	Inability to distribute information to end users quickly.
Vulnerability	To attack or natural disaster.
Environment	Harsh radiation, temperature, debris, etc.
Cost	Both space systems and access to space are expensive.

Developing a Framework

The first element of the framework is a series of definitions (table 10). To construct a generic framework for a space architecture the space, ground, and launch segments—fleshed out with their elements, as defined—make up one axis of a matrix. The second axis is the attributes. The result forms the basis for describing the specific features of an architecture, and thus allows comparison of different architectures. The real-world determinants of requirements, technology, and cost as described later provide additional detail and refinement.

Table 10

Space System Terms and Definitions

Architecture	The overall, grand design for the hardware, infrastructure procedures, and measures of performance of a "system of sytems." A strategic theory for exploiting space and a doctrine for employing space assests are implicit in an architecture, though these things may not be well articulated.
National security space architecture	The architecture assoicated with military, intelligence, and other functions commonly referred to as the "national security" sector.
Segments	Parts of an architecture grouped by their role and environment. The space segment is what remains on orbit for the duration of its mission. The ground segment is employed by space "operators" and "customers" to make the space segment useful to terrestrial operations. The launch segment is concerned with deploying the space segment, though certain kinds of "launch" vehicles may perform other missions.
Elements	The component pieces of the segments; for example, the ground segment would include command, control, communications, processing and distribution, logistics, and supporting infrastructure elements.
Operator	An organization that controls the activity of a space sytem.
Customer	An organization or individual with a need for a space product or service.
Attributes	The desired/required, implied or predetermined characteristics of the elements. For example, survivability (robustness?) is a general attribute, which is determined by a system's size, "hardness," maneuverability, stealthiness, and other properties (subattributes). Some measure of survivability may be required by military necessity and expected threat. The way this is specified will determine parts of other attributes, such as cost or logistics.
Functional area	Force enhancement, force application, space control, space support.
Mission area	A subset the functional areas, such as navigation under force enhancement.
Determinants	Operational requirements, technology, cost.

Elements of a Space Architecture

The challenge is to make the list of elements a useful break-out inclusive of different types of systems but not overly specific. One way to do this is to use general types of elements as described below, rather than listing every possible element of each segment.

The *space segment* consists of the mission payload, the spacecraft, and the constellation. The *mission payload* includes the sensors, transceivers, or other equipment that produce a satellite's capability. Depending on design, this could be either a fairly modular and easily identified element, or it could (in a highly integrated system) merge with the spacecraft element. Normally though, the *spacecraft element* provides support to the payload, power, navigation, control, and maneuvering capability, communications, and structure. The *constellation* is the number of satellites and their orbits. Together, the elements of the space segment determine much of the performance, lifetime, degree of ground support required, and other qualities of a space system.

The *ground segment* is composed of elements that support the satellites in orbit and exploit the information they provide, and can be broken down into telemetry, tracking, and control (TT&C), facilities and infrastructure, and user equipment. TT&C is primarily related to those functions needed to maintain the satellites in orbit and ensure they perform properly. Facilities and infrastructure are buildings, antennas, and other support equipment, but also any intermediate communications or processing capabilities needed to deliver and make the product of the satellites useful to their ultimate customers. This also includes common use equipment, such as the space surveillance network which keeps track of orbiting objects. User equipment could range from things like Global Positioning System (GPS) receivers and special satellite communications (SATCOM) equipment to field-deployable ground stations and tactical dissemination capabilities. The features of the ground and space elements interact strongly and provide many potential areas for trade-offs.

The *launch segment* includes equipment, facilities, and procedures needed to deploy the space segment. These can be divided into the command and control functions and the sites required to physically prepare and launch a vehicle. Of course the vehicle itself makes up the third category of launch segment elements. Although it is called *launch*, this segment would also include other functions, such as orbit transfer, recovery, and deorbit, or even suborbital missions. It may be worth calling this the *transport segment* as (if and when) the United States moves toward a more comprehensive and sophisticated space capability. This segment, though traditionally seen as completely subordinate to the requirements of the spacecraft designers, may in fact hold the key to flexibility in the other segments.[18] Summarizing the discussion above, the basic elements of a space architecture can be listed as in table 11.

Table 11

Space Architecture Elements

Segment	Element
Space	Mission payload
	Spacecraft
	Constellation
Ground	Telemetry tracking and control (TT&C)
	Facilities/infrastructure
	User equipment
Launch/transport	Command and control
	Launch sites/ranges
	Vehicle

By themselves, the elements described above offer only a physical description of a space architecture.[19] Functional characteristics, like data transmission, information processing, and data fusion, are in fact incorporated in the physical elements as are seen in later architectural description. To make value judgments about an architecture and especially to compare alternatives, some qualitative description is needed. For this purpose, the attributes below will help complete the picture.

Attributes of Space Systems

As defined in table 12, the attributes describe the characteristics of each element. These attributes should anticipate design requirements and possibilities, but not predetermine the actual design.

Table 12

Space Architecture Attributes

Attribute	Definition
Performance	Ability to provide a service with necessary detail, precision, and accuracy.
Responsiveness	Ability to deliver the required performance as needed and on time.
Flexibility	Ability to shift functional or geographic focus.
Robustness	The system should not fail catastrophically or become unable to perform its mission satisfactorily in the face of attack or mishap.
Logistics requirements	Quantity and type of support needed.
Reliability/availability	The chance of the system being fully or sufficiently operational day-to-day.
Ease of operations	Degree of specialized training required.
Environment impact	Amount of debris, waste or other pollution or need to construct new facilities.
Cost	Life cycle: research, development, acquisition, operation, and disposal.

The attributes in table 12 reflect the key considerations involved in designing space systems.[20] As with the elements of a space architecture, it is useful to group the attributes into categories rather than deal with specific items separately. The reason for this is that overspecifying the attributes can unintentionally foreclose design choices. For example, the generic attribute of robustness could be achieved in several ways involving the following interrelated (to themselves and to other attributes) qualities:

- survivability through hardening of spacecraft; location (altitude); proliferation/distribution of assets; stealth/ deception/decoys; defense (either organic or with dedicated platforms); and maneuver (this and defense depend on threat detection and assessment);
- ability to augment/reconstitute capabilities through on-orbit spares or rapid launch;
- graceful degradation of individual systems and/or the constellation; or
- reduced vulnerability to attacks on links and ground sites through autonomous satellites; antijam/low probability of intercept/encryption; and hardening, mobility, and/or proliferation of ground equipment.[21]

The attributes are presented without priority or weighting at this point. Adding that level of detail—deciding on the relative importance of the attributes—requires making strategic choices about the nature of the space architecture. Fully describing the elements and making design choices (such as the one on robustness mentioned above) requires both prioritization and application of real-world determinants. The framework is already of some use in describing generic types of architectures. Specifically, it can help illuminate the differences between command- and demand-oriented approaches.

Command and Demand Orientation

The distinction between command and demand orientation is significant because the two types are optimized differently and have different priorities. In this sense, there is a similarity to the debate over centralized versus distributed control of airpower.[22] The two types of architecture also imply significant differences beyond command, control, and organization, namely in the capabilities, design, and deployment of space systems.

A command-oriented architecture is a centralized approach, relying on central direction and control for efficiency and economy of force. In theory, as with the centralized control of airpower, this command-oriented system ensures that the best use is made of scarce yet flexible assets. Because of the nature of space systems (worldwide access) and the potential significance of the functions they perform, this kind of architecture responds first to national and strategic needs, leaving needs at the operational and tactical levels to be satisfied as lower priorities or as by-products of higher-level requests.[23]

Command orientation emphasizes the attribute of performance in specific tasks, which has several consequences. It leads to small numbers of large, complex, high performance, and long-lived satellites with highly specialized mission support infrastructure, and attempts to make long-range forecasts of future space system requirements. To deal with future contingencies, the system must anticipate unknowable demands, which often leads to the inclusion of performance "pads" in the design. The number of launches needed to maintain this architecture is small, though it often uses heavy-lift vehicles. The attributes emphasized here—as with the satellites—are performance and reliability.

Organizationally, a command-oriented architecture (in theory) has a single executive agent for the mission. In practice, however, the value of space systems for various missions and the security/secrecy requirements for "exotic" capabilities can lead to vertically integrated organizations to design, develop, and operate systems specialized along functional lines. Operations within each of these "stovepipes" are centralized, and then an additional element of centralization is added through coordinating or oversight committees. This phenomenon tends to improve the responsiveness of a system to its functional community, but at the expense of making access from outside that community more difficult.

To help visualize the nature of a command-oriented architecture, a matrix combining the elements and attributes of a space architecture can be used to reflect the priorities described above. Of necessity, this will be a rough portrait; it cannot readily incorporate qualitative features (such as the degree to which a spacecraft might need to operate autonomously) without the framework becoming much more detailed. Nor is it easy to portray the relative importance of the different elements in terms of resource allocation without creating confusion. As a first cut at describing an architecture type, as a possible basis for an operations analysis approach, and in preparation for applying the real-world determinants of the next section, this approach has some utility. Using this framework, a command-oriented architecture would look like table 13.

Table 13

Command-Oriented Architecture Priorities

	Space segment			Ground segment			Launch segment		
	Payload	Constel.	Craft	TT&C	Facilities	User	C^2	Sites	Vehicle
Performance	●	●	●	●	●	◗	●	●	●
Responsiveness	◗	◗	◗	●	◗	●	◗	❍	❍
Flexibility	◗	◗	❍	❍	◗	❍	❍	◗	❍
Robustness	●	●	●	◗	❍	❍	◗	◗	❍
Logistics requirements	❍	❍	❍	❍	❍	❍	❍	❍	❍
Reliability	●	●	●	●	●	●	●	●	●
Ease of operations	❍	❍	❍	◗	◗	◗	◗	❍	❍
Environmental impact	❍	❍	❍	❍	❍	❍	❍	◗	❍
Cost	❍	❍	◗	❍	❍	◗	❍	❍	◗

For simplicity, the table uses only three levels of priority, with darker symbols indicating greater relative weight/emphasis of each attribute-element pair in design considerations (● = high, ◗ = medium, ❍ = low). This does not mean that a low emphasis is unimportant, only that it would fare poorly in a trade-off with a higher priority item. Finally, this is an attempt to describe a hypothetical command-oriented architecture, not one that exists in the real world.

In contrast, a demand-oriented architecture is organized around the attributes of responsiveness and flexibility.[24] Again in theory, this type of system would accommodate the needs of any potential user with the priorities determined by a given situation. To support these goals, a demand-oriented architecture would consist of relatively (to command orientation) larger numbers of smaller, more autonomous, specialized, and short-lived satellites deployed in constellations that could be tailored to specific situations. Because of the larger number and more rapid launches that would be required, launch systems would be driven by two primary attributes—responsiveness and cost—and would operate much more like current air transport. Specialized infrastructure—from launch through end user equipment—would be minimized, either by a reduction in infrastructure requirements or through sharing of infrastructure with other systems.

Organizationally, command and control would be decentralized to some extent, for example with fielded units at some level able to directly task as well as receive information from space systems, though overall spacecraft "health and welfare" functions might be performed centrally. The danger that a demand-oriented system presents, if poorly coordinated, is the same as that of decentralized airpower-potentially inefficient, poorly coordinated, and misdirected effort (table 14).

Table 14

Demand-Oriented Architecture Priorities

	Space segment			**Ground segment**			**Launch segment**		
	Payload	Constel.	Craft	TT&C	Facilities	User	C^2	Sites	Vehicle
Performance	◗	●	◗	◗	◗	●	●	●	◗
Responsiveness	◗	●	◗	◗	●	●	●	●	●
Flexibility	○	●	●	●	◗	◗	●	●	●
Robustness	◗	●	◗	◗	○	●	●	●	●
Logistics requirements	○	◗	◗	○	○	●	○	●	●
Reliability	◗	●	◗	◗	◗	●	●	○	◗
Ease of operations	●	◗	●	●	◗	●	◗	◗	●
Environmental impact	○	○	○	○	○	○	○	◗	◗
Cost	●	◗	●	◗	◗	◗	●	●	●

In comparison to the command-oriented architecture, this illustration shows differences in the attributes that are important for particular elements, as well as differences in the priorities of attributes across the architecture. This is particularly noticeable in comparing the priorities for the launch vehicle and in comparing the emphasis placed on the flexibility and reliability of different elements in the two architectures. It also shows that there are more high priorities in the demand-oriented architecture, perhaps an indication of why creating one may be difficult. Although not fully representative of the differences between the architectural types, the chart illustrates the value of building an analytical framework.

The next section explains the priorities and why some of the features described as part of one architecture are not available to the other. For several reasons, pure architecture types cannot exist in the real world. Some of those reasons, which will help to introduce the real-world determinants, can be illustrated by a brief look at our current architecture.

Current Architecture

A thorough description of our current national security space architecture is not possible in an unclassified paper, but the outline of its functions shows that it is primarily command-oriented. The four Joint Chiefs of Staff (JCS) space functional areas are force application, space control, force enhancement, and space support.[25] Except for ballistic missiles, which still have only a strategic nuclear mission, we have no force application capability from or through space. Likewise, we have no space control capability except for the monitoring function of the space surveillance network. The force enhancement mission areas that are currently supported are navigation, communications, missile warning, environmental sensing, and RSTA.[26] Space support consists of launch, satellite control, and logistics.

With few exceptions, the architecture reflects the characteristics of command orientation. Overall, considering the number of missions performed and potential customers, there are a relatively small number of spacecraft.[27] Satellite constellations tend to reflect the coverage needs of the cold war.[28] We also have a small number of operating sites—the primary ones are at Shriever AFB, Colorado, and Sunnyvale, California, and there are only two launch sites.[29] Our launch vehicles and operating procedures are not able to respond rapidly to a crisis.[30] Finally, those who can task, communicate with, or even receive information from a space system directly are relatively few.[31]

Such functions of the current architecture as communications, certain intelligence indicators, and missile warnings are now provided relatively transparently to the ultimate users through such existing channels as the tactical information broadcast system (TIBS), tactical receive equipment (TRE), and tactical related applications (TRAP). These are excellent examples of a push approach, since the transparency of information delivery causes users to often be unaware of the contributions of space systems, or the potential or procedures to get additional information.

In practice, the architecture was designed to respond to the needs of the National Command Authorities and national intelligence centers and to support strategic nuclear missions. It still has these as its top customers and priorities.[32] The architecture has evolved over the past few years, but it has done so by exploiting built-in but underused capability, not by changing its basic orientation.

Of course, there are exceptions. The GPS system is one obvious example with widespread applications. Also, there have been numerous Tactical Exploitation of National Capabilities (TENCAP) initiatives by the services, especially since the Gulf War, to make national systems more useful to theater commanders in chief (CINC) and war fighters. The creation of the Space Warfare Center and space support teams promise to bring in some elements of demand orientation, but these measures do not change the basic characteristics of the architecture. Access, allocation, and priorities are decided centrally, and there are only a few assets to satisfy many needs.

There are many interrelated reasons for this focus. Security has played a major part, since there exists the need to limit knowledge of our most sophisticated capabilities. Security will continue to be a source of tension given a limited number of assets, since any knowledge of their operating procedures could compromise their effectiveness. A lack of well-documented requirements for expanded capabilities and in some cases an inability to articulate requirements from the side of the war fighter remains a factor. Bureaucratic politics have also played a part. Those organizations that in the past successfully pressed a claim to some control over a capability now are reluctant to give up any of it. Technology has certainly been a factor, since for many years our space systems were on the cutting edge and therefore limited by what was deemed possible. Cost, which certainly relates to technology, is often a deciding factor in whether we can do a certain mission and how it will be done. Finally, national politics, whether of the visionary or the pork barrel sort, has affected everything from the direction of space research and development (R&D) to the nature of our spacelift and space access.

Perhaps the bottom line is that our current space architecture was not built as part of a grand design, but rather evolved gradually under the pressure of many influences. Policy makers are now struggling with technical, physical, and bureaucratic inertia, and the various demands of a changed national security environment, shrinking budgets, and an exploding technology base to determine the future of our space architecture. The question for the next section is whether there is a rational way to evaluate these many influences, and what messages this process might hold for the future direction of space systems.

Applying Real World Determinants

Even if "ideal" command or demand architectures do not exist in the real world, it is useful to ask when a bias toward one approach or the other is appropriate. No general discussion can anticipate all the factors that might affect the choice of a system or architecture. In keeping with the theme of a framework for comparison, the study proceeds with a method for applying real-world determinants in the areas of operational requirements, technology, and budget to the framework of space architecture elements and attributes.

The first step is to identify the determinants, describe how these challenge assumptions made in the past, and describe how the determinants interact. Finally, the determinants are applied to the generic framework of command or demand architectures to show how the inherent assumptions and restrictions of each produce different implications.

Real-World Determinants: Requirements

In the real world, requirements are debated endlessly and often have different meaning to different people. Requirements also tend to be focused on specific missions or mission areas, at least when formalized as official documents. Though developing detailed requirements in itself implies some analysis, there are a few generic requirements for future space systems that would seem to apply across the board.[33]

The first is that in the uncertain international environment of the post-cold-war world, we cannot optimize coverage of any particular region for an indefinite length of time. US interests are global, and our potential enemies are both less obvious than the Soviet Union, and more likely to be changing (this year's friend could be next year's revolutionary trouble spot). Compounding this problem is the fact that fewer US forces will be forward based, so that much, if not all, of the ground support equipment we need to exploit space in response to a crisis will have to be deployed from the continental United States.

The second requirement is for capabilities to be available at the earliest possible stages of any crisis. History suggests that a prompt and appropriate response to a developing situation can often obviate the need for a more drastic response later. To make this possible, the United States must have forces, including space systems, that can be on the scene, tailored to the situation, and fully operational in limited time. The question is just how short the reaction time must be; shorter is likely to be better, but at what cost?

The third requirement is that systems be able to function with little strategic warning and, perhaps in the case of space systems, that they provide the strategic warning. In other words, systems must not only have short operational and tactical reaction times (the issue above) but will have to be adaptable to vastly different types of situations.[34] Crises of the future will tend to pop up unpredictably or else suddenly flare up after a long period of dormancy to grab the headlines and demand attention from policy makers. Somalia, the previously repressed nationalist and ethnic conflicts in eastern Europe and the former Soviet Union, and North Korea's nuclear weapons are all recent examples. The dilemma posed by this and the preceding requirement is that the kind of coverage needed for global situational awareness is so massive that it will tax our ability to deploy and operate the systems and assess the information.

The fourth requirement is that capabilities be flexible enough to respond to many different types of crises, from large-scale armored attacks to humanitarian relief operations. Also, the demand for the services of our space architecture is likely to expand suddenly and massively. For example, the desire to limit collateral damage in wartime and the possibilities of precision weapons have opened the door to potentially huge requirements for extremely detailed data on short notice. Worldwide deployments in response to crises could mean great surges in demand for remote, high bandwidth communications capabilities. The dilemma is whether to build capabilities that will be insufficient and then prioritize tasks, build in so much excess capability that unanticipated tasks can be accommodated, or try to augment and update capabilities as required.

The final general requirement is that our systems perform their functions with little or no delay for processing, analysis, and transmission of information. This has been expressed in many ways—real time, near real time, and in time—and implies not just the delivery of a product, but its delivery to exactly the right customers in an immediately useful form. In a future world where transit through space is used for rapid delivery of cargo, people, and weapons, these concerns will apply to the physical as well as the ethereal.

In summary, the future national security space architecture will have to function globally, bring its full capability to bear on an uncertain enemy and situation rapidly, and provide enough of the right kind of service in near real time. Many aspects of this situation favor space systems of any kind, but not without reservation, especially when we must operate in a constrained budget environment as is discussed below.

Real-World Determinants: Technology

This study does not explicitly evaluate all technologies that could contribute to space systems. As in the area of requirements, there are some trends and general issues that merit consideration. The first is the general trend away from the Department of Defense (DOD) leading developments in high technology sectors of the economy to DOD's product cycles trailing far behind those of the commercial world. Arguably, this is a reversal of an historically atypical post-World War II trend, but the implications for development of future systems are profound. As equipment takes longer to produce, it will increasingly include out-of-date components, design practices, and materials. This is true in many militarily significant areas such as microelectronics, though not in certain niche areas such as armor plating and nuclear submarine construction. The question faced is whether space systems are one of those niche areas or not.

A related issue is the current trend favoring dual-use spending for government research and development money. How well do space systems take advantage of this trend? Will a dual-use focus allow the government to continue investing as much as it believes necessary in all the military niches? If not, what are the priorities in technology development, and do they support space system requirements?

In specific technology areas, advances over the past few years have been dramatic. This is particularly true of microelectronics and microprocessors. Not only is their capability today much greater than anything expected when our current space systems were designed, but progress in the near future may be even more rapid. Are military systems in general, and space systems in particular poised to take advantage of this?

Both military and commercial R&D have made possible advances in command, control, and communications. Higher bandwidth links, especially using lasers, new methods of compressing information to fit into less bandwidth, more efficient ways of managing communications channels, the development of more autonomous machine capabilities, and the development of expert systems to reduce human workloads are all examples. Has the space system design kept up?

Several technologies funded by the Strategic Defense Initiative Organization (SDIO) during the 1980s appear close to fruition now. These include miniaturization of sensors, many spacecraft components, and the ability to design and build smart structures that provide strength, rigidity or precise alignment, and vibration control at a fraction of the weight of current designs. Materials technologies, advanced by many different research and development efforts, also offer a chance to reduce weight or increase performance of structures and surfaces.

Both the commercial and to a lesser extent the military sectors of industry have made progress in the related fields of standardization, modularity, and flexible manufacturing.[35] Together, these capabilities allow products tailored to a specific customer's desires to be produced quickly without requiring extensive, costly redesign, testing, and fabrication by hand. How well do space systems take advantage of these capabilities and trends?

On the negative side, there has been relatively little progress in recent years in improving spacelift capability. With minor exceptions, such as the Pegasus small launch vehicle, our systems and operating concepts remain closely tied to the ICBM-derived launchers we have used since the beginning of the space age.[36] Concepts that could radically cut costs and improve access to space would seem to merit high priority, but the efforts and results to date have been paltry.[37] Is this because of technological hurdles or because of a lack of institutional agreement on what is needed? Can space architecture comparisons shed any light on this issue?

In general, reviewing technology and technology trends raises the issue of what are the best choices or combinations for a future space architecture. Does the nature of an architecture affect its ability to apply new capabilities? Do technologies make possible some things thought unworkable in the past?

Real-World Determinants: Budget

No discussion of real-world determinants would be complete without the bottom line. Cost has already been raised as an issue in terms of how much capability we can afford, and what sort of research and development we will be able to pursue, so what are the general outlines of the budgetary determinants?

First, absent a new perceived threat to our national survival, defense budgets likely will continue to decline absolutely and in purchasing power in the near term. In an effort to prevent the current military from becoming a hollow force, the research and procurement accounts of the budget will probably be sacrificed to maintain current readiness. Space systems are no exception: the prospect for new system starts in the near term is poor and getting worse, and the acquisition community seems unable to produce any new answers.[38] Even the development programs in the "black" world, traditionally thought to have almost unlimited budgets to get their job done, seem to be feeling the pinch.[39]

As research, development, and acquisition budgets shrink, there is increasing emphasis on reducing the life-cycle costs of systems, including operations, maintenance, and disposal along with procurement. The catch-22 is that building systems with lower life-cycle costs requires more up-front investment in improved designs. In a worst case, this could mean no options but incremental upgrades to system designs. Again this raises the question, do space architecture alternatives offer any way out of this dilemma?

Finally, the budgetary environment raises the question of whether anything can be done in the national security space business to take advantage of the market forces of the commercial sector. Although this issue has mostly been discussed in terms of the commercial sector providing such services as launch, communications, and even remote sensing, we should ask if there are space architectural options that might be more adaptable to a world in which market forces, not government priorities, drive most investment decisions.

How Do the Determinants Interact?

In discussing the determinants, many of their interactions have already become apparent. Requirements drive a system toward greater capability while budgets place limits on what can be done, whether in terms of numbers, quality, or the amount of research and development. Technology, however, can cut both ways. It can force costs higher while enhancing performance, or it can make a mission possible with fewer resources than before. Sometimes technology can create new missions or capabilities, which are very difficult to quantify.[40]

Generally, the interaction of the determinants produces questions that must be answered by engineering trade studies. Can enough assets be kept on orbit to cover all situations? Conversely, can an augmentation be deployed fast enough to matter? Can the ground support equipment needed to make use of our space assets be deployed in a timely manner? What is affordable? Is there a way to get more capability for the same or less money? What are the priorities? Do we/can we sacrifice missions and reduce manning?

Recognizing the way the determinants interact is crucial, because doing so exposes the steps needed to solve a problem. By way of illustration, consider the process of designing a satellite. If the design process begins with requirements that specify a certain satellite lifetime, those requirements will drive several design features such as the quality of parts, redundancy in the system, and the amount of fuel for orbit maintenance. These features, combined with the mission of the satellite, determine its weight and orbit, hence the launch vehicle required. If access to space is expensive, and the number of satellites being launched is small, requirements and fiscal pressure will drive the designers to add additional capability to each satellite, thus increasing its complexity and weight. In extreme cases, this could force the satellite to be launched from a more capable (and expensive) launch vehicle. At this point, recognizing the amount of money being invested in this single system and the number of requirements it is intended to fulfill, designers will feel pressure to make it even more reliable and longer-lived. This means even higher quality, more redundancy, and so forth. Concurrently, recognizing that the system will be on orbit for many years, designers will need to build in additional performance margin. All of these activities lengthen the time needed to build, test, and deploy the satellite, and increase costs dramatically. The result is a stagnating development system, a dearth of successful new program starts, and a reliance on modifications to proven but often dated designs to keep costs under control. Unfortunately, this is very much the situation that the space research and development community finds itself in today. Figure 2 is a simplified illustration of how the interaction of real-world determinants, through three linked cycles of design, performance, and lifetime raise costs, and how this in turn creates demands for more costly features.

The key to breaking the vicious cycle of space system acquisition and getting more capable satellites on orbit rapidly and

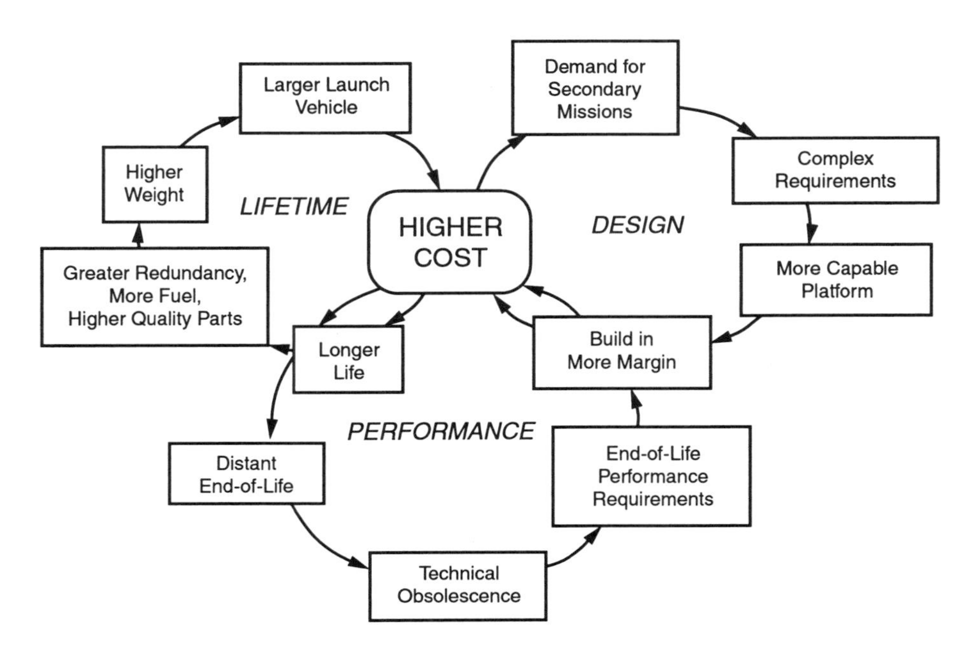

Source: E. Dionneand and C. Daehnick, "The Future Role of Space Experiments," American Institute of Aeronautics and Astronautics (AIAA) Paper 92-1697, presented at the AIAA Space Programs and Technologies Conference, Huntsville, Ala., 24–27 March 1992, 2.

Figure 2. Space System Cycles

affordably lies in understanding the nature and causes of this interaction. Because different types of space system architectures address requirements and take advantage of technology differently, evaluating those architectural approaches may produce some useful insights.

How Do the Determinants Affect Architectures?

Summarizing the determinants in a compact form produces table 15. If it were possible to represent these determinants and the elements and attributes of a space architecture mathematically, the matrix in table 13 or 14 could be cross multiplied with table 15 to produce a complete description of an architecture.[41] Such precision is unlikely to be useful, though, in dealing with qualities that are difficult to estimate and that often involve value judgments. A more subjective and qualitative approach is likely to be more useful.[42]

Two questions need to be answered. What affect do the determinants have on specific elements of the architecture, and how does one apply the determinants to the attribute-element pairs of tables 13 or 14? To illustrate the process, two elements of a space architecture are evaluated: the constellation and the launch vehicle. As should be clear from this and the following section, those elements provide a good representation of the differences between command and demand systems, though a complete picture is only possible if the other elements are incorporated.

Applying the Determinants

The first step is to take a simplified version of the matrix used for tables 13 and 14 (reflecting only one element) and add columns for each of the determinants. Money figures both as an attribute (cost) and a determinant (budget). This is because money is both a characteristic of design choices (better parts cost more) and a sometimes (seemingly) arbitrary restriction imposed for nontechnical reasons. The matrix is used to record qualitative implications (derived from observation) of each of the determinants in table 15 for each attribute of the selected element. A priority column reflects what was assigned

Table 15

Space Architecture Determinants

Requirements	Technology	Budget
Global coverage	DOD ability to drive technology	In decline, especially for research, development, and acquisition.
Early access	Increased emphasis on dual use	Need to reduce life cycle costs
Pop-up crises	Microprocessor revolution	Can market forces be tapped?
Flexible, expandable capabilities	Command, control, and communications improvements	
Rapid throughput	Miniaturization, structures, material	
	Standardization and modularity, flexible manufacturing	

in the previous section, and gives an idea of how to weigh the implications when assembling an overall conclusion.[43] At this level, of course, without discussing a particular mission area, specific requirements cannot be formulated. For now, the differences between command and demand orientation can be illustrated relatively.

Implications for a Command-Oriented Architecture

From the general principles of a command-oriented system—that efficiency or economy of force and therefore centralization are most important—the implications or features of the architecture for each element can be surmised. These are presented in tables 16 and 17 for the satellite constellation and the launch vehicle. It's important to remember that the effects of various determinants are highly interactive throughout the architecture.

A few points about the command-oriented architecture stand out. First, the architecture responds to real-world determinants by building relatively small numbers of highly capable and expensive systems. The space assets are re-

Table 16

Constellation Implications, Command-Oriented Architecture

	Priority	Implications — Constellation		
	Constel.	Requirements	Technology	Budget
Performance	●	Fewer, larger, more capable satellites	Mission-specialized, over-designed, long lead times	Government the sole customer
Responsiveness	◗	Adapt/exploit existing capability	Design to customer spec, leads to "stove-pipes"	Add as many satellites as budget allows
Flexibility	◗	Add multiple functions	Improve C^3, distribution	More satellites?
Robustness	●	Emphasis on individual satellite survival	High mean time between failure, redundancy, best available at tech freeze	Hardening, counter-ASAT (antisatellite) accidents?
Logistics	○	Preplanned launch of spares/replacements	Each satellite unique	Limited incentive to improve
Reliability	●	Likely to need all satellites at all times	Redundancy on each satellite, high reliability parts	Plan for large ground C^2 network
Ease of operations	○	Specialized operators needed	Focus on ground segment upgrades	Limited incentive to try new methods
Environment	○	Boost higher or deorbit	Extra fuel	No money for nuclear
Cost	○	Emphasis on capability, regardless of price	Investment leading to better mission performance	Space segment a large portion of life cycle cost

placed infrequently, and these factors lead to small numbers of launches of relatively high-performance vehicles.[44] In the case of each element shown here, the need for reliability is ensured by building more capability and redundancy into the hardware and the procedures, a practice which achieves the goal but at significant cost.[45] In turn, the cost of keeping the system working strongly affects the ability to invest in radical changes to hardware or operating procedures; these simply don't have the priority to get funded. The result is a relatively slow evolution of capability, limited ability to exploit commercial developments, and ever-increasing operating costs.

Table 17

Launch Vehicle Implications, Command-Oriented Architecture

	Priority	Implications — Launch Vehicle		
	Vehicle	Requirements	Technology	Budget
Performance	●	Driven by large satellite, orbit	Proven designs, upgrades to increase payload	Dual-use fine but government requirements primary
Responsiveness	○	Months of notice	Build on launch pad okay	Minimize infrastructure investment
Flexibility	○	Vehicle tailored to satellite	Limited use of standardization	Whatever is needed to get the job done
Robustness	○	None built-in, need to manage risk	Careful procedures, reduce risk	Better to accept delay cost than have one fail
Logistics	○	Whatever is needed	Proven techniques	Limited incentive to reduce
Reliability	●	Single loss catastrophic	Prefer proven systems	Unlikely to invest in new concepts
Ease of operations	○	Large numbers, contractors needed	Use specialized equipment to meet performance goal	Little incentive to invest in improvements
Environment	○	Performance still key	Expendables, solid boosters acceptable	Only highly toxic additives insupportable
Cost	◗	Need to buy small numbers of expensive vehicles	Refinements such as payload increases, but no radical change	Focus on reducing research, development, and acquisition cost

Implications for a Demand-Oriented Architecture

Demand orientation has responsiveness and flexibility as its overriding goals or principles. To this end, the performance of any individual piece of the architecture is less important than overall capability, with implications as seen below.[46]

The assertion that a demand-oriented architecture will trade off some capability to save money may make some people uncomfortable, and sounds like the claims of the military reformers of the early 1980s that our systems were too complex and expensive to work well.[47] In fact, demand-oriented systems do not try to push the state of the art in technologies, but they do try to take advantage of the most recently available technology and to get it operational faster. This still results in capable systems using advanced technology, but does not require deployment of a system to wait for programmed innovation (tables 18 and 19).

As with the constellation, the need for new investment in launch vehicles appears to be a problem given the real-world budget determinants. One could argue, however, that the type of investments needed by the military closely parallels the type of investments needed for the commercial space launch market and for emerging markets such as rapid surface-to-surface cargo delivery.[48] In general, the demand-oriented system is better positioned to exploit technological advances as they occur regardless of who has sponsored them.

The type of changes called for—improved operability, reduced cost-per-pound to orbit, and more rapid response—

Table 18

Constellation Implications, Demand-Oriented Architecture

	Priority	Implications — Constellation		
	Constel.	Requirements	Technology	Budget
Performance	●	Emphasis on systemic versus satellite measures	Distributed architecture, use most recent technology	Because of the requirement for incorporation of multiple new technologies, need more RD&A money; this is somewhat offset since many of the technologies are being pursued commercially.
Responsiveness	●	Right product available quickly to all users	Tailored systems, rapid build and launch	
Flexibility	●	Adapt to changing situation	Standardization, modularity, C^3, on-board processing	
Robustness	●	Proliferate, degrade gracefully	Autonomy, distribution, C^3, on-board processing	
Logistics	◗	Augment and replenish	Standardization, modularity	
Reliability	●	Backup/swing capability vice individual system	Redundancy, self-healing constellations	
Ease of operations	◗	More systems > need for standardized operations	Autonomy, C^3, processing, expert systems	
Environment	○	Boost or deorbit	Extra fuel, short-life orbits	No money for nuclear
Cost	◗	Trade off some capability for affordability	Technology investment requirements heavy, but dual-use a possibility	

Table 19

Launch Vehicle Implications, Demand-Oriented Architecture

	Priority	Implications — Launch Vehicle		
	Vehicle	Requirements	Technology	Budget
Performance	◗	Less payload needed	Aid rapid access to space	Need for investment in operability of launch systems; requires a shift to a new kind of vehicle while keeping existing capabilities working through a transition
Responsiveness	●	Launch on demand hours/days	Aircraft-like operations	
Flexibility	●	Surge capability	Standard interfaces, reusable vehicles	
Robustness	●	Multiple vehicles/ launch sites	Ability to operate from multiple sites	
Logistics	●	Minimize	Reduce special handling equipment	Reduce expenditures
Reliability	◗	A figure of merit, not a hard fast requirement	Only what is consistent with safety	Gradual approach; improve with practice
Ease of operations	●	No need for contractor support	Ability to operate with reduced support	Build on aircraft experience?
Environment	◗	More launches imply need to reduce impact	Reduce noise, toxins, waste	Avoid cleanup, legal restrictions
Cost	●	Bring down cost per pound to orbit drastically	Reusable vehicles, smaller payloads	Focus on reducing operations costs

would benefit any architecture, but only the demand-oriented architecture requires them. In the command-oriented approach, there is little or no incentive to invest in the new kinds of capabilities mentioned above. The type and number of space systems being deployed, the way in which the architecture responds to new requirements or unexpected events, and the underlying philosophy of what is important all determine the kind of support infrastructure, including launch.

One other point bears mentioning. Smaller payloads may be compatible with reduced cost-per-pound to orbit. This goes against conventional wisdom, since in any aerospace vehicle the greater the payload, the more the costs can be spread out. However, just as airlines do not operate 747s on every passenger route, there is a limit to economies of scale through size.

First, the vehicle must be purchased and large systems will cost more. This drawback is compounded by the need to spread larger development costs over a (generally) smaller production run. In operations, if an airline cannot fill the large vehicle, it will not get all of the benefit of that vehicle's lower operating costs per pound of payload. Finally, in the case of space-lift systems, range (which tends to favor large air vehicles) is not a factor since almost all of a launch vehicle's energy is used to raise its speed. The benefits of large structures (like wings) are reduced because there is no cruise regime, and the penalties are increased (all the mass minus fuel must be accelerated to the final velocity). Although the trade-offs are complicated, the implication is that designing to a specific payload size is a poor way to build a space-lift system.[49] Maximizing operability and minimizing life-cycle cost is better. If access is cheap enough, payloads will be redesigned to fit.[50]

Narrowing the Focus

The above examples are somewhat general and certainly not as rigorous as possible, and improving them requires additional detail. It may be possible to compare the performance of space architectures in detail across all mission areas at once, but that is beyond the scope of this study. Comparing the advantages and disadvantages of command versus demand systems for a single mission area should illustrate the process and provide some additional insights for the big picture.

Architectural Comparison for the Theater Reconnaissance, Surveillance, and Target Acquisition Mission

As presented to this point, the framework for architectural comparison does not say much about when command- or demand-oriented architectures are preferable. Adding a specific mission focus is the next step.

This section describes the general outlines of theater RSTA requirements, shows how these affect (or are affected by) the other determinants of technology and budget, and applies them to the elements and attributes of the competing space architectures. This illustrates the method and produces some useful insights about architectural choices.

Why Examine Theater RSTA?

RSTA is an expansion of the traditional reconnaissance and surveillance missions.[51] Theater RSTA is an essential part of space support to the war fighter since it supports the theater CINC or his forces.[52] Although the emphasis on theater-level operations may change, currently it serves as the basis for most force planning and strategy discussions.[53]

Theater RSTA is a good example, despite limitations on unclassified discussion, because it provides the full range of design responses—upgrades, diversions to different platforms, or architectural change—to shortcomings identified from operational experience.[54] Further, it combines the significant issues relating to space architectures with a mission important enough to highlight the consequences of making poor space choices.

RSTA Mission Description

The theater RSTA mission involves providing the United States and the theater CINC awareness, flexibility, and information needed to respond to actual or potential crises. This ability must be available throughout all phases of an evolving situation, from precrisis indications and warning through hostilities to postconflict monitoring. Theater RSTA includes a wide variety of specific tasks determined by the forces involved and the information they require, and these tasks are not solely military, especially in those phases of the situation that do not involve armed conflict. Table 20 summarizes these issues conceptually as a prerequisite to determining mission requirements.

Table 20 focuses on the specific contributions RSTA can make to the theater mission. RSTA provides information in a way that accommodates each phase of the crisis and adapts to potential enemy action. This adaptation can result in such tasks as augmentation and reconstitution. The table does not

Table 20

Theater RSTA Description

Phase	Function	Meaning/Tasks
Precrisis	Monitoring	Global basic awareness (framework system)
Emerging crisis	Access	Quick reaction augmentation for theater of interest - improved synoptic coverage - gather additional detail; intelligence preparation of the battlefield - limited war-fighting capability if needed
War	Exploit the "high ground"	Theater-level situational awareness Timely location of enemy forces, description of their activity Reduce effectiveness of camouflage, concealment, and deception Detect and characterize "indicators," aid in identifying centers of gravity Find specific targets; report information to "shooters" "in time" Augmentation as appropriate Replenishment and/or reconstitution as needed
Postcrisis	Drawdown redeploy, but maintain awareness	Monitoring as necessary - unobtrusive; noninvasive if appropriate Deactivation/redeployment when no longer needed - or replenishment and augmentation for continuing mission

Source: Air Force Space Command, "Space Primer," preliminary draft, February 1995, and personal experience in discussing the requirements for future RSTA systems with personnel at Air Force Space Command and Air Combat Command in 1992–1993.

extend to such derived capabilities as deterrence based on the enemy's knowing that his adversary is watching and can react. Nor should the table imply that only space forces can perform the theater RSTA mission. A space system that performs or supports this mission will have the elements of the architecture already presented, though many of those elements will support other missions as well.

Questions for Architectural Comparison

Each part of the RSTA mission raises questions about the type of architecture needed. In the past, the desire of the United States to monitor and anticipate crises in "important" parts of the world, coupled with fiscal constraints, has meant that some theaters were much better covered than others.

Keeping in mind the generic requirements of the previous section, space planners must ask if the national security environment of the future will permit the United States to maintain the disparity between theaters, or if something like global situational awareness or global presence is needed.[55] If the United States needs expanded capability, how can our RSTA forces achieve it, and what can we afford? Likewise, should the United States continue to place most space RSTA investment in systems that provide highly detailed coverage of relatively small areas of interest?[56] At the same time, as precision weapons delivery capabilities improve and the national leadership's and American public's desire for economy of force and lack of collateral damage demands ever more accurate target information, do RSTA systems not also need to provide more highly detailed information of more types than ever before? The goal of architectural comparison is to illustrate the trade-offs involved and suggest answers to these questions.

RSTA Mission Requirements

Table 20 shows a need for a time-phased mix of presence, persistence, and access to respond to an emerging geopolitical crisis. By their nature, space systems will usually be first "on the scene" and will provide the initial RSTA functions. Depending on the situation, US objectives, and the means available, additional capabilities to augment the RSTA architecture in the theater of interest could be deployed.[57]

A natural example of a RSTA mission is the detection, location, tracking, and targeting of theater ballistic missile launchers.[58] Briefly, RSTA assets must be able to aid intelligence preparation of the battlefield by gathering information on bases, operating areas, order of battle, and so forth prior to any hostilities, keep this information updated as the crisis evolves, and determine the location of as many launchers as possible at all times and provide sufficient information for targeting. Once hostilities have begun, RSTA systems will need to locate as many missiles and launchers as possible before they can do any damage to friendly forces, keep track of the missiles' movements, and provide timely targeting updates to weapons platforms. Qualitatively, an RSTA architecture will have to include the features listed in table 21.

Table 21

Theater RSTA Qualitative Requirements

Quality	Requirements
Access	All parts of the theater, unrestricted by enemy defenses
Coverage	Wide area synoptic plus ability to focus on specific areas
Revisit time	Allowable gaps in coverage will depend on target; days for fixed sites with little activity, hours or minutes for mobile forces
Spectra	Sufficient to penetrate weather, camouflage, and foliage, and to aid in target discrimination and identification
Resolution	Consistent with requirements for target identification and status determination
Geolocation	Sufficient to cue other sensors, provide adequate target data to weapons
Information dissemination	Ability to provide enough information of the right kind to all customers in a timely manner as often as necessary

These requirements present three challenges to a theater RSTA architecture. The first concerns sensors. Some of the requirements are impossible for a single sensor to satisfy simultaneously, for example, the need for both wide area coverage and high-resolution information.[59] It is also impractical to put sensors that cover all the relevant spectra—spanning at least radar to infrared wavelengths—on a single platform.[60] It may be impractical, depending on cost and employment constraints, to deploy some of the sensor types ideally used in a given situation.

The second challenge is that the type, quantity, and timeliness of RSTA information needs vary considerably among customers. Aircrews planning missions will need the most current threat information for ingress, egress, and the target area; details on aim points; and sufficient information to acquire the target with onboard sensors and place a weapon "in the basket." Campaign planners will need detailed information on particular targets: hardness, extent, dispersal, other physical characteristics, and the targets' use and interaction with other aspects of the enemy system. In general, planners need any information that will help determine the importance of a target to the enemy's war effort, achieve our campaign objectives, and assess the target's vulnerability to attack. Assembling enough information and performing this kind of analysis will take time so planners can usually live with somewhat less reporting timeliness.

In assessing effects, timing, timeliness, and detail are all important.[61] Senior military leaders will want a broad overview of events in the theater so that they can try to judge if events are unfolding according to plan. Although to some extent this overview can be synthesized from detailed information, that approach risks missing the forest for the trees. Policy makers may want to use RSTA to look for indicators of enemy intentions; thus, they may need to examine in detail areas of little use to other RSTA customers. Finally, events may force a diversion of RSTA to address a task because of its political or strategic, as opposed to operational, import.[62] All of these demands for information will have to be accommodated by an RSTA system.

The third challenge is an outgrowth of the first two. Given the competing and sometimes conflicting demands for information, how does the architecture respond? Has the architecture been set up to accommodate all users? How can capabilities be augmented? Can national-level capabilities be dedicated to a theater, and under what circumstances might they be recalled? Who "owns" theater RSTA assets? How can they be used most efficiently and effectively? Although the theater war fighter is intended to be the focus of theater RSTA, can the involvement of national-level agencies be avoided? These questions address the issue of who sets priorities for use of limited assets and on what they base those decisions. In establishing an architecture, how many requirements can be anticipated, and who is best at determining these—a central operator or the customer?

Space systems will make a contribution to meeting the theater RSTA requirements. How they do this, what kind of capabilities they will have, and what other issues they raise will depend on the choice of architecture.

The Command-Oriented Approach to RSTA

The premise of a command-oriented space architecture is that the national-level capabilities will provide the first reliable indications of a crisis. These capabilities will then be apportioned to some extent to support the theater, but the need to monitor other situations around the world will force compromises.[63] Whatever support can be made available will be allocated by a central authority from a fixed pool of assets, and as a rule, there would be no augmentation of space capabilities that hadn't long since been planned. If one were to design a space architecture to support theater RSTA requirements using this approach, the general outlines would appear as shown in table 22.

Table 22

Command-Oriented Architecture for RSTA Requirements

	Implications of RSTA Requirements		
	Space Segment	**Ground Segment**	**Launch Segment**
Performance	Emphasis on strategic needs, tactical met as collaterial function	High throughput; centralized tasking	Ensure that standing capabilities are kept on orbit; replenishments launched on schedules determined years in advance
Responsiveness	Change orbits	Provide central direction to shift assets	
Flexibility	A byproduct of built-in overcapacity	Ability to produce multiple products centrally; deploy some functions forward	
Robustness	Defend, harden	Defend, distribute	Keep CONUS sites operational
Logistics	Prepare to augment for a long conflict	Deploy comm links, specialized ground stations	Prepare to augment for a long conflict
Reliability	Essential	Have skilled technicians to fix on-orbit problems	Near 100% necessary
Ease of operations	Secondary to reliability	Deploy specialists	Secondary to reliability

Table 22 is divided by segment for simplicity and to give an overall view of the architecture. In general, the effects of applying the requirements of a specific mission are apparent at this level, though to see the effect on specific design choices would require a further breakout.

Technology and budget determinants, as illustrated previously, are to a large extent already included in the above table. Some of their key impacts on the command-oriented theater RSTA space architecture are reliance on a relatively small number of large satellites, emphasis on ground-based versus onboard processing of information, and the channeling of information through central locations.

The command-oriented architecture leans toward centralization for doctrinal and physical reasons. Small numbers of satellites mean there is little or no slack in the system to respond to a surge in demand, so a central clearinghouse for tasking is established. This central authority is distant from the theater, both physically and in terms of organizational hurdles, since it spends most of its time responding to national-level requests for information.

Centralization also results from the hardware design. Development and production of large and complex satellites takes years, and designers often cannot anticipate changes in technology with any certainty.[64] Coupled with a lack of standard interfaces and operating systems on satellites, this makes it extremely difficult to insert the latest capabilities. The result is that the processing electronics on a spacecraft will be several generations behind what can be put in a ground station; consequently, designers tend to put minimal processing capability into the satellite.[65] This results in high data rate downlink requirements and, in turn, means that a ground station equipped to receive and process the signals must have significant hardware capability (often peculiar to the system) and highly trained personnel. Neither of which is conducive to easy and rapid deployment to the theater.

The type of information disseminated is also affected by both doctrinal and budget concerns. Because the satellite provides data in only one form, the ground station must convert it into something useful. Again, because of the high level of skill required to do this (and potentially because the product is subject to interpretation), the command-oriented system favors centralizing production and working to assemble the kind of product each user needs. The disadvantage is that this takes time, especially when data is coming in quickly and there are many requests for products. Further, it requires that the end users understand the system's capabilities and limitations to ask for the right product; this often means adapting the theater's operating methods to fit the needs of what is supposed to be a supporting function.

Observations on the Command-Oriented Approach to RSTA

Command-oriented space RSTA systems will be best suited for detecting and responding to the concerns of their primary customers—national-level authorities. These systems are capable of producing highly detailed and customized products, and centralized control should ensure that the space systems on orbit are used efficiently but not over tasked. A command-oriented architecture, because it is intended to have virtually its full capability on orbit at all times, may provide the maximum available global coverage and situational awareness.

A command-oriented architecture, unless it is unconstrained by funding, will respond poorly to surges in demand, especially if those surges occur in parts of the world that do not have optimum coverage or call for sensors that are not deployed. In other words, the effectiveness of a command-oriented system depends on its designers' ability to anticipate specific requirements. Command-oriented systems will suffer from untimely information distribution, again especially at times of increased demand, and probably will use standardized products and request formats. These characteristics will result in a lack of flexibility and produce frustration among the customers.

Because the same systems serve all users, the capabilities of the systems are determined by the most pressing national needs. Since there are only a few available assets, the products of a command-oriented space RSTA architecture will have to be carefully protected. Security is necessary to ensure that enemies, including those not engaged with us at the moment, do not learn so much about our capabilities that they can develop effective countermeasures. Unfortunately, these security requirements will restrict access to the information. Allies and even many of our own troops may not have sufficient "need to know" to get access to the best information available.

Finally, a command-oriented architecture gives the theater CINC little, if any, control over space assets. This does reduce the decision-making burden on the CINC's staff, but it also leaves open the possibility that support from the space systems will not be provided when needed most.[66]

Two further observations are necessary. The characteristics of a command-oriented theater RSTA space architecture as described above are nearly identical to the characteristics of our current space RSTA systems. One example is in the area of flexibility. The Defense Support Program (DSP) early warning satellite, although not intended to support theater war fighting, took advantage of certain built-in capabilities in excess of what was needed for its strategic warning mission to cue other sensors and weapons in a limited way during the Gulf War. It's also apparent that the disadvantages of a command-oriented space architecture are very close to the commonly perceived disadvantages of space systems in general, as discussed previously. A comparison with a demand-oriented approach should help answer the question if these are in fact generic to space systems.

Responses to the Command-Oriented Approach to RSTA Shortfalls

There are three ways of responding to a shortfall in capability—by improving or upgrading existing assets, by diverting missions allocated to one type of system to another one, or by using the same types of systems in a different architectural framework. All three have merits, and the first two are being vigorously pursued to enhance our theater RSTA capabilities.[67] Architectural change, which has received less attention, may offer the greatest long-term payoff in providing better space RSTA.

Upgrades or improvements to the current US RSTA architecture involve speeding up processing times, making more information on system capabilities and limitations available to users in the field, pushing more information to the field (including changing rules on classification), producing better data fusion, working to eliminate system-specific equipment and to provide common terminals and ground stations, and reducing the number of barriers between theater users and those who actually control the systems.[68] Fundamentally, though, the architecture remains the same. Assets are still centrally controlled, and the improvements are a matter of degree, not a matter of kind.

For example, the custom product network allows users at forward locations to create such custom imagery products as mosaics of pictures.[69] This provides the users the additional flexibility of enhancing their intelligence preparation of the battlefield, but these mosaics are limited to material that has been archived or is being sent forward.[70] Users must also consider whether a mosaic picture of the battlefield with each piece possibly taken at a different time or by different sensors is sufficiently accurate for their purposes. For some applications, this mosaic may be accurate enough, but not for others. In other words, while a step forward, this solution is only a partial answer.

Another category of solutions is diversion. In the case of theater RSTA, this means using airborne sensors to cover the gaps that perhaps are too difficult for space sensors to fill. Airborne sensors also offer such advantages as the ability to loiter over a particular area for hours or even longer. By employing unmanned vehicles, reducing payload (and hence vehicle) sizes, and employing low observable features, such platforms can provide coverage over otherwise denied areas, thus incorporating some of the advantages of space systems.[71]

Aerial vehicles also have drawbacks. They require fairly regular launch and recovery operations, and because they are limited to atmospheric speeds, they will have to be based in theater or fairly close to avoid lengthy transits and the sacrifice of loiter time. These considerations mean aerial vehicles will have a considerable logistics tail, much of which will have to be deployed in the theater. Not only does this add to the number of things that are high priority for immediate delivery, but it could complicate the basing and scheduling of other aircraft. Finally, there is the overflight issue. Even though the probability of detection may be low, a political decision is necessary before taking the risk of overflying sensitive territory.

Aerial vehicles offer tremendous possibilities. In some cases, they may be the only way to get close enough to a target to obtain the right kind and quality of information.[72] The best solution to the theater RSTA problem will undoubtedly incorporate aerial vehicles, but should they necessarily address all the disadvantages of a command-oriented system?

Before discussing the demand-oriented architectural approach to theater RSTA, there is one other avenue to improve today's capabilities that should be recognized. This might be called a hybrid of diversion and architectural change, since it involves using systems that are not under direct military control to augment military capabilities. For example, the French *Systeme Probatoire pour l'Observation de la Terre* (SPOT) imaging satellite and the LANDSAT multispectral sensing satellite provide useful products which have been incorporated into databases in the past. One might argue that in the future, as the market for earth observation products grows, the theater CINC could have a variety of commercial sources from which to obtain information.[73] This idea also has merit, but with several caveats. First, depending on the political situation, the availability of those products to us, our allies, and our adversary is questionable.[74] Second, the timeliness of the information is far from ideal.[75] Third, the data may arrive in a format incompatible with the rest of the theater RSTA architecture, requiring additional processing and further delays. All of these concerns indicate that although commercial augmentation may be a valuable addition to RSTA capabilities, it may not be one to rely on in a time of crisis.

The Demand-Oriented Approach to RSTA

The main premise of a demand-oriented architecture is that those individuals responsible for the theater, aided by national capabilities, will have the first indications of a crisis and be able to identify additional RSTA needs. A truly demand-oriented system will respond by surging and augmenting capability, including deploying additional space assets tailored to the situation in terms of orbit and mission payload.[76] Table 23 summarizes the demand architecture.

Table 23

Demand-Oriented Architecture for RSTA Requirements

	Implications of RSTA Requirements		
	Space Segment	Ground Segment	Launch Segment
Performance	Emphasis on operational-level needs	Information on demand to any user	Ability to surge number of launches
Responsiveness	Additional capability available in days	Tasking at low operational level	Orbit satellites within hours of need
Flexibility	Add new satellite types; autonomous platforms	Interoperable with all RSTA assets	Deploy different satellites to different orbits
Robustness	Proliferation; replace and reconstitute as needed	Proliferation of equipment	Operate from multiple sites
Logistics	Reduce number of unique parts	Minimize unique equipment	Reusable (or cheap expendable) vehicles
Reliability	High for functions; lower for individual satellites	On par with other computers/electronics	High for function, not individual vehicle
Ease of operations	Access and control easy for users	Standardized equip and procedures	Rapid turn times

In an extreme form, a demand-oriented architecture would have a minimum essential capability in place for worldwide strategic monitoring. On identification of a crisis, the theater CINC would activate a plan to augment space RSTA capabilities. These augmenting assets would be tailored to the theater's needs and would be tasked directly by theater forces.

Because these RSTA capabilities would not be at the national level, the security requirements would be less stringent and the information distribution broader. The space systems themselves would primarily be small, relatively short-lived systems so it would neither be a major commitment of resources to deploy them nor a significant setback if one failed to work. This implies an extremely responsive space-lift capability, one that can surge to place potentially dozens of satellites in orbit over a few days, and then sustain a launch rate to augment or replace satellites as needed. For crises of indefinite duration or for a world in which the need to augment capabilities occurs on a regular basis, a reusable launch system that provides access often and inexpensively offers clear advantages over an expendable system.

Observations on the Demand-Oriented Approach to RSTA

A demand-oriented architecture offers clear advantages to the theater CINC in terms of responsiveness to tasking and the ability to tailor the coverage to the situation. Because of the proliferation of satellites, it offers the ability to get close to continuous observation of the theater from space. In a future situation in which an enemy could threaten some or all aspects of our space architecture, the demand-oriented approach is clearly more robust than a command-oriented one. It presents the enemy with a proliferated and distributed target set both in space and on earth. Each of the targets is relatively small and insignificant in itself, and there are no critical nodes that will cause the whole system to cease functioning. Because of proliferation, the architecture is also less vulnerable to accident or natural disaster.

On the other hand, the distributed architecture will have a less capable initial configuration than the command-oriented system, and augmenting it will take a finite amount of time.[77] Also, because of the smaller size of the spacecraft, there are missions they will not perform as well as the satellites of a command-oriented architecture. Perhaps most critically, the demand-oriented architecture requires a satellite building and launching capability that the United States does not currently possess, but which is attainable.[78]

Assembling a Workable Theater RSTA Architecture

Satisfying theater RSTA requirements takes more capability than any one class of solution can bring to the table. The above discussion explains how the current, command-oriented US space architecture cannot meet all the requirements.

It also shows that each of the potential enhancements to that system has drawbacks, making it unlikely to solve the problems alone. Because the eventual solution will take the form of a hybrid architecture, there will be many challenges to ensuring the entire network of systems is interoperable in terms of communications, ground stations, databases, geographic coordinate references, and so forth. There are numerous initiatives under way to make the best use of the United States considerable assets.

The class of solution that has received the least attention to date is the potential to build a space architecture that both takes advantage of the unique characteristics of space and provides the theater user the control and responsiveness he or she needs—in other words, a demand-oriented architecture. This type of solution offers considerable potential, but will require work in both technological and doctrinal areas. By helping to identify the key issues, a framework for architectural comparison may advance the process.

Utility of A Framework for Comparing Architectures

Previous sections have built and elaborated on a framework for describing and comparing space architectures. This section answers some of the questions raised in the process and consolidates observations in four areas. First, the distinguishing characteristics of the command- and demand-oriented approaches are reviewed. Second, these characteristics are used to define which perceived disadvantages of space are inherent and which are a result of design choices. Third, the key efforts needed to overcome those design disadvantages are identified, and finally, the contribution that having a framework makes to this process is discussed.

Distinctions between Command- and Demand-Oriented Architectures

Command- and demand-oriented architectures can be distinguished by physical, temporal, and philosophical differences. The physical ones are the most obvious—satellite size and number and launch vehicle size and type. Command-oriented architectures will have fewer, larger, and more complex space systems. Those systems will individually be more capable than those of a demand-oriented system. Space lift will be an infrequent activity that is scheduled far in advance and optimized to lift the maximum weight into orbit per launch. In demand-oriented systems, satellites might be built to perform a specific mission and would be more likely to use off-the-shelf components than custom-designed ones. Although less complex overall, demand-oriented satellites will be "smarter." More of the navigation, system management, communication, and information processing capabilities will be on orbit than in a command-oriented architecture. Information from a satellite is more likely to be broadcast or directly downlinked to end users than in the command-oriented system.

The temporal differences are of two types. First, systems for the command-oriented architecture will take longer to design, build, and deploy. Technology will be inserted more slowly than in a demand-oriented system, and satellites will be designed to last longer. Demand-oriented satellites could be built on short notice and for relatively short (months instead of years) duration missions. The second type of difference is in the response of the system to new situations. A demand-oriented architecture can be reconfigured rapidly and is designed from the beginning to provide the fastest possible response to its users. The command-oriented architecture changes slowly, if at all, in response to changing situations on the ground, and tends to sacrifice timeliness of information for precision.

Underlying the physical and temporal difference are the philosophical or doctrinal ones. In a command-oriented system, efficiency and economy of force are the driving principles, which lead to centralization both of command and control and of information distribution. In a command-oriented architecture, the designers assume that a central authority not only has the greatest ability to control tasking and distribution but is also best postured to decide upon priorities. In contrast, a demand-oriented architecture is built around the principles of flexibility and responsiveness, under the assumption that the end users of a system's product are best able to determine priorities and control tasking. Both control and dissemination of information (or execution) are distributed and decentralized, leading to a more robust but somewhat anarchical and inefficient system.

Command-oriented architectures are inherently better suited to missions having long-term and reasonably predictable requirements. Demand-oriented architectures in contrast are best suited to those missions that are unpredictable, may involve sudden changes in the capabilities needed, and could conceivably be of short duration. A command-oriented architecture emphasizes efficiency by making best use of assets in place, but this requires excellent long-range planning. A demand-oriented architecture relies on the ability to react and adapt quickly to new situations, even if this means accepting less than optimum use of assets. Paradoxically, for unpredictable situations, the demand-oriented architecture may be the most efficient one. By allowing rapid changes in capability, it prevents the architecture from having to be overdesigned initially and, by providing a more rapid and tailored response, it may preempt the need for a larger or longer term commitment.

The Inherent Disadvantages of Space

Here is a recap of the list of perceived space disadvantages from table 9.

- Distance
- Predictability of movement (to the enemy)
- Poor continuity, meaning
  - — lack of dwell time (a low-earth-orbiting satellite has a point on the ground in view for only about 10 minutes out of every orbit)
  - — gaps in revisit time (the ability to have a specific mission capability over a specific point on the earth with sufficient frequency)
- Poor responsiveness
  - — (strategic) to crises the systems were not designed for
  - — (operational) to theater requirements

- Inflexibility for retasking
- Unsatisfactory timeliness/distribution of information
- Vulnerability to attack or natural disaster
- Need to operate in a harsh environment
- Cost

Are there any lessons from the foregoing architectural comparisons concerning which of these disadvantages are inherent and which are design dependent?

Distance is an intrinsic disadvantage. There are some missions where proximity is needed, and to be in space a platform must be in a sustainable orbit, generally accepted as at least 93 miles above the surface of the earth.[79] As a result, these are the closest approaches a space system can make.[80] Other practical concerns may push this minimum higher, but to keep things in perspective, our airborne systems will sometimes have to operate at similar ranges from the target to see deep with reasonable security and have to look through more air in the process.[81] The other aspect of this drawback is that this distance is vertical. Considering the high speeds needed to achieve orbit, a large increase in kinetic and potential energy is needed to enter space from earth. As a result we have not yet been able to deploy space systems routinely and inexpensively.

Predictability is also inherent in orbital systems, but with significant caveats. First, orbital mechanics is not deterministic; knowing a satellite's location precisely requires frequent observations. Unless the satellite is in a synchronous orbit, orbital perturbations will change its path from what has been predicted in a relatively short period of time.[82] This phenomenon is compounded if the satellite has the ability to perform even small maneuvers (especially if out of sight of enemy sensors).[83] With the addition of decoys, a satellite could leave even relatively sophisticated adversaries uncertain as to the actual time that they were being overflown.[84] Maneuvers cost fuel, which in turn shortens the life of the satellite, but this could be overcome either by deliberately designing the rest of the satellite to last a short time (and thus saving money) or possibly by refueling the satellite on orbit. In other words, although maneuver of an orbiting body is difficult, relatively small maneuvers can have substantial effects. The disadvantages of predictability can largely be overcome.

Continuity is clearly a function of design, and usually becomes an issue because of cost constraints. If enough satellites are built cheaply and if launching them and maintaining them in orbit is affordable, there is no physical reason why space systems could not provide continuous coverage over any part or all of the globe. This is due now with GPS (24 satellites), Iridium (66 satellites), and Teledesic (approximately eight-hundred satellites). Communications constellations will also provide coverage from much lower orbits. The problem is that in lowering the altitude to overcome the distance problem the number of satellites needed rises considerably, which complicates the cost and command and control issues. Use of advanced technologies such as standardization and modularity, flexible manufacturing, and autonomous spacecraft operations make the problems manageable; however, these approaches clearly fit more closely into a demand-oriented architecture.

Responsiveness and flexibility are strongly architecture dependent. An architecture that can be tailored to an emerging situation on the ground would provide good strategic flexibility. An architecture that can be rapidly augmented to provide additional capabilities will fulfill the needs of operational users. A command-oriented system by its nature is ill-suited to these kinds of adjustments; however, and our lack of a responsive spacelift capability makes the problem worse. Even if we had a launch-on-demand system, a command-oriented architecture would not be flexible and responsive. This is due to the long lead times for satellite building, the requirement that they last a long time, the cost of each satellite, and our operating procedures that put most of the maintenance and control functions on the ground, and thus require large numbers of operators.

Less obvious, but equally important, is the fact that the architectural philosophy strongly influences the responsiveness of spacelift systems. Although it is desirable to be able to launch satellites rapidly, a command-oriented architecture may hinder the development of rapidly responsive launch capabilities. Since the command-oriented architecture is ill-suited to take advantage of these capabilities, it provides little incentive to justify the development costs of a revolutionary kind of space access which we only use a few times a year.[85]

Timeliness also is a function of design. Even a command-oriented system, given sufficient motivation, can push information through its bottlenecks quickly on occasion. By removing the choke points of centralized processing and control, a demand-oriented system can respond rapidly to almost any request.

Vulnerability, whether to enemy action or to mishap, can also be greatly reduced by design. The more systems there are on orbit, the smarter those systems are, the more pathways available for information transmission, the more launch sites, and the more ground sites capable of tasking the satellites, the less chance there is that loss of any one component will seriously affect the system. Because this is the case, each component could be designed and tested to less stringent specifications, which would make the cost of proliferating systems more bearable.

The need to operate in a harsh environment is an inherent feature of space systems. But what is the answer to this problem? Long system life is seen as a plus, but this seems to tackle an environmental disadvantage head on. It also adds to the cost and technological backwardness of spacecraft by increasing their weight and complexity and thus lengthening the time needed to design, build, test, and deploy them. Much cheaper systems could operate perfectly well for more limited periods, so that in the long run responsiveness could be improved at no greater cost. This disadvantage of space shows that alternative architectural approaches can mitigate even inherent limitations of the operating environment.

Cost is a drawback for space systems, but it may not be an insurmountable barrier. Cost is driven by many things, but two stand out—the satellite itself and the cost of access. Many current satellites are extremely expensive, but so are aircraft carriers and B-2 bombers. Just as not all ships and aircraft are as expensive as those premier systems, neither are all satellites. In fact, by taking advantage of new technologies and methods as described throughout this study, individual satellites could be produced relatively inexpensively.[86] As for launch, the combination of demand for responsiveness, robustness, and low operating costs with the potential of improved design and construction capabilities could finally provide the incentive to pay the up-front costs of developing a radically improved spacelift system. There will always be a price to pay for having an advanced capability, but it need not be exorbitant.

From the discussion above, the intrinsic disadvantages of space systems appear to be distance and harsh environment. The rest are essentially a function of two things: the design choices made in developing the space architecture, and the difficulty (cost and slowness) of access to space which is also related to architectural design choices. Even the intrinsic disadvantages of space are no worse than the intrinsic disadvantages of air or the ocean in the sense that properly designed systems will always be needed to exploit the environment.

The Key Factors Needed to Overcome Design-Driven Disadvantages

The features of a demand-oriented architecture enable space system designers to make better use of rapidly evolving technology. At the same time, that technology may make some of the features of a demand-oriented architecture possible. For example, by taking advantage of a standardized and modular spacecraft design, including operating system software, future military space operators could draw from a stock of subsystems and essentially "bolt" together a satellite in a few hours.[87] Since the design and interfaces would be standardized, a minimum amount of testing would be needed before the system was deployed. Using an advanced internal architecture, the basic components could be assembled to support several different missions.[88] Sensor or other mission payload packages could likewise be designed to fit the standard interfaces.

By assembling a satellite shortly before launch, operators could incorporate the latest in processor and storage technology.[89] Just-in-time assembly would produce better performance on orbit and would cut down on the stockpiles of parts that would have to be continually maintained.[90] With improved on-orbit performance, satellites could do more for themselves including autonomously navigating, maneuvering, and monitoring their own health and status. Except for emergency situations, the amount of ground support needed by the satellite could be reduced to almost zero, making the additional support costs of proliferated constellations small. The additional processing capability would also allow the satellites to provide different kinds of products directly to the users, for example, on-board target detection processing and cueing other sensors or combat forces with target type and location. At the same time, the satellite could pass a full image of an area already annotated with detected targets to theater headquarters for correlation with other sources and send unprocessed raw data to a central location for evaluation of the satellite's performance. This ability has the potential to make RSTA systems much more decentralized and available to multiple users.[91]

An architectural approach resulting in a proliferation of autonomous and potentially maneuvering satellites on orbit and increasing transits of space will strain existing ability to monitor activity in space. At some point, an enhanced space surveillance capability will need to be deployed, one that can adapt to this new environment.[92] This will probably include space-based sensors and a new concept of space traffic control.

Technology trends seem to favor the demand-oriented approach. As more advanced capabilities become part of the commercial sector, particularly in electronics and software but also in materials, design, and manufacturing, our national security space architecture will be challenged to adopt the capabilities ever more rapidly. In many cases, the technological advantage among military forces will go not to the country that can develop the best technologies, but to the one that can best exploit new technologies regardless of the source. In a similar vein, technological advantage is becoming more perishable, and it is ill-advisable to lock into a specific technology for 20 plus years. The key to exploiting the trends in technology is to change design philosophy to one that not only accepts early obsolescence but plans for it. This will undoubtedly prove uncomfortable to some (as it does to any personal computer buyer who finds a cheaper and better system on the market a month later), but it is also the key to adapting to a rapidly changing environment and to making swift progress, thus staying ahead of rivals.

Inevitably, the concept of a demand-oriented architecture will depend on the performance and cost-effectiveness of small satellites, sometimes called *lightsats*. Space systems can be built rapidly and affordably only if size, weight, and complexity are reduced.[93] Lightsats have numerous critics as well as advocates.[94] Because of progress in miniaturizing components and subsystems, there is substantial reason to believe that performance goals can be achieved, especially if those performance goals are realistic, that is, based on need rather than want.[95] A more debatable proposition is whether satellites can ever be cheap enough to be routinely considered for short duration (three to six months) or special purpose missions.[96]

Taking advantage of commercially available equipment and relatively large production runs and using smaller, more modular spacecraft would dramatically reduce the cost of each satellite.[97] This would be of little use, however, if launch costs were not also dramatically decreased.[98] The preceding sections should make it clear that rapid, affordable space lift is a critical issue in moving toward the advantages of a demand-oriented space architecture.[99] Advocates of improved space capability for the United States should make the development of greatly improved spacelift an overriding priority.

How Does an Architectural Comparison Framework Help?

The Department of Defense has recognized the need for a coordinated approach to developing future space capabilities and has designated a deputy defense undersecretary for space acquisition and technology programs.[100] Building on this concept, the Air Force has proposed seven strategies to improve the space capabilities of the United States.[101] Both actions imply a recognition that future space capabilities must be considered in a holistic sense. Addressing problems or pursuing opportunities in just one area—satellite size, launch systems, command and control, or operational concepts—at best leads to suboptimization and at worst leads to poor conclusions by assuming away key issues or ignoring interdependencies. To avoid this pitfall, some method of describing what a space architecture is and how its components interact is needed.

The framework for architectural comparison presented in this paper is a foundation for thinking about the real differences between alternative approaches to designing space systems. With considerable expansion, it could serve as the basis for quantitative comparisons between different architectural designs. At this point, however, its main value is in providing some qualitative boundaries to the debate and helping to frame questions about both space system hardware and the underlying operational philosophy or doctrine. By highlighting the nature and extent of the differences between command and demand orientation, their respective advantages and disadvantages, and the relationship of the architectural features to real-world determinants, the framework also opens the way to discussion of what doctrinal or physical changes are desirable in future space systems.

Summary and Implications

This study has explored the possibility of using the concepts of command and demand orientation to describe not only the way information might flow in a space system but also to encompass the nature of the entire space architecture including hardware, facilities, and operational procedures. To give this expanded definition of command and demand orientation meaning, the study presented a framework for describing and ultimately comparing space architectures in terms of their physical elements, operational and other attributes, and the real-world factors determining how an architecture will look and perform.

This approach appears to have some utility. Although the distinctions that the study draws between command- and demand-oriented architectures are largely qualitative, they are real and unambiguous. Command orientation manifests itself in a centralized system with extensive ground control; a relatively small and fixed number of large, highly capable on-orbit assets; and a spacelift capability driven by reliability and the need to launch on schedule. In contrast, a demand-oriented architecture will be decentralized and distributed with more processing and control functions in orbit, a relatively large number of smaller satellites that can be augmented in a short period of time, and an emphasis on system-wide (constellation) performance as opposed to the performance of an individual satellite. Demand-oriented systems, unlike command-oriented systems, favor satellites that can be built quickly and affordably using the latest off-the-shelf technology. Demand-oriented architectures will tend to allow more users at a lower organizational level to communicate directly with the satellites and will delegate authority to task space systems to the operational level. The spacelift capability required by a demand-oriented architecture must be responsive and be able to surge to augment on-orbit capabilities.

An architecture with demand-oriented characteristics also appears to be the only viable alternative if in the future an enemy contests the right to use space freely. The demand-oriented approach is inherently more robust, able to absorb damage, and able to respond with augmented or replenishment capability. A command-oriented architecture, because of the centralization and the criticality of each asset, is less robust in the face of attack or mishap. Although the concept of wars extending to space is not yet widely accepted, space superiority has not yet become an issue in a conflict, and at present the threat seems minimal in both a strategic and tactical sense similar to the threat of aircraft to surface ships before World War II—this is an issue that space planners should not ignore.

The contrast between command- and demand-oriented architectures reveals that many of the disadvantages traditionally associated with space systems—among them predictability, inability to provide continuous coverage, and poor responsiveness and flexibility—are more a function of architectural design characteristics than the inherent limitations of the environment. With the right approach to space system design, much more can be achieved in space than expected. On the other hand, failure to separate fundamental or intrinsic limitations from those imposed by design choices can lead to less overall capability by producing faulty assumptions.

It should come as no surprise that the current US national security space architecture is predominantly command oriented with the associated advantages and disadvantages. Since this is the only architecture that most space planners are familiar with, it is also unsurprising that many planners think of those advantages and disadvantages as intrinsic to space systems. Furthermore, the United States tremendous investment in existing systems and the undeniably superb capabilities of those systems encourage the development of incremental improvements rather than wholesale changes in equipment or concepts of operations.

Although this situation has valid historical and technical roots, changes over the past several years in the geopolitical environment have reshaped basic assumptions about requirements and budgets. Budgetary reality means that the United States cannot respond to additional needs by deploying more of the existing kind of systems; it's simply not affordable. Emerging requirements clearly point to the need for a more demand-oriented architecture, at least to augment the command-oriented "backbone" in times of crisis, if not to provide the bulk of the space support to regional or theater CINCs. Providing a fundamentally different architecture is the best way to be responsive to the needs of the theater war fighters, to convince those war fighters that space assets will always be there when needed, and to encourage full use of the advantages of space.

Demand-oriented architectures require not only the development of new procedures, but the exploitation of new technology. Technological trends—in computing power, structures, materials, and design techniques, such as standardized interfaces and modular construction, enable a break from the vicious cycle of increasing cost and complexity that space systems are now in and allow for more rapid deployment of modern systems. Harnessing the capabilities in those technologies will make possible smaller, less expensive, but still highly capable space systems. The same trends also make it imperative that we assess our current doctrine and practices, since failure to recognize and adapt to a changing environment could allow an enemy to leap ahead of us in capability.

To take advantage of the potential of demand-oriented systems, some obstacles must be overcome. These obstacles are two-fold. First, planners are saddled with outdated assumptions about what space systems can and should do. The United States lacks a coherent strategy for controlling and exploiting space that could help shape military doctrine and direct system development efforts. Second, the ability to change the nature of space architectures depends heavily on creating a much more responsive and inexpensive spacelift capability. Unfortunately, the nature of the current national security space architecture does not produce sufficient incentives to develop that new type of spacelift.

The United States needs to move toward more rapid exploitation of technological opportunities vice comprehensive, dedicated leading edge development. This process needs to be made routine so that a space architecture can adapt more rapidly when hit with surprises or opportunities. Since it's impossible to anticipate everything, flexibility must be built into future space assets—but at the "architectural" level, not by creating heroically capable individual systems. Perhaps the ultimate test of a space architecture is whether it encourages or retards this flexibility.

Recognizing that the United States cannot, and should not, completely abandon the type of space systems that have served it well for over 30 years, there are still some useful steps to be taken. Strategy and doctrine for decentralized space operations keyed to supporting the theater CINCs should be developed from a clean sheet of paper, with no preconceptions as to what is possible. A detailed analysis of the potential for demand-oriented systems to respond to requirements and technological opportunities in several mission areas should be conducted. Development of a spacelift system that provides rapid, reliable, and inexpensive access to space must be given the highest priority, with the payload such a system carries treated as a measure of merit, not specified in a requirement. To demonstrate the possibilities of new types of space systems, DOD should promote a design-to-cost competition for small satellites to perform various missions, and should encourage the development of modular designs and standard interfaces. If such steps are taken in parallel—and it is important to recognize that strategy, doctrine, technology, cost goals, and perhaps above all, the ability to get useful payloads into orbit as quickly and as often as the strategy demands, are linked—the next few years could see the emergence of a new space architecture.

Fundamentally, the choice of an architectural approach in developing future space systems matters. We first need to recognize that there are viable alternative approaches. To understand the nature and implications of those alternatives, we need a common basis for discussion and comparison. The framework presented in this study is a first step in that direction.

In the words of Air Force Gen Henry "Hap" Arnold in 1945, "National safety would be endangered by an Air Force whose doctrines and techniques are tied solely on the equipment and process of the moment. Present equipment is but a step in progress, and any Air Force which does not keep its doctrines ahead of its equipment, and its vision far into the future, can only delude the nation into a false sense of security."[102] The words are as relevant for space forces as they have been for the Air Force, and as applicable today as ever.

Notes

1. This study is concerned with national security as opposed to the commercial or civil sector of space operations and with the near-earth region (both terms as defined in Joint Pub 3-14, Joint Doctrine, Tactics, Training, and Procedures [JDTTP] 3-14, *Space Operations*, 15 April 1992) because these are the areas of immediate concern to the US military, and because the ability to exploit near-earth space will be the foundation for any future, wider ranging endeavors. This focus does not preclude discussion of issues that overlap with other sectors, such as the relevance of the commercial sector to our launch and technology development.

2. The terms *command oriented* and *demand oriented* were used in Air University's 1994 SPACECAST 2020 study. Traditionally, those terms have been used to describe the information flow in a system, but they seem to have implications for every aspect of a space architecture. In part, this paper arose from a need to more fully explain and justify those terms and ideas.

3. This environment has a single major adversary, a focus on strategic-nuclear intelligence needs, an overriding national priority that demands performance with little regard for cost or operational difficulties, and an

organizational and security structure with "stovepiped" systems with little interoperability or connectivity to conventional military forces. Historian R. Cargill Hall, in "The Eisenhower Administration and the Cold War: Framing American Astronautics to Serve National Security," unpublished essay, January 1994, 2, says that our "astronautical enterprise" was "impressed with a near-indelible Cold War seal" by its origins. See, for example, the *Blue Ribbon Panel of the Air Force in Space in the 21st Century, Executive Summary* (Washington, D.C.: the Panel, 1992); Gen Charles A. Horner, commander in chief, US Space Command, "Space Seen as Challenge, Military's Final Frontier," prepared statement to the Senate Armed Services Committee, 22 April 1993; Gen Charles A. Horner, "Space 1990 and Beyond . . . The Turning Point," presentation to the US Air Force Today & Tomorrow conference; Lt Gen Thomas S. Moorman, "Space: A New Strategic Frontier," undated essay, 14–15; and John T. Correll, "Fogbound in Space," *Air Force Magazine*, January 1994, 22–29.

4. I. B. Holley, "Of Saber Charges, Escort Fighters, and Spacecraft," *Air University Review*, September–October 1983, 10.

5. I am more accustomed to the term *capabilities* and *limitations*. Also I use *advantage* to avoid confusion with joint publications which use the word *capability* to mean the ability to perform a specific type of mission.

6. For example, altitude (a characteristic of air or space systems) offers the possibility of seeing farther, but it requires being able to see far enough (meaning resolving those things of interest from the background) and overcoming or living with obscuration.

7. Old methods can be perpetuated beyond their useful limits. For a discussion of the persistence of the horse cavalry and the lessons for space forces, see Brig Gen Bob Stewart, USA, Retired, untitled address to the Space Support to the Warfighter Conference, Peterson AFB, Colo., 15 December 1993.

8. Joint Pub 3-14, III-2 to III-5. This definition appears to imply that space systems offer no unique capabilities, so it doesn't help in deriving space "advantages."

9. Ibid., chap. 2.

10. Platform speed and weapons range are based on the environment (gravity, composition, etc.) and make the extent of the environment irrelevant in some cases. Does it matter how far a satellite travels to perform its mission? No, what matters is the amount of time it takes.

11. Air Force Space Command, "Space Primer," preliminary draft, February 1995.

12. Ibid., 7.

13. *SPACECAST 2020 Executive Summary* (Maxwell AFB, Ala.: Air University, June 1994), i.

14. An example is a prohibition on basing weapons of mass destruction in orbit per the 1967 Outer Space Treaty, "The Treaty on Principles Governing the Activities of States in the Exploration and Use of Outer Space, including the Moon and Other Celestial Bodies"; and Joint Pub 3-14, A-5. In

other words in international law there is no legal basis for "closing" space or restricting access to it, even if it were physically possible. The continuity of the medium ensures that once in space a spacecraft has access to the entire planet's surface. Of course, space could become a battleground in the future, as the high seas have been in the past.

15. A low-earth-orbiting satellite has a point on the ground in view for only about 10 minutes out of every orbit. *Revisit time* is the ability to have a specific mission capability over a specific point on the earth with sufficient frequency. Vulnerability of a satellite requires some elaboration. Although it can be difficult to physically reach a deployed satellite, some, especially in low orbits, could be vulnerable to directed energy weapons, ECM or physical attack. Weight is at a premium, and spacecraft themselves are usually quite "soft." Many ground stations are relatively soft targets. Launch facilities for current systems are also vulnerable to terrorist or other attack or to natural disaster.

16. This can go the other way also, as seems to be the case with the Global Positioning System largely replacing terrestrial navigation systems.

17. Examples of this in the commercial sector are the Motorola Iridium and Microsoft/McCaw Teledesic communications satellite concepts which, because of a desire for global coverage and other requirements, use relatively large numbers of low-earth-orbiting constellations rather than traditional geosynchronous platforms. Although neither is yet deployed, Iridium is approaching deployment on or ahead of schedule. William B. Scott, "Iridium on Track for First Launch in 1996," *Aviation Week & Space Technology*, 3 April 1995, 56–61. Teledesic is pressing ahead with launch options and other advance planning.

18. Referred to as the "tyranny of the payload." The Honorable Sheila E. Widnall, secretary of the Air Force, address to the National Security Industrial Association, Crystal City, Va., 22 March 1994, 3.

19. This study differs from AFDD-4, *Space Operations Doctrine*, 1 May 1995, which lists three elements: space, ground, and the "link," which primarily means communications. The identification of space, ground and launch segments, and the breakdown into more specific elements, is both more in line with traditional space system descriptions and more useful in forming an architectural comparison. Dr. James R. Wertz and Dr. Wiley J. Larson, eds., *Space Mission Analysis and Design* (Dordrecht, The Netherlands: Kluwer Academic Publishers, 1991), 9–10.

20. I base this on my experience in conducting the operational analysis of the SPACECAST 2020 study, on my education as an astronautical engineer, and on various texts and short courses on space system design including TRW Space and Technology Group, *TRW Space Data Book*, 4th ed. (Redondo Beach, Calif.: TRW S&TG Communications, 1992); Wertz and Larson; and James R. French and Michael D. Griffen, "Spacecraft Systems Design and Engineering," short course presented as part of the American Institute of Aeronautics and Astronautics (AIAA) Professional Studies Series, 14–15 February 1990.

21. Steven R. Petersen, *Space Control and the Role of Antisatellite Weapons* (Maxwell AFB, Ala.: Air University Press, May 1991), 72; and Mark H. Shellans and William R. Matoush, "Designing Survivable Space Systems," *Aerospace America*, August 1992, 38–41.

22. Stephen J. McNamara, *Air Power's Gordian Knot: Centralized versus Organic Control* (Maxwell AFB, Ala.: Air University Press, August 1994).

23. An example is the "push" orientation of certain intelligence information, meaning that everything available that meets certain broad parameters is forwarded. The system makes little attempt to respond to the "pull" of specific requirements from the field.

24. An imperfect but helpful analogy would be to contrast Marine Corps air organization and doctrine, particularly for close air support, to that of the US Air Force. McNamara.

25. Joint Pub 3-14, III-5.

26. Based on functional descriptions in Air Force Space Command, "Space Primer."

27. Unclassified sources typically list about 60 national security satellites, a third of which are in the Navigation Satellite Timing and Ranging (NAVSTAR) GPS constellation, to perform missions worldwide. Paul B. Stares, *Space and National Security* (Washington, D.C.: The Brookings Institution, 1987).

28. This assertion can only be inferred indirectly from comments in unclassified literature, For example, Adm William O. Studeman in "The Space Business and National Security: an Evolving Partnership," *Aerospace America*, November 1994, 27, says that the United States is "living off the resources of the past. The space inventory in orbit today is generally less capable than that in orbit during Desert Storm."

29. This does not include sites dedicated to operating classified systems, or the numerous sites of the Air Force Satellite Control Network. The former are examples of stovepiping in specific functional areas, and the latter are manifestations of a command-oriented architecture, not of a decentralized or distributed command and control system, as is explained in the sections which follow. Since some orbits can only be reached from one of these launch sites, in effect we have no backup in case that site should become unavailable.

30. The United States would need two to three months to replace a satellite that failed unexpectedly. Some satellite and launcher combinations would take nearly a year to replace. Horner, "Space 1990 and Beyond," 11.

31. The exception to this, of course, is the GPS constellation, since it is a broadcast system.

32. According to Martin Faga, budget cuts "translate into less service for policymakers and intelligence consumers." Bill Gertz, "The Secret Mission of the NRO," *Air Force Magazine*, June 1993, 60–63. See also Horner, "Space 1990 and Beyond," 5; and Studeman, 29.

33. The general nature of the requirements is apparent in *A National Security Strategy of Engagement and Enlargement* (Washington, D.C.: The

White House, July 1994), and current US National Security Strategy, and the Department of the Air Force, *Global Presence 1995* (Washington, D.C.: Government Printing Office [GPO], 1995).

34. As opposed to the cold war situation, in which the predominant strategic concern was a large-scale conflict with the Soviet Union, the indicators of an emerging crisis could be developed and refined over years, and systems could even be built specifically to look for certain indicators.

35. An example is interfaces (physical and electrical) of software or of certain key components. The personal computer and automobile businesses offer numerous examples. Standardization should not be taken to extremes. One size (or color or model) will not fit all customers.

36. These exceptions are minor since there are no operational (as opposed to research and development) DOD satellites suitable for Pegasus launch. The space shuttle is different of course, but it is not really a player in the national security space architecture for political reasons (DOD payloads were phased out of shuttle launches following the *Challenger* explosion in 1986), operational concerns (there are few shuttle flights every year and manifesting a major payload on the shuttle must be done years in advance and requires extensive coordination, testing, safeguards, and integration work) and cost (even by conservative estimates, launching a shuttle costs more than any other system). See the *Space Launch Modernization Study Executive Summary* (Washington, D.C.: Department of Defense, 18 April 1994), 13.

37. See Maj William W. Bruner III, "The National Security Implications of Inexpensive Space Access," (master's thesis, School of Advanced Airpower Studies [SAAS], June 1995) for a discussion of the strategic and operational implications of a vehicle that provided rapid, reliable, and low cost access to space.

38. A perfect illustration of this is our attempt to replace the DSP early warning system, which is basically a 1960s design. Since the early 1980s we have attempted to develop the Boost Surveillance and Tracking System, the Advanced Warning System, the Tactical Warning/Attack Assessment System, and the Follow-On Early Warning System. All were canceled and then reborn under a new name, the latest of which is alert, locate and report missiles (ALARM). The reasons for these repeated failures are complex, but the root of the problem may be that our observe-orient-decide-act (OODA) loop is too slow. We take so long to specify requirements for a system, design and test it, and field it that the requirements and budgetary ground shifts out from under us, leaving the system with no support. It goes without saying that this cycle adjusts poorly to rapidly changing technology. For an idea of the complexity of the political and bureaucratic issues involved, see United States Congress, House Committee on Government Operations, Legislation and National Security Subcommittee, *Strategic Satellite Systems in a Post-Cold War Environment* (Washington, D.C.: GPO, 2 February 1994).

39. Studeman, 27; and Gertz, 63. Trying to satisfy requirements no longer seems to be an option, and much of the space community seems resigned to just doing the best they can with the money they're given. As Martin Faga is quoted in the Gertz article: "It isn't the requirements. The requirements are infinite."

40. Low observable, or stealth, technology is a good example of the latter two points.

41. As in linear algebra.

42. Though perhaps still one that would lend itself later to numerical weighting and operations analysis.

43. This is done to compare the command- and demand-oriented approaches. For a different sort of comparison, the weighting would obviously reflect different criteria.

44. This becomes a problem because the small number of launches means that each one must succeed, yet the need to push performance to the maximum forces costs up and reliability down, According to W. Paul Blase, "The First Reusable SSTO Spacecraft," *Spaceflight*, March 1993, 91, as you approach the limits of performance, every 10 percent increase in performance doubles the cost and halves the reliability.

45. In the case of the launch vehicle, everything may be sacrificed for performance, so that reliability can be ensured only by intensive review and highly involved checks and cross-checks.

46. For example, the ability of a constellation of electrooptical sensors to have continuous coverage of an area of interest and timely reporting of that information may be more important than the resolution of an individual sensor.

47. These assertions were at least challenged by performance in the Gulf War.

48. Especially if Iridium, Teledesic, and similar concepts come to fruition. Bruner.

49. For more on this, see *SPACECAST 2020 Final Report*, vol. 1 (Maxwell, AFB: Air University, June 1994), specifically the section entitled, "Spacelift: Suborbital, Earth to Orbit, and On-Orbit."

50. For this assertion to be true, cheap enough means that savings on launch costs are greater than the cost of redesigning the satellite. In some cases, this may mean redesign at the architectural rather than the satellite level. For example, it may not make sense to launch a large geosynchronous communications satellite in pieces and assemble it on orbit, but the same mission could be accomplished by several smaller, lower-capacity communications satellites, either in geosynchronous or lower orbits, at a lower life cycle cost.

51. RSTA recognizes the expanded utility of integrating the more intelligence-oriented aspects of the mission with those directly supporting operations, with the ultimate goal of creating (as the Russians call it) a reconnaissance-strike complex that will detect, locate, identify, and attack enemy forces much faster than they can react.

52. The phrase, "support to the war fighter," has cropped up in briefings attended by the author on everything from developmental systems to Space Command organization. It appears in virtually every article on military space written since the Gulf War, and, in the form of "support to military operations" has even become a raison d'être for parts of the national intelligence community. See, for example, Studeman, 26.

53. An example is the DOD Bottom-Up Review which is based on the ability to fight two major regional contingencies. The theater CINC focus was, of course, promoted by the 1986 Goldwater-Nichols Act, and reinforced by experience in Desert Shield/Desert Storm.

54. These are primarily discussed as lessons learned from Desert Storm. See, for example, Moorman; Lt Gen James R. Clapper Jr., "Imagery—Gulf War Lessons Learned and Future Challenges," *American Intelligence Journal,* Winter/Spring 1992, 13–17; and Kevin H. Darr, "DIA's Intelligence Imagery Support Process: Operations Desert Shield/Desert Storm and Beyond," *American Intelligence Journal,* Winter/Spring 1992, 43–45.

55. Department of the Air Force, *Global Presence.*

56. One of the major shortcomings identified in the Gulf War was the inability of RSTA systems to provide synoptic coverage of the theater. Studeman, 26. Actual capabilities of our space RSTA systems are classified, but it is fair to surmise from their cold war mission of monitoring the USSR that they are optimized to collect highly detailed information on fixed targets, not to provide near-real-time coverage of a dynamic situation. As partial confirmation of this, many recent TENCAP projects have focused mainly on intelligence preparation of the battlefield functions. Chief of naval operations (CNO), briefing, N-632, "JCS TENCAP Special Project 95 Night Vector, Project Summary," April 1995. There are exceptions, and there have been efforts to allow our current space systems to provide more direct support (Talon Sword, Radiant Ivory), but I believe in general that the statement in the text is true.

57. The United States will employ systems to fill in gaps in coverage or add a type of sensor not normally present.

58. RSTA was one of the foci of Exercise Roving Sands 1995, in which the author took part as a member of the air operations center.

59. D. Brian Gordon, "Use of Civil Satellite Imagery for Operations Desert Shield/Desert Storm," *American Intelligence Journal,* Winter/Spring 1992, 39.

60. Covering all relevant spectra is impractical because of inevitable space, weight, and power constraints; the possibility of physical or electromagnetic interference among different sensors; the engineering drawback of having to design the platform to suit the most demanding of sensors (e.g. providing far more stability than most sensors need for the one that does require it); and the volume of data that would result, which would either require multiple data links on one platform or a link of extremely high capacity.

61. These factors make battle damage or combat assessment one of the most consistently difficult tasks for intelligence.

62. This importance is similar to the Scud hunt during the Gulf War. Of course this doesn't effect RSTA alone, but also the conventional forces who must deal with the problem

63. For example, it may not be possible to adjust the orbit of a satellite to optimize coverage over a given region, either because it would use up too much of the satellite's maneuvering fuel (and hence its operational life), or because coverage of some other theater requires a compromise orbit.

64. The "technology freeze date" for a major space system is typically five or more years before the first launch. Since that satellite is likely to operate for several years, the onboard electronics could be 10 to 15 years behind what is available on the ground toward the end of the satellite's life.

65. Additionally, processors use power and produce heat. Both of these phenomena add weight to the satellite design.

66. Similar concerns have been raised in the past about virtually all types of forces—air combat forces, airlift forces, and naval forces. The solutions worked out to date have generally given the CINC control of any forces while they are operating in his area of responsibility, a compromise that is somewhat problematic with regard to space systems.

67. CNO briefing; "Space Warfare Center," *Space Tactics Bulletin* 1, no. 1, June 1994. On the air side, the Defense Airborne Reconnaissance Office is planning to spend $9.23 billion over the next five years. Maj Gen Kenneth R. Israel, "An Integrated Airborne Reconnaissance Strategy," *Unmanned Systems* 12, no. 3 (Summer 1994): 17–32.

68. Since the current system is primarily command oriented, it offers good examples of how to improve a generic command-oriented architecture. For example, the Vista project and the joint force air component commander (JFACC) joint situational awareness system (JSAS) described in the CNO briefing.

69. CNO briefing.

70. Additional coverage can be requested, and this system makes it easier to know what to ask for, but the theater user is still at the mercy of someone else's priorities.

71. Maj Stephen P. Howard, "Special Operations Forces and Unmanned Aerial Vehicles: Failure or Future?" (master's thesis, SAAS, June 1995) describes how unmanned aerial vehicles (UAV) with enhanced capabilities could contribute to the special operations forces missions.

72. An example is the use of a laser imaging, detection, and ranging (LIDAR) or other active sensor to discriminate real targets from decoys, or performing highly detailed battle damage assessment.

73. See, for example, Dr. Brian McCue, "The Military Utility of Civilian Remote Sensing Satellites," *Space Times*, January–February 1994, 11–14; and Dr. Ray A. Williamson, "Assessing U.S. Civilian Remote Sensing Satellites and Data," *Space Times*, January–February 1994, 6-10.

74. John G. Roos, "SPOT's 'Open Skies' Policy Was Early Casualty of Mideast Conflict," *Armed Forces Journal International*, April 1991, 32.

75. According to Joint Pub 3-14, B-15, requests for LANDSAT data can take from several days to several months to fill. Because of the orbit and the limitations of the sensors, a given area on the ground is imaged only every 16 to 22 days.

76. This implies that the demand-oriented system might be overlaid on an existing, command-oriented architecture in times of crisis.

77. Arguably, though, in a world of multiple, ongoing crises there may be considerable residual capability on orbit that could be chopped from CINC to CINC as appropriate.

78. On building satellites quickly, see Leonard David, "Faster, Better, Cheaper: Sloganeering or Good Engineering?" *Aerospace America*, January 1995, 28–32, and "New Techniques Allow 22-Day Satellite Assembly," *Aviation Week & Space Technology*, 3 April 1995, 57. There are many opinions on ways to radically improve space lift; three examples are John A. Copper, et al., "Future single stage rockets: Reusable and Reliable," *Aerospace America*, February 1994, 18–21; John R. London III, *LEO on the Cheap: Methods for Achieving Drastic Reductions in Space Launch Costs* (Maxwell AFB, Ala.: Air University Press, October 1994); and *SPACECAST 2020 Final Report*, vol. 1. The Advanced Research Projects Agency has also spearheaded efforts to provide more user-responsive space assets. One example is the DARPASAT program.

79. Joint Pub 3-14, 1–2. The US Air Force awards astronaut wings for flights above 50 miles. The thermosphere—the boundary above which the atmosphere provides virtually no protection from ultraviolet radiation—begins at about 55 miles. Wertz and Larson, 194. In legal terms, there is no authoritative definition of where national airspace ends and international space begins.

80. This not to say that weapons or other payloads coming from or through space could not perform missions now done, for example by aircraft, only to note that even with the increased speed of an orbital or suborbital system there will be a delay before the payload arrives. If the target is immobile, located where an aircraft cannot reach, or too well hardened to be destroyed by a nonnuclear aerial weapon, a space solution may be attractive.

81. This makes a difference for some missions, since atmospheric absorption and disturbance is greater at some wavelengths in the electromagnetic spectrum than others. Notes from Air Force Institute of Technology Course Physics S21, Space Surveillance, Summer 1989.

82. Geosynchronous, sun-synchronous, and Molniya orbits are examples. William E. Wiesel, *Spaceflight Dynamics* (New York: McGraw Hill, 1989), chap. 3 and 66–67. Over a period of time, the lack of precise knowledge of initial conditions and the effect of various perturbations on the satellite's orbit grow into a large enough positional uncertainty that a viewer

no longer has a track on the satellite. Regular observations are needed to update a satellite's orbital elements.

83. These maneuvers can be on the order of meters per second of velocity change, compared to orbital velocities on the order of 10 kilometers per second.

84. This hypothetical conclusion is based on the difficulty of predicting the track of a given space object in the absence of sufficient observations. If our operational procedures force us to use a particular, more predictable orbital path, this advantage is nullified. History shows that this liability is not confined to space systems, witness the often predictable pattern of air operations in Vietnam.

85. Since satellites are large, heavy, and complex; require a long time to build; last a long time; and are extremely expensive, they are not likely to be stockpiled. Performance and reliability of satellites and launch systems are emphasized above all else, and in general the command-oriented architecture depends on having adequate capabilities in place for crises, not on augmentation or reconstitution and replenishment.

86. Certainly the backers of Teledesic are betting that this is true.

87. Although we cannot yet do this, trends in technology clearly point in this direction. "New Techniques Allow 22-Day Satellite Assembly," 57, discusses reducing satellite assembly and testing time from years to days and uses the analogy of the revolution Henry Ford brought to automobile assembly. Continuing that analogy, there is every reason to suspect that as assembly automobile lines of today are far more efficient and flexible than they were in Henry Ford's day, future satellite assembly—with the aid of type certification to reduce individual satellite testing, standardized modules, and so forth—will also be much simplified. It bears remembering that the Iridium satellites are both fairly complex and designed for five- to seven-year lifetimes. The satellites needed for augmenting a demand-oriented architecture could be much simpler.

88. The architecture could be like the open systems architecture used by the personal computer industry.

89. This advantage does not minimize the difficulties of using commercial electronic equipment in space, but it can be done with intelligent design, suitable redundancy and without expecting 100 percent reliability or exceptionally long lifetimes.

90. This just-in-time delivery would presume reliable suppliers.

91. Certainly a tremendous amount of onboard processing, perhaps even intelligence, will be needed to produce an RSTA architecture that could broadcast information—as GPS does—without overwhelming most users with unnecessary data. This kind of concept also requires that each receiver be able to correlate, fuse, and act on information from the multiple off-board sensors that it may be receiving. Although these seem like difficult problems, they also seem like issues that must be tackled if we are not to depend on centralized processing nodes and a few high-capacity data links. In other words, these problems must be solved if the US military is going to

prepare to fight in a high-tempo operating environment against enemies who understand our reliance on information and have the means to attack.

92. *SPACECAST 2020 Final Report*, vol. 1, section D, "Space Traffic Control, The Culmination of Improved Space Operations."

93. There is no consensus on the definition of a *lightsat*, though a lightsat is usually a satellite of 1,500 to 2,000 pounds or less. The Teledesic satellites fall into this category. In recent years, there has been increasing talk and development work on *microsats* weighing as little as a few pounds. Theresa Foley, "Tiny Satellites Aim to Please the Bean Counters," *New York Times*, 5 March 1995, 10F.

94. For example, Donald C. Latham, "Lightsats: A Flawed Concept," *Armed Forces Journal International*, August 1990, 84–86. S. Roy Schubert, James R. Stuart and Stanley W. Dubyn, "LightSats: The Coming Revolution," *Aerospace America*, February 1994, 26–29, 34; "Fitting the Small to the Infinite," *The Economist*, 12 October 1991, 87–88.

95. The Clementine lunar mapper and asteroid rendezvous mission showed many of the reasons for pursuing this course. For $80 million, far less than any previous mission outside of near-earth orbit, a satellite weighing just over five hundred pounds (without fuel) was able to provide valuable and in many cases unique scientific information. The mission was not completed due to a software defect, which highlights a danger of building satellites "faster and cheaper." Given time, additional testing, and considerably more money such problems can be avoided, but the point of the Clementine demonstration was to make acceptable some risk to gain rapid and affordable response. Col Pedro L. Rustan, "Clementine: Mining New Uses for SDI Technology," *Aerospace America*, January 1994, 38–41; and Rustan, "Clementine: Measuring the results," *Aerospace America*, February 1995, 34–38. It's worth mentioning that doing things the conventional way, with substantial testing and a much larger budget, is no guarantee of success either. Witness the loss (for causes still not precisely known) of NASA's Mars Explorer mission. For the ideas of one of the pioneers, see William E. Howard III, "Cheaper by the Dozen?," US Naval Institute *Proceedings*, February 1989, 70–74.

96. An example is a constellation dedicated to providing near-continuous RSTA coverage of a particular theater.

97. How much costs can be reduced is a difficult question. How much they should be reduced is perhaps a better question. In other words, what cost per satellite would the military find acceptable for an important mission for which the hardware was expendable? I suggest that the cost goal for a satellite to enable this approach should be in the neighborhood of what a cruise missile costs, around $1 million.

98. How much can launch costs be decreased? A better question may be, how low do launch costs have to be to make a responsive launch system cost effective? Based on research and numerous discussions during the SPACECAST 2020 study, I believe the threshold is about five hundred dollars per pound, so that the total lift cost for a nominal 1,000-pound

satellite would be five hundred thousand dollars. To achieve this kind of cost level, the launch vehicle would almost certainly have to be reusable. This compares to current launch costs of roughly four thousand dollars to $12,000 per pound on relatively nonresponsive launch systems. United States Air Force Space Command, *Space Launch Modernization Plan (Moorman Study) Executive Summary* (Peterson AFB, Colo.: Department of Defense, 18 April 1994). The Pegasus light launch vehicle is currently the cheapest dedicated ride that is not piggybacking on another satellite's larger launch vehicle. Pegasus can attain orbit at about $12 million for a payload under 900 pounds, but at a cost-per-pound of nearly $15,000. London, 5.

99. Another valid question is, "Given a responsive space launch capability, how does one ensure that it will have enough capacity in times of crisis without huge overhead costs (either for an inventory of expendable vehicles or an idle fleet of reusable ones) during noncrisis periods?" The answer is to build reusable vehicles that can have multiple missions, perhaps including surface-to-surface cargo transport, weapons delivery, or even reconnaissance. In *SPACECAST 2020 Final Report*, vol. 1, the chapter "Spacelift: Suborbital, Earth to Orbit, and On Orbit" discusses how this could be done with a transatmospheric vehicle using a modular (or containerized) payload approach.

100. William B. Scott, "'Architect' to Reshape Defense Space Policy," *Aviation Week & Space Technology*, 20 February 1995, 50.

101. Briefing, "Seven Strategies for Space—The Way Ahead," Col C. A. Waln, USAF Space and Missile Systems Center; Col H. E. Hagemeier, Air Force Space Command, and Mr. Darrell Spreen; USAF Phillips Laboratory, January 1995.

102. AFM 1-1, *Basic Doctrine of the United States Air Force*, 1984, 4–7.

Bibliography

Books

Bate, Roger R., Donald D. Mueller, and Jerry E. White. *Fundamentals of Astrodynamics.* New York: Dover Publications, Inc., 1971.

Baucom, Donald R. *Clausewitz on Space War: An Essay on the Strategic Aspects of Military Operations in Space.* Maxwell AFB, Ala.: Air University Press, June 1992.

Campen, Alan D., ed. *The First Information War.* Fairfax, Va.: AFCEA International Press, 1992.

London, John R., III. *LEO on the Cheap: Methods for Achieving Drastic Reductions in Space Launch Costs.* Maxwell AFB, Ala.: Air University Press, October 1994.

McNamara, Stephen J. *Air Power's Gordian Knot: Centralized versus Organic Control.* Maxwell AFB, Ala.: Air University Press, August 1994.

Ninth National Space Symposium, Proceedings Report. Colorado Springs, Colo.: United States Space Foundation, 1993.

Petersen, Steven R. *Space Control and the Role of Antisatellite Weapons.* Maxwell AFB, Ala.: Air University Press, May 1991.

Stares, Paul B. *Space and National Security.* Washington, D.C.: Brookings Institution, 1987.

TRW Space and Technology Group, *TRW Space Data Book.* 4th ed., Redondo Beach, Calif.: TRW S&TG Communications, 1992.

Wertz, Dr. James R., and Dr. Wiley J. Larson, eds. *Space Mission Analysis and Design.* Dordrecht, The Netherlands: Kluwer Academic Publishers, 1991.

Wiesel, William E. *Spaceflight Dynamics.* New York: McGraw Hill, 1989.

Articles

Berkowitz, Bruce D. "More Moon Probe for Your Money." *Technology Review,* April 1995, 24–31.

Blase, W. Paul. "The First Reusable SSTO Spacecraft." *Spaceflight,* March 1993, 90–94.

Buenneke, Richard H., Jr. "The Army and Navy in Space." *Air Force Magazine,* August 1990, 36–39.

Canan, James W. "Normalizing Space." *Air Force Magazine,* August 1990, 12–14.

Clapper, Lt Gen James R. Jr., "Imagery—Gulf War Lessons Learned and Future Challenges." *American Intelligence Journal,* Winter/Spring 1992, 13–17.

Copper, John A., et al. "Future Single Stage Rockets: Reusable and Reliable." *Aerospace America,* February 1994, 18–21.

Correll, John T. "Slipping in Space." *Air Force Magazine,* October 1993, 2.

——— "Fogbound in Space." *Air Force Magazine,* January 1994, 22–29.

Darr, Kevin H. "DIA's Intelligence Imagery Support Process: Operations Desert Shield/Desert Storm and Beyond." *American Intelligence Journal,* Winter/Spring 1992, 43–45.

David, Leonard. "Faster, Better, Cheaper: Sloganeering or Good Engineering?" *Aerospace America,* January 1995, 28–32.

Deshmukh, Anton R. "Privatization of Civil Remote Sensing—A Win-Win Proposition." *Space Times,* January-February 1994, 1–2.

Dougherty, Vice Adm William A. "Storm from Space." US Naval Institute *Proceedings,* August 1992, 48–52.

"Fitting the Small to the Infinite." *The Economist,* 12 October 1991, 87–88.

Foley, Theresa M. "Tiny Satellites Aim to Please the Bean Counters." *New York Times,* 5 March 1995, 1 OF.

———. "Zooming in on Remote Sensing Markets." *Aerospace America,* October 1994, 22–27.

Gertz, Bill. "The Secret Mission of the NRO." *Air Force Magazine,* June 1993, 60–63.

Gordon, D. Brian. "Use of Civil Satellite Imagery for Operations Desert Shield/Desert Storm." *American Intelligence Journal,* Winter/Spring 1992, 39–42.

Gray, Colin S. "Vision for Naval Space Strategy." US Naval Institute *Proceedings,* January 1994, 63–68.

Hall, R. Cargill. "The Origins of U.S. Space Policy." *Colloquy,* December 1993, 5–6, 19–24.

Hamel, Lt Col Michael A., and Lt Col Michael L. Wolfert. "Space Strategy and the New World Order." *Proceedings of the Twenty-Ninth Space Congress, Space Quest for New Frontiers,* 1992.

Holley, I. B. "Of Saber Charges, Escort Fighters, and Spacecraft." *Air University Review,* September/October 1983, 2–11.

Howard, William E., III. "Cheaper by the Dozen?" US Naval Institute *Proceedings,* February 1989, 70–74.

Howard, William E., III and Owen K. Garriot. "Can You See Ships From Space?" US Naval Institute *Proceedings,* December 1989, 89–94.

Israel, Maj Gen Kenneth R. "An Integrated Airborne Reconnaissance Strategy." *Unmanned Systems* 12, no. 3 (Summer 1994): 17–32.

Kronz, James C. "NAVSTAR GPS: A New Era in Navigation." US Naval Institute *Proceedings.* December 1986, 91–93.

Latham, Donald C. "Lightsats: A Flawed Concept," *Armed Forces Journal International,* August 1990, 84–86.

McCue, Dr. Brian. "The Military Utility of Civilian Remote Sensing Satellites." *Space Times,* January–February 1994, 11–14.

Moore, George M., Vic Budura, and Joan Johnson-Freese. "Joint Space Doctrine: Catapulting into the Future." *Joint Forces Quarterly,* Summer 1994, 71–76.

"Navigation: Finding the Future." *The Economist,* 6 November 1993, 115.

"New Techniques Allow 22-Day Satellite Assembly." *Aviation Week & Space Technology,* 3 April 1995, 57.

Noor, Ahmed K., and Samuel L Venneri. "Perspectives on Future Space Systems." *Aerospace America,* February 1994, 14–18, 5.

Parrington, Lt Col Alan J. "US Space Doctrine: Time for a Change?" *Airpower Journal,* Fall 1989, 51–61.

"Rethinking Space: An Interview with Norman R. Augustine." *Technology Review,* August/September 1991, 33–40.

Rip, Michael R. "How Navstar Became Indispensable." *Air Force Magazine,* November 1993, 46–49.

Roos, John G. "SPOT's 'Open Skies' Policy Was Early Casualty of Mideast Conflict." *Armed Forces Journal International,* April 1991, 32.

Rustan, Col Pedro L. "Clementine: Mining New Uses for SDI Technology." *Aerospace America,* January 1994, 38–41.

———. "Clementine: Measuring the Results." *Aerospace America,* February 1995, 34–38.

Ryan, Terry. "Imagery Intelligence Reform: Is It Time?" *American Intelligence Journal,* Winter/Spring 1992, 19–23.

Schubert, S. Roy, James R. Stuart, and Stanley W. Dubyn. "LightSats: The Coming Revolution." *Aerospace America,* February 1994, 26–29, 34.

Scott, William B. "Military Space 'Reengineers'." *Aviation Week & Space Technology, 15* August 1994, 20–21.

———. "'Architect' to Reshape Defense Space Policy." *Aviation Week & Space Technology,* 20 February 1995, 50.

———. "Iridium on Track for First Launch in 1996." *Aviation Week & Space Technology,* 3 April 1995, 56–61.

———. "Space-Based Radar Pushed as Anti-Terrorist Weapon." *Aviation Week & Space Technology,* 17 April 1995, 56–57.

Shellans, Mark H., and William R. Matoush, "Designing Survivable Space Systems." *Aerospace America,* August 1992, 38–41.

Smith, Bruce A. "Military Space Establishment Moves toward Major Changes." *Aviation Week & Space Technology,* 20 March 1989, 121–23.

"Space Almanac." *Air Force Magazine,* August 1994, 44–59.

Studeman, Adm William O. "The Space Business and National Security: an Evolving Partnership." *Aerospace America,* November 1994, 24–29.

Szafranski, Richard. "When Waves Collide: Future Conflict." *Joint Forces Quarterly,* Spring 1995, 77–84.

"The Uncontested Arena." *The Economist,* 26 January 1991, 78.

Wilkins, D. E. B. "Mission Control in 2023." *Spaceflight* 38, no. 1 (January 1994).

Williamson, Dr. Ray A. "Assessing U.S. Civilian Remote Sensing Satellites and Data." *Space Times,* January–February 1994, 6–10.

Worden, Col Simon P., and Lt Col Jess M. Sponable. "Management on the Fast Track." *Aerospace America,* November 1994, 30–39.

Government Documents

Blue Ribbon Panel of the Air Force in Space in the 21st Century, Executive Summary. Washington, D.C.: The Blue Ribbon Panel, 1992.
A National Security Strategy of Engagement and Enlargement. Washington, D.C.: The White House, July 1994.
AFDD-4. *Space Doctrine,* 1 May 1995.
AFM 1-1. *Basic Doctrine of the United States Air Force,* 1984.
AFM 2-25. *Space Operations,* initial draft, 29 March 1991.
Air Force Space Command. "Space Primer," preliminary draft, February 1995.
Army Space Institute. *Space Support in Mid-Intensity Conflict,* 1990.
Chief of Naval Operations. "JCS TENCAP Special Project 95 Night Vector, Project Summary," Briefing, N-632. April 1995.
DeKok, Brig Gen Roger. "Air Force Space—Then and Now," undated. Briefing.
Department of the Air Force, *Global Presence.* Washington, D.C.: GPO, 1995.
Horner, Gen Charles A. "Space Seen as Challenge, Military's Final Frontier." Prepared statement to the Senate Armed Services Committee, 22 April 1993.
———. "Space 1990 and Beyond: The Turning Point." Address delivered to the US Air Force: Today and Tomorrow Conference, October 1993.
Joint Doctrine, Tactics, Training, and Procedures (JDTTP) 3-14, *Space Operations,* 15 April 1992.
Report, Vice President's Space Policy Advisory Board, *The Future Light U.S. Space Launch Capability,* November 1992.
SPACECAST 2020 Executive Summary. Maxwell AFB, Ala.: Air University, June 1994.
SPACECAST 2020 Final Report. Vol. 1. Maxwell AFB, Ala.: Air University, June 1994.

SPACECAST 2020 Operational Analysis. Maxwell AFB, Ala.: Air University, June 1994.

Space Launch Modernization Plan (Moorman Study) Executive Summary. Peterson AFB, Colo.: Department of Defense, 18 April 1994.

Space Warfare Center. *Space Tactics Bulletin* 1, no. 1 (June 1994).

Waln, Col C. A., Col H. E. Hagemeier, and Mr. Darrell Spreen. "Seven Strategies for Space—The Way Ahead." Briefing. January 1995.

Webster, Maj Mark N. "Lightsat Operational Control: Who's in Charge?" Naval War College paper, DTIC Document no. AD-A249 844, 13 February 1992.

Widnall, Sheila E. secretary of the Air Force. Address to the National Security Industrial Association, Crystal City, Va., 22 March 1994.

United States Congress, House Committee on Government Operations, Legislation and National Security Subcommittee. *Strategic Satellite Systems in a Post-Cold War Environment.* Washington, D.C.: GPO, 2 February 1994.

Unpublished Sources

Air Force Institute of Technology "Physics 521, Space Surveillance." Course Notes, Summer 1989.

Bruner, Maj William W., III. "The National Security Implications of Inexpensive Space Access." Master's thesis, School of Advanced Airpower Studies, June 1995.

Canavan, Gregory H., and Simon P. Worden. "Military Space in the Next Century," unpublished article.

Dionne, E., and C. Daehnick. "The Future Role of Space Experiments." AIAA Paper 92-1697, presented at the AIAA Space Programs and Technologies Conference, Huntsville, Ala., 24–27 March 1992.

French, James R., and Michael D. Griffen. *Spacecraft Systems Design and Engineering.* Short course presented as part of the AIAA Professional Studies Series, 14–15 February 1990.

Hall, R. Cargill. "The Eisenhower Administration and the Cold War: Framing American Astronautics to Serve National Security." Unpublished essay, January 1994.

Howard, Maj Stephen P. *Special Operations Forces and Unmanned Aerial Vehicles: Failure or Future?* Master's thesis, School of Advanced Airpower Studies, June 1995.

Johnson, Dana J. "Issues in United States Space Policy." RAND Project Memorandum PM-141-AF/A/OSD, June 1993.

Moorman, Lt Gen Thomas S. "Space: A New Strategic Frontier." undated.

Stewart, Bob. Untitled address to the Space Support to the Warfighter Conference, Peterson AFB, Colo., 15 December 1993.

US Air Force Academy. "Astrodynamics Short Course." Undated.

Wiesel, William E. *Advanced Astrodynamics.* text from an Air Force Institute of Technology Class, December 1989.

———. *Modern Methods of Orbit Determination.* text from an Air Force Institute of Technology Class, 1990.

PART II

Sanctuary/Survivability Perspectives

Chapter 4

Safe Heavens: Military Strategy and Space Sanctuary

David W. Zeigler

Undoubtedly the most provocative subject in any discussion of the future of space is the subject of space weapons and the likelihood of their use. Here I am referring to the broadest categories: space-based lasers to shoot down hostile intercontinental ballistic missiles, space weapons that attack other satellites, or weapons released from space platforms that destroy terrestrial targets. Today these kinds of systems clearly break the current thresholds of acceptability and introduce Anti-Ballistic Missile Treaty issues, as well as social and political reservations. But the 21st century could well see a change.

—Gen Thomas S. Moorman Jr.

Today, as they have since the 1950s, American leaders are debating the efficacy of US space weapons. In military circles these discussions frequently gravitate to issues of technology, legality, cost, and the military employment of the weapons themselves. Such a focus—one that predominantly concerns itself with how space weapons can be deployed—inevitably overshadows the question of what happens if they are deployed. This result jeopardizes the foundation of knowledge from which Americans will judge the merits of space weapons. Decision makers may be forced to act without a complete and rigorous analysis of the compatibility of space weapons with national strategy.

When Basil H. Liddell Hart succinctly defined strategy as "the art of distributing and applying military means to fulfill

This work was accomplished in partial fulfillment of the master's degree requirements of the School of Advanced Airpower Studies, Air University, Maxwell AFB, Ala., 1997.
Advisor: Maj Bruce DeBlois, PhD
Reader: Dr Karl Mueller

the ends of policy," he correctly subordinated a nation's force structure and doctrine to its national policy objectives—they are inextricably linked.[1] As a result, militarily promising weapons and doctrines can still prove incompatible with higher policy objectives. Three historical examples illustrate this idea, beginning with the Allies' choice of weapons against Germany in the Second World War.

During World War II, the Allies developed proximity-fuzed antiaircraft shells used with great success against German V-1 missiles. Undoubtedly these same weapons would have brought the Allies better performance against the Luftwaffe in combat over France and Germany. Allied commanders banned the weapon from that region, however, fearing that if the Germans manufactured their own from a captured specimen they might use it with devastating effectiveness against Allied bombers in the crucial combined bomber offensive (CBO).[2] Although deploying the shells to continental Europe offered military advantages, those advantages were incompatible with the CBO's central role in Allied strategy.

President James Earl "Jimmy" Carter's rejection of the neutron bomb offers an example of higher national policy ruling out a promising weapon system still in the conceptual stage. The president's complete repudiation of these weapons rested not with their ineffectiveness—they were well-suited for stopping a Soviet offensive while preserving Europe's infrastructure—but rather with the incompatibility of the bombs with broader American strategy. That strategy motivated the United States to internationally maintain the moral high ground, preserve the North Atlantic Treaty Organization (NATO) coalition, and promote arms control.

American deliberations over chemical weapons provide the most contemporary illustration of the potential clash between military expediency and national policy objectives. In April 1997 the US Senate formally ratified the Chemical Weapons Convention by obligating America to forsake future development, production, acquisition, transfer, stockpiling, and use of chemical agents. The treaty was controversial in that such historical American adversaries as Russia, Libya, and Iraq refused to sign it.[3] Treaty critics preferred, instead, to preserve

America's freedom to retaliate with chemical weapons against adversaries who used such weapons against American troops. They accurately asserted that lacking such freedom weakened the ability of the United States to control conflict escalation. As with the case of the neutron bomb, however, the United States elected to forgo the military benefits of a chemical deterrent in deference to higher political objectives. US leaders calculated that America's reputation as a responsible superpower and its commitment to arms control were better served by formally renouncing the American chemical arsenal.

Military policy makers for space find themselves treading similar waters. Today, space weapons are becoming increasingly practical in terms of military promise and associated costs. Yet in the context of higher military and national strategy, the decision to deploy them is complicated by related social, political, economic, and diplomatic factors. As in the past, military missions like "space control" and "space force application" cannot be decoupled from broader national strategy. Though they may promise military advantages, space weapons are desirable only if they prove to be compatible with policy at the national level.

There is no question that Department of Defense (DOD) officials fully appreciate the subordination of military space operations to America's civilian-led national strategy. In February 1997 the commander in chief, US Space Command (CINC USSPACECOM), Gen Howell M. Estes III, emphasized that decisions to develop space-based weaponry are not made by the military. "We . . . support whatever decisions our elected leadership may arrive at with regard to space control and the weapon systems required," he remarked.[4]

As the elected leadership moves closer to these decisions, military strategists should work now to consider the issue of space weapons from every angle, including potential arguments against their development. A quick review of today's defense literature, however, reveals that this is not happening. While there is much written in support of space weapons and their attendant missions, attempts to understand the counterarguments against deploying space weapons are scarce. Few strategists, if any, are testing the conventional wisdom of space weapon proponents with any rigor. For example, military planners and strategists are silent on the evidence of some 40 years of American cold war space policy—a history that shows US national interests ultimately being served by preserving a space sanctuary relatively free of American space weapons. This should not be the case. There must be a disciplined consideration of why cold war space operations developed the way they did and the relevance (or irrelevance) they have today. Instead, some advocates for space weapons continue to see sanctuary thought as a form of "unstrategy," viewing its proponents as "making head-in-the-sand plans."[5] This perspective only serves to undermine useful debate. It leads to a situation in which everybody interprets the universe of possible strategies to include only those they are already predisposed to. As a result, even the most ardent space weapon advocates find themselves at a disadvantage when crafting strategy. They compromise their ability to implement a weapons program that still incorporates, to the extent possible, useful features of sanctuary thought. They forfeit the opportunities, afforded by another point of view, to fairly appraise and ameliorate any weaknesses associated with space weapons.

Regardless of their initial convictions, strategists must strive for totally objective thought. They should take apart every conviction and recast it to optimally fit the current situation. They must explore all avenues of approach to a problem and ranges of possible solutions. Hence the purpose of this study. It endeavors to develop a better understanding of the arguments against space weapons by asking the question: Could pursuing a space sanctuary in the near future benefit the national interest? The product—the space sanctuary argument articulated here in the strongest reasonable terms—offers military strategists a counterpoint to round out the pro-weapons literature on their shelves. Since its purpose is to challenge mentally and not to persuade, the question of whether space should or should not be weaponized is left unanswered. Instead, strategists are invited to put the sanctuary perspective in their cognitive "toolboxes" as but one of many tools required to decide the future of space weapons.

In laying out the sanctuary perspective, basic concepts essential to any discussion of sanctuary thought are first clarified. An underlying premise is emphasized: that US military strategy—especially one associated with space—cannot be divorced from broader national strategy. Since that is true, President William Jefferson "Bill" Clinton's 1996 US national security strategy is used to give the phrase "national strategy" greater substance. The clarification of basic concepts concludes with definitions for "space weaponization" and "space sanctuary."

Having established a framework for discussion, the study turns to America's history with space weapons. Any treatment of contemporary military space policy must at least consider where the nation has been in the past. Although most of America's space history is indelibly colored by the cold war—a geopolitical environment far different from that of 1997—it nevertheless bears some relevance for policy today. The restrained manner in which the United States pursued antisatellites (ASAT) through the end of the 1980s is a classic example of sanctuary concepts in action.

Contemporary American space policy remains relatively consistent with that of the cold war. Domestic support for operational space weapons is growing, however. After transitioning from the past to the present, fundamental convictions driving the arguments of American space weapon advocates today are explored. These convictions are then challenged with sanctuary counterarguments. The case for a sanctuary policy is further bolstered with rationale independent from the convictions of weapon advocates. No attempt is made to critique the weaknesses of the sanctuary argument presented—further acknowledgment that this study merely aims to give sanctuary thought its full day in court. It is left to the reader to balance the space weapon and space sanctuary perspectives.

With the sanctuary argument complete, the conclusion calls upon military strategists to embrace the complex debate over national military space strategy. It encourages strategists to consider military space policy from every perspective in search of the very best strategy. Strategists are also challenged to disregard the idea that sanctuary thought leads to a passive national strategy. Instead, examples illustrate how sanctuary tenets demand coordinated action of all national instruments of power. They also show how sanctuary thought remains relevant even if there is an eventual US decision to deploy space weapons.

Definitions

The United States is a spacefaring nation—it operates some two hundred military and civilian satellites with a combined value of $100 billion.[6] As impressive as these statistics appear, they do not reflect the additional billions of dollars and millions of American lives influenced every day by space communication, navigation, weather, environment, and national security satellites. Space is big business and is inseparable from US economic strength. It attracts international attention and therefore diplomatic power. It is absolutely crucial to American military operations. Since the "high frontier" underpins almost every facet of US national power, American strategists must consider space from a perspective broader than pure military concerns. To do so, however, they must define "broader perspective." In that regard, *A National Security Strategy of Engagement and Enlargement* (February 1996) provides a solid point of departure and conveys the president's priorities for formulating and conducting national policy. "The nature of our response must depend on what best serves our own long-term national interests. Those interests are ultimately defined by our security requirements. Such requirements start with our physical defense and economic well-being. They also include environmental security as well as the security of our values achieved through expansion of the community of democratic nations."[7] Subsequent use of "national interests" in this study is meant to connote the four most basic security requirements arranged by the White House: physical defense, economic well-being, environmental security, and the expansion of the community of democratic nations.

The rudimentary framework provided by the 1996 publication prompts military strategists to evaluate space strategies across the full spectrum of national interests. Before that occurs, however, strategists must clearly understand the space strategies themselves. Therefore, the specific ideas conveyed by "space weapon" and "space sanctuary" must be explicitly defined.

A space weapon is defined as any system that directly works to defeat space assets from terrestrial- or space-based locations or terrestrial-based targets from space. Space weaponization is distinct from the extensive militarization of space that began in the late 1950s. Since that decade, nations have launched thousands of military satellites into space to support surveillance, reconnaissance, communications, navigation, and military research.[8] Today, these satellites make important but indirect contributions to the final defeat of targets. Space weapons, if ever employed, will directly attack and defeat targets via mechanisms ranging from physical destruction to spoofing.

Significantly, the definition adopted for space weapons leaves out two categories of weapon systems that routinely operate in space—ballistic missiles and antiballistic missiles (ABM). Although ballistic missiles traverse space en route to their targets, they are more accurately appraised as surface-to-surface systems. In addition ballistic missiles are well established in strategic thought and provide national security with a deterrent function that has long since been accepted. Considering ballistic missiles as space weapons, then, would inordinately complicate the debate with no apparent gain.

The same is true of the second notable exclusion from the definition for space weapons, the ground-launched ABMs. Including ABM systems in the context of the space sanctuary debate would cloud the central issues related to weapons that attack targets in space and weapons that attack targets from space. Note, however, that ABM systems modified to perform ASAT missions are not excluded. In that event, the modified system clearly becomes a space weapon.[9]

Understanding what is implied by the concept *space sanctuary* is as important as defining space weapons. In the strictest sense, space is a sanctuary when it is completely unthreatened by terrestrial- or space-based weapons. This definition, however, is impractical on two counts. First, such a sanctuary has not existed for decades and realistically never will again. It therefore becomes a rather inflexible construct for a serious policy discussion. Second, even when a nation sincerely believes a sanctuary exists, other nations may disagree. Consider that starting in 1981 the Soviets strenuously objected to the American space shuttle as an ASAT because of its capability to "snatch" satellites from space.

A second, more flexible, definition for space sanctuary might see it in light of national intentions. By this reckoning, a space sanctuary would exist even where nations possessed space weapons, so long as they truly intended never to use them. Again, however, the construct becomes problematic. Good intentions notwithstanding, no nation as a practical matter can accept an armada of adversarial space weapons on the faith they would never be used. Instead of continuing to search for a conceptual definition of space sanctuary in absolute terms, then, this study seeks a more pragmatic approach linked to current realities.

Today, the number of operational space weapons is unchanged from that of a decade ago. In fact the number is actually down from cold war peaks discussed in the next section. The international community, therefore, lives with a degree of space weapons that is stable. Nations are not fielding new weapon systems and the operational systems that already exist are extremely limited in capability. As support builds for American space weapons, however, US decision makers are rapidly approaching a crossroads—a point of decision. This study asserts that any US strategy advocates a space sanctuary if it endeavors to cap the current level of space weaponization *where it stands today*. In other words, a sanctuary exists today given the present equilibrium.

Introducing new space weapons would violate that sanctuary. If the threshold for viewing space as a sanctuary is set at current levels of weaponization, then the strategist ought to know the history that generated those levels. The next section describes past space weapons and elucidates the drivers behind America's space weapons policy during the last 50 years.

Space Weapons and the American Experience

The cold war was a tense affair. For 40 years, two global superpowers stood toe-to-toe, eye-to-eye poised for a war that promised devastation for both. Amidst this tension, the impetus for superiority was so strong and the level of mutual distrust so powerful, that America's nuclear arsenals were built to levels far beyond what some assert were ever useful. The global confrontation also drove innovation and modernization of American conventional forces. United States policy makers never deliberately allowed the Soviets to achieve favorable asymmetries in major weapon systems except antisatellite weapons. Many caution that the cold war fostered geopolitical conditions so unlike today's that its lessons are totally irrelevant. In her book *Rational Choice in an Uncertain World*, Robyn Dawes notes that "a great deal of thinking is associational, and it is very difficult indeed to ignore experience that is associationally relevant, but logically irrelevant."[10] Correspondingly, one might assert that while today's weapon races appear to be comparable to those of the cold war, the unique bipolar tension of the cold war makes any comparison of the two logically flawed—what worked in the cold war may fail in today's multipolar world. That hypothesis, however, is more true for some weapon systems than it is for others. In the case of space weapons it is suspect.

The American cold war experience with space weapons presents a bit of a conundrum. Despite the pressure for relative military parity, if not US superiority, the Soviets finished the cold war with an operational ASAT while the United States possessed none. Significantly, this asymmetry cannot be traced to greater Soviet technological prowess. Instead, its roots lie with American restraint. Unilateral arms restraint during the cold war, however, runs counter to the prevailing sentiments of that period. If the United States did in fact deliberately opt against pursuing an aggressive ASAT program, it must have been to advance interests beyond simple military effectiveness.

American cold war space policy, therefore, is highly relevant for space sanctuary advocates in 1997. The sanctuary argument proposes the very restraint observed in that era. It suggests that broader national strategies can preempt even the strongest justifications for space weapons just as occurred during the cold war maelstrom. For this reason, the argument for a space sanctuary strategy should consider the history of cold war space weapons.

Two Historical Themes

This section briefly describes America's historical experience with space weapons. From the 1950s to the start of the 1990s, two general themes emerge.

First, although space weapon technologies matured over the years, any long-term US commitment to a vigorous space weapons program was constrained by perceived American vulnerabilities in space. When operational US ASATs did appear, they were in direct response to the Soviet threat of orbiting nuclear weapons. Second, in spite of their reluctance to develop space weapons, US policy makers consistently "hedged their bets" with the technological insurance of space weapons research.

Protecting American Vulnerabilities through Restraint

Historical US space policy consistently embraced American restraint in the deployment of space weapons. Policy makers were motivated to legitimize and protect other US space missions from attack. On two occasions, US policy makers ordered ASAT systems to go operational. In both cases, the systems were motivated by Soviet involvement with orbiting nuclear weapons.

By the mid-1950s, the United States was engaged in a cold war of atomic proportions. The perceived adversary was a monolithic Communist movement adroitly led by the Soviet Union—a conviction reinforced by the confrontation with the Soviets over the blockade of Berlin, the 1950 Sino-Soviet Pact, and the Korean War. The technology was nuclear and the introduction of relatively lightweight hydrogen bombs now meant intercontinental ballistic missile (ICBM)-launched warheads were feasible.[11] Assessing the situation in 1954, President Dwight D. Eisenhower observed that "modern weapons have made it easier for a hostile nation with a closed society to plan an attack in secrecy and thus gain an advantage denied to the nation with an open society."[12] His observation hastened the first military space program, Project Feedback, a study recommending that the United States develop satellite reconnaissance as a matter of "vital strategic interest to the United States."[13] By July 1954 Program WS-117L (advanced reconnaissance system) was approved.[14] It was the first step in a long-term American commitment to satellite reconnaissance.

The first serious US discussions of space weapons were prompted by the Soviet launch of sputnik in October 1957. Already that year, Gen Bernard A. Schriever, US Air Force, had stressed the need for "space superiority," predicting that in decades to come the decisive battles would be fought in space.[15] Sputnik inflamed such convictions—even the public soon shared the concern over a perceived "space weapons gap" with the Soviets.[16] This public climate led defense officials to be more specific in their calls for American space weapons. Gen James Gavin, US Army, urgently recommended that Americans "acquire at least a capability of denying Soviet overflight—that we develop a satellite interceptor."[17] In November 1957 his service proposed two ASAT solutions: a modified Nike Zeus ABM and a "homing satellite" carrying a destructive charge.[18]

Despite the mounting pressure to weaponize space, President Eisenhower resisted. He believed it was more imperative that the international community embrace the legitimacy of the satellite reconnaissance mission.[19] In his estimation, jumping out to a lead in ASATs would undermine the credibility of America's efforts to promote space for "peaceful" purposes and encourage the Soviets to redouble their own ASAT efforts. By 1958 Eisenhower articulated this policy in National Security Council (NSC) 5814/1, stating the United States should "in anticipation of the availability of reconnaissance satellites, seek urgently a political framework which will place the uses of U.S. reconnaissance satellites in political and psychological context favorable to the United States."[20]

By the early 1960s, President John F. Kennedy was forced to reassess Eisenhower's sanctuary strategy when Soviet statements and actions indicated they might develop orbiting nuclear bombs. Kennedy feared such weapons could blackmail Americans in a crisis and knew waiting to counter the threat, after it appeared, might embarrass his administration later.[21] So in May 1962, Secretary of Defense (SECDEF) Robert S. McNamara ordered the Army to modify the Nike Zeus ABM for a future ASAT role. The modified system, Program 505, was based at Kwajalein Atoll in the Marshall Islands. Each missile carried a nuclear warhead capable of destroying satellite targets.[22]

As evidence of Soviet efforts to deploy orbital bombs continued to mount, so did pressure for a long-range American ASAT. In 1963 President Kennedy approved Program 437—a ground-launched ASAT system based on the Thor intermediate-range ballistic missile (IRBM)—stating that the United States should "develop an active antisatellite capability at the earliest practicable time, nuclear and non-nuclear."[23] Program 437 was eventually based at Johnston Island in the Pacific. Like Program 505 it carried a nuclear warhead.[24]

Both Programs 505 and 437 went operational in May 1964.[25] Program 505 was quickly phased out by May 1966 in deference to Program 437's longer range.[26] Four factors indicate that these programs were simply emergency stopgaps against a specific nuclear threat and did not signal an American priority to deploy a general-purpose ASAT against other types of satellites. First, after the United States conducted the Starfish Prime series of space nuclear tests in 1962, American policy makers clearly understood that nuclear ASAT detonations would cripple friendly satellites as well as hostile ones.[27] Second, any use of Programs 505 and 437 would have violated the Partial Test Ban Treaty signed only one day before President Kennedy approved Program 437.[28] Third, both systems were hamstrung by their single remote bases. Operating from fixed locations severely limited the number of satellites vulnerable to each system. Satellites that were periodically vulnerable would often be out of view for days.[29] Finally, more flexible systems for targeting general purpose satellites across the spectrum of conflict—nonnuclear ASATs—were never produced despite President Kennedy's directive. DOD considered several projects, but each failed to win administration endorsement.[30]

President Lyndon B. Johnson's administration completed the ASAT programs started by Kennedy, sharing the view that any US ASAT program was principally a hedge against Soviet orbital weapons. An administration report stated that "an anti-satellite capability (probably earth to space) will be needed for defense of the United States. . . . Current high priority efforts should be continued and extended as necessary in the future."[31] Significantly, that same report considered using American ASATs against "space targets in time of war whether or not the orbital nuclear delivery vehicles were introduced." It also proposed that US ASATs could "enforce the principle of noninterference in space."[32] When it came to these additional missions, however, the Johnson administration reiterated Eisenhower's conclusions—targeting Soviet satellites invited retaliation and the United States was more dependent on its space assets. As the report stated, "the usefulness to the United States of observation [satellites] . . . as a means of penetrating Soviet secretiveness is obvious. The value to the USSR may be less clear; indeed, the value is probably much lower."[33] As a result, the Johnson administration proved ambivalent to ASATs, and little was done to replace the limited capabilities of Program 437.[34] That decision was complemented by Johnson's broader space policy: "We should continue to stand on the general principle of freedom of space. We should actively seek arms control arrangements which enhance national security. We should pursue vigorously the development and use of appropriate and necessary military activities in space, while seeking to prevent extension of the arms race into space."[35] President Johnson's policy was another example of America's traditional inclination for sanctuary thought and a key contributor to international acceptance of the 1967 Outer Space Treaty. The treaty's signatories agreed "not to place in orbit around the earth any objects carrying nuclear weapons or any other kinds of weapons of mass destruction, install such weapons on celestial bodies, or station such weapons in outer space in any other manner."[36] America's ASAT posture and policy remained rooted in the sanctuary perspective through 1977. As a case in point, Program 437 was terminated on 1 April 1975, leaving the United States with no operational ASAT capability.[37] This termination is particularly striking in light of the Soviet involvement with ASATs during the same period.

The Soviets began testing their co-orbital ASAT in 1967.[38] The tests' prevailing pattern involved the launch of a target satellite followed by the launch of a "killer satellite" boosted into a coplanar orbit. Typically within two orbital revolutions, the killer satellite would be maneuvered to detonate near the target satellite, destroying it in a cloud of shrapnel.[39] Although these tests often failed, when the initial series of Soviet tests ended in December 1971, they had demonstrated the ability to intercept US photoreconnaissance, electronic intelligence, weather, and TRANSIT NNSS (US Navy navigation satellite system).[40]

President Richard M. Nixon's national security advisor, Henry A. Kissinger, reacted to the Soviet ASAT tests by calling for a "quick study" of possible US responses in 1970.[41] Remarkably, the lack of urgency was such that the report was not submitted until 1973. By that time détente, including the Strategic Arms Limitation Talks (SALT) I treaty and the Soviet hiatus in ASAT testing, had diverted interest from the subject of ASATs.[42]

Détente aside, the report's findings are further indication of US reluctance to deploy space weapons—even when provoked. It recommended steps to reduce the vulnerability of US satellites to attack but explicitly argued against a US ASAT program in response. The rationale was reminiscent of previous administrations. A US ASAT was "not an area where deterrence works very well because of dissimilarities in value between US and Soviet space systems."[43]

By 1977, however, three developments gave new impetus for a renewed US ASAT effort. The first was a series of government panels expressing concern over the growing vulnerability of US satellites. The second was the blinding of US satellites over the Union of Soviet Socialist Republics (USSR) and the resumption of Soviet ASAT testing. The third was a president concerned about the obvious cold war asymmetry in ASAT capability.

In 1975 President Gerald R. Ford's advisors convened the Slichter Panel to review the military applications of space. The panel focused on satellite reconnaissance and tactical communications concluding that "the US dependence on satellites was growing and that these satellites were largely defenseless and extremely soft to countermeasures."[44] This warning was the catalyst for a second panel convened to specifically analyze these vulnerabilities and consider the need for an American ASAT program.[45] The Buchsbaum Panel determined that an ASAT would not enhance the survivability of other US satellites—deterrence was ineffective given the heavy American dependency on space. The Buchsbaum Panel did recognize, however, that while the United States was more dependent on space than the Soviets, the Soviet dependency was increasing. In this regard, the panel believed an American ASAT possessed at least some utility against Soviet intelligence and radar ocean reconnaissance satellites. This utility could also strengthen ASATs as a negotiation chip in future arms control discussions.[46]

Anxiety over the vulnerability of US satellites was heightened by the blinding of US satellites over the USSR and the resumption of Soviet ASAT testing. On three occasions in 1975, US satellites were saturated with intense radiation from sources in the Soviet Union.[47] These incidents reinforced reports that the Soviets were rapidly progressing in directed energy weapon technologies.[48] To aggravate matters further, the Soviets resumed testing of the co-orbital ASAT. In 1976 alone, there were four such orbital tests.[49] The net effect of these developments was a subtle shift in US ASAT policy presaged at the end of 1976 by comments from the Director of Defense Research and Engineering Malcolm Currie. "The Soviets have developed and tested a potential war-fighting anti-satellite capability. They have thereby seized the initiative in an area which we hoped would be left untapped. They have opened the specter of space as a new dimension for warfare, with all that this implies. I would warn them that they have started down a dangerous road. Restraint on their part will be matched by our own restraint, but we should not permit them to develop an asymmetry in space."[50]

Subsequent policy statements continued to emphasize restraint and space as a medium for nonaggressive purposes, but in January 1977 President Ford released National Security Decision Memorandum (NSDM) 345 ordering DOD to develop an operational ASAT.[51]

President Carter inherited Ford's NSDM 345 weeks after it was signed. Elected on a platform of arms control and reduced military spending, however, Carter returned the nation to its tradition of working to stabilize space as a sanctuary. He continued with the ASAT initiative principally on the grounds that it would strengthen arms negotiations as a bargaining chip. If arms control succeeded, the American ASAT would never become operational. President Carter's 1978 Presidential Directive on Space Policy stated that "the United States finds itself under increasing pressure to field an antisatellite capability of its own in response to Soviet activities in this area. By exercising mutual restraint, the United States and the Soviet Union have an opportunity at this early juncture to stop an unhealthy arms competition in space before the competition develops a momentum of its own."[52] In line with this policy, the Carter administration opened ASAT arms control talks with the Soviets in June 1978.[53] The negotiations stalled over a number of issues, however, and finally collapsed with the Soviet invasion of Afghanistan in December 1979.[54]

By the time President Ronald W. Reagan assumed office in 1981, America's ASAT program was in an advanced stage of development.[55] Specifically, the miniature homing vehicle (MHV) ASAT—a direct ascent, air-launched missile designed to home in on and collide with satellites—was approaching the point of operational testing.[56] In contrast with Carter's perspective on space weapons, Reagan unabashedly accelerated the program stating at the beginning of his first term "the United States will proceed with development of an antisatellite (ASAT capability), with operational deployment as a goal. The primary purposes of a United States ASAT capability are to deter threats to space systems of the United States and its allies and, within such limits imposed by international law, to deny any adversary the use of space-based systems that provide support to hostile military forces."[57]

In further contrast to his predecessor, Reagan pressed on with the MHV ASAT effort even as the Soviets called for a space weapons treaty. In 1983 Foreign Minister Andrey A. Gromyko proposed to supplement the Outer Space Treaty so as to outlaw the use of force in space to include a prohibition on "any space based weapons intended to hit targets on the Earth, in the atmosphere, or in space." Significantly, the Soviets underscored the sincerity of their calls by imposing a unilateral moratorium on their own ASAT testing in the same year.[58] Nevertheless, Reagan categorically rejected all Soviet offers citing various weaknesses in the proposed treaty drafts.[59]

In spite of President Reagan's strong support, the MHV ASAT program faced congressional opposition. The Soviet overtures for a space weapons treaty were well received by legislators and many viewed the MHV as an unnecessary start to an arms race in space.[60] As a result, Congress passed a law in 1984 that banned further US ASAT testing. Only a short lapse between this ban and its successor permitted a September 1985 test to occur. On 13 September 1985, an F-15 launched an MHV ASAT at a US satellite collecting scientific data in space. Seconds later, the MHV struck the satellite shattering it into several hundred pieces.[61] The success belied the program's future. In March 1988 congressional test restrictions and budgetary limitations killed the ASAT program before it went operational.[62]

Although President George W. Bush was handed a dead ASAT program in 1989, Reagan's Strategic Defense Initiative (SDI) remained very much alive. Ironically, the Bush administration deemphasized any push for an operational US ASAT effort because of SDI. The administration believed ASATs were destabilizing and above all a threat to the sophisticated ballistic missile defense satellites planned for the future. Addressing the question of stability, President Bush's National Security Advisor Brent Scowcroft observed that "all scenarios involving the use of ASATs, especially those surrounding crises, increase the risks of accident, misperception, and inadvertent escalation."[63]

The vulnerability of the expensive SDI space architecture to ASATs was also recognized early in its development. The government's Defensive Technologies Study Team found in 1984 that "survivability is potentially a serious problem for the space-based components. The most likely threats to the components of a defense system are direct-ascent antisatellite weapons; ground- or air-based lasers; orbital antisatellites, both conventional and directed energy; space mines; and fragment clouds."[64] The technologists designing the SDI architecture would echo the same thoughts in subsequent years. According to the director of the Lawrence Livermore National Laboratory in 1986, "if extensive strategic defenses are deployed, the ASAT and counter ASAT picture changes completely. This is particularly true if space-based weapons are developed and deployed. Under such circumstances, all space assets, whether needed for defense or offense, for warning or other purpose, would have to operate in a very hostile environment."[65]

President Bush, then, returned the nation to a familiar ASAT policy. President Eisenhower had rejected operational ASATs because of the US's dependency on reconnaissance satellites. Subsequent administrations rejected operational ASATs because of the US's growing dependency on satellites of all types. President Bush rejected operational ASATs, in part, because of a predicted US dependency on ballistic missile defense satellites.

The fact that Bush elected not to deploy an operational ASAT does not mean he dismissed ASAT work altogether. In 1989, a year after the MHV was canceled, all three military services remained engaged in ASAT research.[66] This approach to ASATs is patently American and represents a second consistency in the history of US space weapons. US policy makers have consistently "hedged their bets" with the technological insurance of space weapons research and development (R&D) programs.

Technological Insurance through ASAT Research

As the first president to adopt a sanctuary policy for space, Eisenhower nevertheless authorized the Advanced Research Projects Agency (ARPA) and all three of the military services to conduct space weapon research. NSC 5802/1 called for a "vigorous research and development program" to consider weapons against "satellites and space vehicles."[67] Consistent with his broader policy, however, Eisenhower disapproved the services' requests for more advanced stages of system development.[68] A B-47-launched ASAT missile tested in the Bold Orion program and the satellite interceptor (SAINT) program were two notable R&D efforts during Eisenhower's presidency.[69]

In the course of congressional hearings in 1962, Director of Defense Research and Engineering Dr. Harold Brown acknowledged that the Kennedy administration would follow Eisenhower's precedent of pursuing ASAT R&D as insurance. Brown stated that "we must, therefore, engage in a broad program covering basic building blocks which will develop technological capabilities to meet many possible contingencies. In this way, we will provide necessary insurance against military surprise in space by advancing our knowledge as a systematic basis so as to permit the shortest possible time lag in undertaking full-scale development programs as specific needs are identified."[70]

Technology associated with the X-20 Dynasoar, a manned hypersonic space glider, is perhaps the most well recognized military space R&D program during this era.[71] That program, as well as the Manned Orbiting Laboratory, lasted well into the Johnson years.[72] The United States continued to consider vigorous R&D as sufficient insurance against future space weapons threats even as the Soviets demonstrated their co-orbital ASAT. President Nixon's NSC recommended that the United States respond to the Soviet demonstrations with an R&D effort aggressive enough to permit quick turnaround of an operational ASAT system.[73] The MHV ASAT program eventually fulfilled this R&D requirement for both the Ford and Carter administrations.

Measuring national commitment to ASAT R&D after 1983 is very difficult due to President Reagan's SDI. The line between ASAT and ballistic missile defense (BMD) weapons is so blurred as to often make it impossible to distinguish between the two. Indeed, some opponents regarded SDI as little more than cover for a "bloated ASAT development effort."[74] While that assertion is undoubtedly inaccurate, it correctly appreciates that defensive capabilities against ballistic missiles can equate to offensive capabilities against satellites. Since this is so, it is reasonable to assert that the United States continued to pursue ASAT technologies through the R&D associated with

SDI and President Bush's subsequent global protection against limited strikes (GPALS).

In the two years after President Reagan's Star Wars speech in 1983, SDI became the Pentagon's largest single R&D program.[75] Reagan's planned SDI architecture included space-based missile warning satellites, traditional ground-based ABMs with conventional warheads, and constellations of space-based interceptors—hundreds of satellites, each equipped with small rockets to destroy ICBMs. Over the long-term, SDI intended to replace this architecture with various directed-energy weapons deployed on the ground, in the air, and in space.[76]

The 1972 ABM Treaty clearly influenced SDI's research and test methodology. Since the traditional interpretation of that treaty only allowed for testing of sanctioned ground-based ABM systems and their components, the Reagan administration declined to conduct SDI space experiments in the ABM mode.[77] As a result, active space experiments were always conducted against other "space objects," not missile components, underscoring the tenuous distinction between BMD and ASAT R&D.

With the end of the cold war, President Bush reoriented SDI to GPALS. Since the Soviet threat was now replaced by that of rogue nations with rapidly developing ballistic missile programs, GPALS emphasized more mature technologies suitable for theater and tactical defenses.[78] In addition to the traditional warning satellite and ground-based ABMs, Brilliant Pebbles—an improved space-based interceptor—became the critical space weapon in GPALS. Brilliant Pebbles would consist of hundreds of small interceptors deployed in orbits 400 kilometers above the earth. These interceptors would maneuver to collide with any detected ballistic missiles.[79]

Although the concepts for SDI and GPALS never matured to operational systems, they fostered significant advances in space weapon technologies. For example, ground ABM tests showed significantly improved probabilities for intercepting ballistic missiles from long ranges;[80] a high-intensity particle beam irradiated a miniature reentry vehicle in 1986;[81] space experiments collected data on target signatures in space;[82] a neutral particle beam was fired in space from a satellite;[83] and in 1991, SDI Office officials unveiled a chemical laser with practical potential to be an effective space-based weapon.[84]

Conclusions Regarding the Historical Trend

In summary US space policy has a strong sanctuary tradition behind it. Since the 1950s and through eight US presidential administrations, Americans significantly restrained their deployment of space weapons. Policy makers recognized that acting otherwise invited international counterefforts that, in turn, would jeopardize satellites viewed as essential to American national security. In place of operational space weapons, US decision makers opted for research designed to maintain technological parity in space weapons in case production was required to meet new threats. History shows the US government deployed operational ASATs only when the Soviets directly threatened the continental United States with nuclear space weapons, and the utility of these ASATs was quite limited.

Undoubtedly, the United States's sanctuary policies were instrumental in limiting the degree to which space weapons proliferated. Today, space remains relatively unweaponized—defying more than 40 years of a superpower arms race in land, sea, and air weapons. It would be impossible to guess with any precision how things might have turned out had the United States opted to aggressively weaponize space.

Are US space policies of the past relevant for today's decision makers? That question has no simple answer because historical contexts never precisely repeat themselves. Nevertheless, history provides a powerful case study of space sanctuary policy. Understanding the sanctuary perspective in its strongest form requires one to fully appreciate the implications of the historical record. If contemporary US leaders elect to weaponize space today, that decision will stand in marked contrast to almost all US space policies of the past. It would be viewed, domestically and internationally, as a significant discontinuity in US national strategy.

Contemporary US Policy on Space Weapons

The United States is committed to the exploration and use of outer space by all nations for peaceful purposes and for the benefit of all humanity. "Peaceful purposes" allow defense and intelligence-related activities in pursuit of national security and other goals. The United States rejects any claims to sovereignty by any nation over outer space or celestial bodies, or any portion thereof, and rejects any limitations on the fundamental right of sovereign nations to acquire data from space. The United States considers the space systems of any nation to be national property with the right of passage through and operations in space without interference. Purposeful interference with space systems shall be viewed as an infringement on sovereign rights.

—President Clinton's National Space Policy
19 September 1996

Today, US space policy continues to reflect the sanctuary tradition of the past. Like so many of his predecessors, President Clinton opposes aggressive weaponization of space.

President Clinton is being challenged by space weapon advocates around the defense community and in Congress. As that debate unfolds, the United States persists with a familiar course of action—space weapons research and development to a point short of operational deployment.

Space Weapons and the Clinton Administration

While President Clinton tacitly accepts the military missions of space force application (the projection of firepower against surface targets from space) and space control, he clearly has reservations about space weapons. The White House's National Space Policy directs the DOD to "maintain the capability to execute the mission areas of space support, force enhancement, space control, and force application."[85] A more pointed statement remarks later on that "consistent with treaty obligations, the United States will develop, operate, and maintain space control capabilities to ensure freedom of action in space, and, if directed, deny such freedom of action to adversaries."[86] These policy statements cannot be construed to mean President Clinton emphatically endorses space weapons. His administration has consistently demonstrated an aversion to such systems.

When President Clinton assumed office in 1993, he acted to prune space weapons from two high-profile defense initiatives. First, he redirected the Ballistic Missile Defense Office's agenda to emphasize local theater missile defense (TMD) at the expense of a more global national missile defense architecture.[87] Reflecting a stricter adherence to traditional interpretations of the 1972 ABM Treaty, this new approach to ballistic missile defense substituted ground-based defenses for space-based weapon systems.[88] Specifically, the Brilliant Pebbles interceptors central to President Bush's global protection against limited strikes was conceptually replaced by the Patriot advanced capability, the upgraded Aegis radar, and the theater high-altitude area defense (THAAD)—all ground-based ABM systems. The only space systems to survive the rearchitecture were satellites designed for passive surveillance.[89]

President Clinton's aversion to space weapons is communicated in his ASAT policy, as well. After his inauguration, he marked for termination President Bush's kinetic energy (KE) ASAT initiative.[90] He has yet to propose a budget with funding for that system.[91]

The Convictions of American Space Weapon Advocates

Growing elements of Congress and the defense community are resisting the president's position, however. Since 1994 the Senate has sustained the KE ASAT program with unrequested funds.[92] In the fiscal year 1997 budget, for example, Congress unilaterally added $50 million to develop this antisatellite system.[93] An analyst for the Congressional Research Service notes that on the subject of ASATs, "the current Congress is certainly more supportive than the last several congresses."[94]

Congress, supported by senior defense leaders, believes its actions are consistent with national security requirements. Their case is built around two basic convictions. First, proponents believe space is too central to America's power to remain unprotected. They view the US space infrastructure as a center of gravity. Soon after assuming command of the US Space Command, Gen Howell M. Estes III, noted that, "we are the world's most successful space-faring nation . . . , one of the major reasons the United States holds its current position in today's league of nations. But, we are also the world's most space-dependent nation, thereby making us vulnerable to hostile groups or powers seeking to disrupt our access to, and use of, space. For this reason, it is vital to our national security that we protect and safeguard our interests in space.[95] The ability of our potential adversaries to affect our advantage in space is growing. We, in military space, are just now beginning to consider and deal with these threats."[96]

Senior DOD leaders particularly highlight America's growing dependence on space systems for economic and military prowess. In February 1997, the Deputy Under Secretary of Defense for Space Robert V. Davis underscored the economic vulnerability of satellites that pass extensive electronic commerce through space.[97] That same month, CINC USSPACECOM cautioned that DOD space systems also present adversaries with lucrative targets. He observed that "in purely military terms, the national dependence on space-based systems equates to a vulnerability. History shows that vulnerabilities are eventually exploited by adversaries, so the United States must be prepared to defend these systems."[98] Recognizing these vulnerabilities, many policy makers see space combat and weapons as inevitable. "The United States will . . . eventually fight from space and into space," remarked Gen Joseph W. Ashy, CINC USSPACECOM at the time of interview.[99] "We are developing direct-force applicators," he emphasized on another occasion. "They can be delivered by terrestrial [means], as well as from aircraft, shooting [targets] in the air or in space."[100] Secretary of the Air Force Sheila Widnall allowed that these direct-force applicators might range from shooting down satellites to less obtrusive interference with an adversary's signals.[101]

As a second basic conviction, US space weapon proponents believe that adversaries will unilaterally develop space systems in pursuit of greater relative power. Proponents are concerned about hostile space intelligence surveillance, and reconnaissance, information (ISR) satellites, as well as hostile space weapons. They recommend the deployment of US space weapons to counter these international developments.

US advocates of space weapons decry the improving ISR space posture of our potential adversaries. At the end of 1995, some 31 nations or international ventures had at least one such satellite payload in orbit.[102] Gen Robert S. Dickman, the DOD's space architect, predicts that in the next decade more than 20 nations will field space systems that "will have some ability to influence the battlefield."[103] Such systems will put US soldiers at risk, as adversaries take advantage of the force multiplication offered by their own satellites. In the words of the deputy undersecretary of defense for space, the United States must begin to prepare for adversaries that "will be able to use space to [their] advantage the same way we use it for ours. . . . I guarantee, in the near future, that threat will emerge; it's only a matter of time."[104] Vice Chief of Staff of the Air Force Gen Thomas S. Moorman Jr. sees this development as unacceptable. "Just as it would be unthinkable in a future conflict to permit an adversary to use an aircraft to reconnoiter our battle lines for intelligence and targeting, so is it equally unacceptable to allow enemy reconnaissance satellites free and unhindered flight over US military positions. An operational ASAT capability designed to eliminate an adversary's space capabilities must be considered an integral part of this country's force structure."[105]

General Moorman's message is winning support on Capitol Hill, where some lawmakers worry about enemy reconnaissance satellites and commercial satellites. "There is concern in this Congress over the proliferation of imagery" from commercial satellites that can be used for military purposes, said a Congressional Research Service policy analyst. The DOD is sensitive to similar concerns. In March 1997, for the first time, the Army publicly linked its eight-year-old ASAT development with the threat of foreign space-based remote sensing. Specifically, the Army Space and Strategic Defense Command acknowledged it needs rapid development of an ASAT to combat the growing "spread of space-based photography" that has led to concerns that "hostile reconnaissance could be used against the United States and allied military forces in the future."[106]

In addition to the threat posed by proliferating ISR satellites around the globe, advocates of space weapons are wary of foreign ASATs. Senior DOD officials acknowledge that the facilities and launch pad for Russia's co-orbital ASAT are still in place.[107] Many strategists also point to the likelihood that others will follow suit. One such strategist logically points out the attractiveness of ASATs to America's competition. "We should expect interest in anti-satellite weapons (ASATs) to proliferate. . . . ASATs may represent a particularly attractive weapon, because the problems posed by a hostile satellite may be most effectively banished by attacking a single target in space rather than numerous and dispersed Earth-bound targets. The United States has concentrated its space functions on a small number of satellites, meaning that the loss of one or more systems in the midst of hostilities could have fatal repercussions."[108]

Motivated by convictions that space is a US center of gravity and that foreign military competitors will exploit space systems of their own, weapon proponents are successfully impacting today's plans and budgets. For the first time since President Reagan's SDI, a draft National Security Space Master Plan endorses the creation of an offensive space capability against "surface, space, and airborne targets" as US national policy.[109] Consistent with this master plan, the Pentagon is requesting some $84 million for RTD&E under budget lines for "space and electronics warfare," "advanced materials for weapons systems," "advanced weapons technology," and the "DOD high-energy laser facility."[110] This money would be in addition to the congressional funding for a KE ASAT.

Thoughts on Departing the Traditional Sanctuary

In summary, while President Clinton resists deployment of space weapons, other senior policy makers continue to argue for their utility. These policy makers see space weapons as inevitable guardians of US access to space—access fundamental to national power. In addition, advocates promote space weapons as a counter to proliferating foreign ISR and ASAT technologies.

It is interesting that these convictions were just as true during the cold war as they are today, if not more so. Then,

US leaders also recognized that space played a central role in US national security. The threat posed by Soviet ISR satellites and ASATs was considerable during the cold war. In fact, both the threat and its implications were arguably far graver than those posed by potential adversaries today. Yet, US officials restrained themselves from more than token weaponization of space during that conflict.

How contemporary US decision makers would distinguish their situation from that of cold war strategists is a lengthy debate in itself. Perhaps today's looser association of space with the nuclear "sword of Damocles" permits greater freedom to act aggressively there. Then again, perhaps technology has matured to the point where cost-effective weapon concepts are feasible. The proliferation of ballistic missiles to the third world and a heightened US sensitivity to casualties might make those cost-effective space weapons particularly attractive.

Whatever the differences between the eras, some US decision makers believe those differences now make space weapons necessary. Indeed, they may be absolutely correct—this study in no way attempts to belittle their concerns. Nevertheless, decisions addressing space weapons should be postponed until strategists seek out and understand all sides of the debate. This is the goal of the next section. It seeks to round out the debate by articulating a contemporary argument against space weapons today.

The Sanctuary Argument

This section strives to articulate the strongest possible case against weaponizing space further in the immediate future. It works to capture the essence of what sanctuary advocates might argue given their "day in court." The basic premise of this sanctuary argument is that US interests are better served by preserving the present equilibrium in space weapons. It cannot be overemphasized that the case presented here does not propose that the United States should *never* introduce space weapons, but rather that it should *postpone* weaponization until current conditions change.

No attempt is made here to rebut the sanctuary argument. Rather, this section aims to present space weapon advocates with a counterargument to round out the debate. Indeed, the section will be written with a parochial edge to emphasize that counterargument.

The sanctuary argument is presented in two parts. First, it challenges the two basic convictions of space weapon advocates previously summarized. In some cases, that means asserting the basic convictions are incorrect. Where the convictions are incontestable, it means offering policy alternatives to space weapons. Second, the argument makes a positive case for a contemporary sanctuary strategy independent of the two basic convictions—with the goal of connecting such a strategy to broader national interests.

Challenging Weapon Advocates' Basic Convictions

As a first conviction, weapon advocates propose that space is central to US power and must be protected as a center of gravity (COG). This conviction rests on the fundamental assumption that in guarding against exploitation of a presumed US space Achilles' heel there is no alternative but to protect it with space weapons. Military history offers many examples of similar dilemmas solved by eliminating the COG rather than protecting it. In the 1960s, US military credibility rested heavily on bombers and land-based ICBMs. These systems constituted a friendly COG. Improved Soviet nuclear strike capabilities eventually rendered these COGs vulnerable. The principal US response was not to protect their land-based forces by active defenses designed to defeat inbound Soviet missiles. Instead, the United States mitigated its vulnerability by reducing the extent to which the ICBMs and bombers themselves were COGs. The development of submarine-launched ballistic missiles devolved part of the nuclear mission to a third medium—the sea. US strategic vulnerability was reduced. A similar approach is open to policy makers concerned about the exposure of US space assets.

Strategists must recognize that space communication, surveillance, reconnaissance, and navigation systems are not COGs because they are in space; they are COGs because they are centralized communication, surveillance, reconnaissance, and navigation systems. Options exist, however, to share these missions with other terrestrial systems and pursue a widely distributed space architecture. This decentralization would not only reduce US vulnerability in space but might do so without degradation of mission performance. Significantly, as the vulnerability is reduced, the case for space weapons weakens. Protection is accomplished through decentralization and diversification rather than through active defenses.

Current technology hints that this approach to national security is reasonable. Unfortunately, the possibility is masked by the past successes of centralized space assets. Operations such as Desert Storm continue to foster a paradigm that space is now and must always be the principal medium for DOD command, control, communications, computers, and intelligence (C^4I) systems. An overwhelming 90 percent of the coalition's intertheater communications and 60 percent of their intratheater communications were carried by satellites in that conflict. These statistics downplay the fact that 40 percent of the intratheater communications were successfully carried through terrestrial communication links. Microwave, tropospheric, and switched network communications quickly established operational connectivity and began to replace point-to-point satellite communications at both the intertheater and intratheater levels.[111]

The statistics from Desert Storm also understate the vulnerability of satellite communications (SATCOM) to jamming, interception, monitoring, and spoofing. The Iraqis were known to have at least four Soviet-made ultrahigh frequency jammers capable of shutting down up to 95 percent of the wartime communications to and from the US Navy.[112] Such vulnerability led the cochair of a Defense Communication Agency review of the Gulf War to emphasize the need for alternatives to SATCOM.[113] Some of the more promising alternatives that permit this are maturing at a blistering pace.

Fiber-optic technology is one example and is already routinely used by the commercial sector. A single optic fiber exceeds the entire carrying capacity of current satellite designs. In fact, the international demand for fiber-optic paths has prompted trans-Atlantic cables boasting 60,000 channels each. The performance and cost-effectiveness of fiber optics presages its rapid growth in the future.[114] In addition to fiber optics, technologies employing microwave, millimeter wave frequency, infrared, and laser communications also offer enormous broadband capabilities.[115]

General Dickman, the DOD space architect, recently advanced another alternative to present SATCOM architectures. Citing that one of his biggest challenges was getting the military and national security space communities to accept "a different way of looking at space," Dickman proposed communication packages be carried aboard unmanned aerial vehicles (UAV).[116] The military is on the verge of being able to field such a capability. For example, by the end of 1997, the United States was scheduled to build two Global Hawk UAVs capable of line-of-sight data link communications. These vehicles can be launched from ranges up to three thousand nautical miles and still loiter over a target area for 24 hours at altitudes greater than 60,000 feet.[117] With launch bases closer to the theater, loiter times approach 48 hours. The communications payload built for the Global Hawk is equally impressive. It essentially equals the communications capacity of a defense satellite communication system (DSCS) satellite, making the Global Hawk a viable and extremely cost-effective satellite surrogate.[118] The current DOD contract fixes the average unit price of the Global Hawk at $10 million.[119] This contrasts dramatically with the $140-million price tag of a DSCS satellite and its $86-million Atlas booster.[120]

In addition to their contributions to communications, systems such as the Global Hawk are strong candidates to perform reconnaissance and surveillance missions traditionally dominated by satellite platforms. The Global Hawk carries an advanced suite of ISR capabilities. The data from these sensors is processed by the equivalent of an onboard supercomputer before downlink—a system that allows coverage of a geographic area the size of Illinois in just 24 hours at three-foot resolution.[121] It is also capable of spot images with one-foot resolution.[122] No wonder a summary of UAV contributions reads like that of satellites: "responsive and sustained data from anywhere within enemy territory, day or night, regardless of weather, as the needs of the warfighter dictate."[123] Significantly, the UAV provides these capabilities within an architecture that is easily reconstituted. It is less expensive and far simpler to replace a downed UAV than a satellite lost on orbit. The last major satellite mission area is that of navigation. No discussion of the Gulf War can overlook the significant contribution of the global positioning system (GPS). By the end of the war, close to 10,000 receivers guided ships, aircraft, tanks, and infantry soldiers through deserts with no distinguishable landmarks.[124] GPS is even more valuable today. DOD is basing the guidance of a new generation of precision-guided munitions on space-based data. This trend leads advocates of space weapons to posit that GPS satellites warrant protection from attack or interference. Nevertheless, the better solution might be to shift navigation capability back to terrestrial systems. Inertial navigation systems, for example, free navigation from external data links and are rapidly improving. Not only are inertial navigation systems becoming more accurate, they are also becoming more portable, as the military recognizes. Between 1996 and 1999 the Pentagon plans to triple its investment in micromechanical systems with an emphasis on miniaturized inertial measurement, distributed sensing, and information technology.[125] A concerted emphasis on these kinds of technologies could not only build a military relatively insensitive to attack on its space navigation assets or jamming of its signals but also might allow the United States to deny less-developed adversaries access to free GPS data when the shooting starts.

Shifting space missions to terrestrial mediums is one way to minimize US vulnerabilities in space. Another way is to evolve today's centralized space architecture to one that is more distributed and decentralized. Not only would this further mitigate the potential US vulnerability in space but system performance might actually improve. Lt Col Christian C. Daehnick, in the previous chapter of this book, determined that a space architecture with smaller, distributed satellites "more directly responds to the needs of today's primary users and can adapt more readily to changes in both requirements or technological opportunity."[126] Others are reaching the same conclusions.

The National Reconnaissance Office (NRO) revealed it will downsize its national security satellites to a maximum of "½ their current size, and in some cases ¼ of the current weight," while making them more capable than today's spacecraft.[127] Similarly, the Air Force's improved space and missile tracking system will eventually launch 12 to 24 681-kilogram satellites into a distributed constellation.[128] In the future, the space community may consider even these satellites overly large and centralized. The Phillips Laboratory will begin space-based testing of miniaturized components that could lead to grapefruit-sized smart satellites within a decade.[129]

As US space assets shrink in size and weight, "clouds" of small satellites will foster survivability by eliminating single point failures in mission capability. The smaller satellites also enhance survivability by allowing more economical launch systems to replenish satellite constellations. In anticipation of this, the US Air Force is considering a reusable launch vehicle (RLV). The RLV technology, developed in the National Aeronautics and Space Administration (NASA) programs, promises to reduce today's $4,500-per-kilogram costs for low Earth orbit payloads to some $450 per kilogram. NASA administrator Daniel Goldin predicts the RLV will also bring a tenfold improvement in launch reliability.[130]

In summary, advocates of space weapons are correct in their diagnosis, but misguided in their cure. The degree to which the United States has centralized its communication, surveillance, reconnaissance, and navigation systems in space translates to a potentially serious US vulnerability. Rather than introduce weapons to defend these assets, however, the systems themselves could be decentralized and diversified across the air, land, and sea mediums. In this way, the American COG in space could be defended by eliminating it. Note that this does *not* mean the United States should work to abandon space. Instead, it means finding a balance between reliance on space and terrestrial systems, between centralization and decentralization, so as to mitigate the value of US space assets as a COG and obviate the requirement for space weapons for defense.

As a second conviction, space weapon advocates postulate that the US's international competitors will unilaterally move to exploit and control space. More specifically, this conviction assumes that adversaries will develop effective ISR space platforms. Next, it presumes that adversaries will not stop with ISR space systems but will strive to weaponize space as early as possible—with or without provocation from similar US actions. The significance of the first assumption and the accuracy of the second are debatable. For the first, it is disputable whether foreign ISR satellites should significantly alter US military effectiveness. Even if they did, the United States would find it very difficult to target them without recrimination. The commercial and international character of satellites present the targeteer with troublesome sensitivities. Evidence against the second assumption asserts that, unless provoked by extensive US space weaponization, the US's adversaries will not be inclined to pursue space weapons.

Some proponents of space weapons believe foreign ISR satellites—particularly reconnaissance—warrant weapons for preemptive strikes. There are other ways to defeat ISR systems without incurring the costs and risks associated with space weapons. Consider that an opponent being as "blind" as the Iraqis were during the Gulf War is a historical anomaly and not a prerequisite for victory. In World War II, for example, the United States prevailed over adversaries who possessed ISR assets nearly equal to those of the Allies. Allied techniques like concealment, communications security, deception, and operations security proved to be effective countermeasures to enemy ISR capabilities. In this respect, Americans would do well to recall the effectiveness with which the North Koreans, Chinese, North Vietnamese, and Afghani mujahideen operated against superpower militaries. These superpowers possessed space and air superiority—accessing at will any spot in the theater with ISR capabilities. Repeatedly the superpowers were frustrated by their opponents' low-tech countermeasures. December 1950 offers one telling example. In that month, a surprise Chinese offensive drove the US Eighth

Army back into southern Korea. To support the Eighth Army, the Fifth Air Force was ordered to locate precisely the Chinese forces on the other side of the front. Robert F. Futrell notes that 10 days of unspared aerial reconnaissance and 27,643 reconnaissance photographs revealed nothing in front of the Eighth Army's position. What the all-out reconnaissance effort missed were 177,018 troops of the Chinese Fourth Field Army—true masters of camouflage and operations security.[131]

Although US countermeasures will not render enemy ISR satellites totally benign, US military effectiveness is far from lost. Seeing US forces is one thing, attacking them is another. The United States employs a formidable array of defensive technologies designed to prevent enemy penetrations of all types. Even the troublesome ballistic missile threat is well on its way to being thwarted by maturing US theater ballistic missile defense systems. The United States also possesses the world's most effective offensive forces, capable of destroying an enemy's terrestrial links to ISR satellites. So while the adversary's satellite may not be blind, the data is nevertheless lost. For example, during the 1991 Gulf War, Iraqi access to Arabsat telecommunication satellites was severed when a coalition air attack destroyed the Arabsat earth station in Baghdad.[132]

In summary, then, the United States is neither compelled nor limited to countering enemy ISR satellites with space weapons. US military effectiveness can be preserved through operational security, defensive technologies, and attacks on the key terrestrial nodes supporting the enemy space systems.

US strategists still bent on augmenting passive countermeasures with preemptive attacks on foreign ISR satellites face the challenging task of distinguishing between military and commercial systems. Writing from the Centre for Defence Studies and Space Policy Research Unit in Great Britain, Alasdair McLean notes that "all remote sensing satellites relay data on the area of the earth's surface they observe. If, within that area, lie sites of military interest, the data thus obtained is of military value. Likewise, communications satellites, even if not specifically dedicated to military use, can be used for such purposes, whether by normal commercial contracts, or by special agreement in time of crisis or conflict."[133]

The Meteosat-4 satellite, operated by the European Space Agency, illustrates McLean's contention. That satellite transmits signals every 30 minutes to any user with proper receiving equipment. During the Gulf War, a Plymouth College professor built his own homemade receiver and was surprised to see that he could detect troop concentrations in the Gulf area from the weather imagery. Clearly this shows the "undoubted military potential of the most innocent civilian satellite."[134] The high-resolution imaging capabilities of the French *Systeme Probatoire pour l'Observation de la Terre* (SPOT) made it less innocent in the context of the Gulf War. Fortunately for the United States, SPOT Image agreed not to sell its photoreconnaissance outside the coalition. During the same conflict, however, the US-based company that operates Landsat insisted on selling imagery to noncoalition countries, arguing it had a legal obligation to do so.[135] Such uncooperative civilian and commercial systems present military planners with dubious if not provocative targets. Aggressors against these systems must carefully balance military necessity with collateral damage. They must also recognize that allies may be users of the targeted systems. This is precisely what happened in the Gulf War. Iraq had access to civilian-run Intelsat, Inmarsat, and two regional Arabsat telecommunications satellites.[136] Such arrangements will immeasurably complicate future efforts to attack satellites.

Whereas foreign ISR satellites are a reality, foreign space weapons are not. Today there is little to suggest that another nation with the economic, technological, and space expertise required to pursue space weapons is inclined to do so. This includes Russia, Europe, Japan, and China.

Except for the United States, Russia is the only nation to have demonstrated any historical interest in ASAT technologies. In November 1991, the Russians announced that their co-orbital ASAT remains "operational" today. Although this Russian ASAT does threaten certain US space assets, its effectiveness should be kept in context. First, in 29 tests of the system between October 1968 and June 1982, there were 12 failures.[137] Second, the most recent test was conducted 12 years ago.[138] Third, tests were only conducted across orbital inclinations of 62 to 66 degrees and altitudes of six hundred to 1,000 miles.[139] Most of the US's satellites are at altitudes greater than 1,000 miles and well outside the tested inclinations. The performance of the Russian co-orbital ASAT is limited by other operational constraints as well. Days are often required to achieve the orbital conditions that allow a successful launch and intercept. In addition, the nature of the co-orbital intercept provides advance warning of hostile intentions, thus allowing evasive actions on the part of the target. In David Lupton's words: "US terrestrial assets are more vulnerable to numerous threats (including terrorist acts) than are space systems threatened by the Soviet ASAT."[140] Reportedly the Russians have also experimented with other forms of ASAT weaponry. Starting in the 1970s, Russia extensively pursued high-powered, ground-based lasers and microwave weapons. A more conventional ASAT program, very similar to the US F-15 air-launched ASAT, was also kicked off in the late 1980s.[141] Although it is unclear what these efforts finally achieved, there are no indications that any of the concepts matured to become operational systems. Nor is it likely any of the concepts will do so, given the current fiscal condition of the Russian space program. In January 1997, Russian Space Agency (RSA) Director Yuri Koptev warned that without increased funding, Russia would be unable to maintain even a skeleton space program. He acknowledged that of 20 nations active in space research and satellite launches, Russia ranked second to last. Only India spent less. In 1996 this meant that only 11 of the RSA's 27 planned civil missions were actually launched. The RSA's woes are affecting its personnel, as well. Since 1989 half the engineers and technicians have left the RSA as Russian spending on space programs fell each of the previous eight years.[142] Money is so scarce that Russia risks losing its place in the highly visible international space station program. Vice President Albert Gore warned in 1997 that Russian participation would be jeopardized if Russia failed to release millions of rubles withheld from time-critical contracts.[143]

Less information is available on Russia's annual military space budget, but requests for 1995 reveal planned expenditures roughly equal those of the RSA.[144] This indication of dramatically reduced spending on military space systems is corroborated by other evidence. In 1996 there were no Global Navigation Satellite System (GLONASS) navigation satellite launches despite the fact that three GLONASS satellites stopped transmitting signals in that year.[145] Consider also that between 1962 and 1994, the Russians averaged more than two photoreconnaissance spacecraft on orbit. During that same period, there was never a gap in coverage.[146] Today, although it had planned to keep at least one imaging system operational, Russia has no imaging reconnaissance satellites in orbit—a Russian first that stands in stark contrast to the five imaging satellites the United States currently has aloft.[147] As yet another example of deep spending cutbacks, the Russians postponed the December 1996 launch of a new missile warning satellite "to conserve carrier and spacecraft."[148] In light of this and the other operational and fiscal constraints noted above, a concerted Russian effort to develop space weapons appears unlikely in the near future.

While Russia struggles to regain its footing in space, Europe is pursuing strategies for cooperation in the civilian sector. Joint European endeavors in military programs like the Helios reconnaissance satellite are clearly the exception and not the rule.[149] Consistent with this position, European nations continue to rebuff US initiatives to cooperate in ballistic missile defense technology developments. Hence, Alasdair McLean's conclusions on Europe and space weapons: "no evidence exists for any real enthusiasm for European nations to develop active space-based weapon systems."[150]

Any analysis of Japanese ambitions to weaponize space must ultimately consider Japan's constitutional prohibition against offensive military capabilities. Since 1945, Japan has severely constrained its defense expenditures in deference to public support for that prohibition and the military security already provided by US forces.[151] Japan's national sentiment fosters budget woes for the Japanese Defense Agency. Plans for a missile warning satellite were scrapped in favor of the short-term solution of buying US airborne warning and control system (AWACS) aircraft instead.[152] On a related note, Japan recently declined to participate in a joint venture to develop an operational theater missile defense. This evidence indicates that Japan is not inclined to weaponize space.

In terms of space programs, China is Asia's most visible nation. Recently, however, Chinese energy has been devoted to securing the cooperation of the United States and Europe in aerospace ventures. New Chinese initiatives into the next century include an improved booster, technology work geared to a Chinese manned space presence, new imaging spacecraft, and many new communication satellites. Analysts see the Chinese willingness to cooperate as China's admission that it is falling behind its Asian neighbors, such as India and Japan, which are already cooperating with the West.[153] A series of booster failures confirms that there may be cause for Chinese concern. The August 1996 explosion of a Long March 3 rocket pushed China's launch failure rate to more than 30 percent and is the sixth failure in less than four years.[154] In contrast, the January 1997 failure of a US Delta 2 at Cape Canaveral represents an anomaly for a program that enjoys a 98 percent success rate even after the accident.[155] In total, then, it is reasonable to conclude that the Chinese desire to encourage cooperation with the West and the Chinese struggle for reliable space technology will discourage near-term pursuit of advanced space weapons—as long as they do not feel threatened.

In summary, any assertion that the United States should aggressively pursue weaponization to beat adversaries already rushing in that direction is questionable. While it is true that potential adversaries continue to perfect ISR spacecraft, US responses are not limited to shooting those spacecraft down. Time-tested techniques with passive countermeasures and attack of terrestrial choke points offer alternative solutions. Since these options remain effective, the United States should shun provoking potential adversaries by unilaterally employing space weapons. In addition, a close examination of the principal actors in space today indicates that the nations pursuing ISR spacecraft do not appear to be inclined to weaponize space. A depolarizing world headed toward widespread democracy, tight military budgets, mission failures, and flat out disinterest in weapons currently motivate these principal actors to put aside space weapons development. Therefore, contrary to the view of a world racing to weaponize space, the world seems poised to follow the US lead. Today, foreign interest in space weapons may hinge entirely on US restraint or weaponization.

Independent Arguments for a Sanctuary Strategy

Simply refuting the basic convictions of space weapon advocates shortchanges the strongest possible argument for a sanctuary strategy. Sanctuary strategists should also attempt to prove their concepts best serve US national interests on other grounds. These interests are broader than the military objectives that support them. White House policy makers clearly convey these broader interests in the 1996 National security strategy. That document states that "the nature of our response must depend on what best serves our own long-term national interests. Those interests are ultimately defined by our security requirements. Such requirements start with our physical defense and economic well-being. They also include environmental security as well as the security of our values achieved through expansion of the community of democratic nations."[156]

As a starting point to extending the sanctuary argument, it is reasonable to postulate that physical security, economic well-being, and democratic expansion depend on the quality of American international relations. If that is accepted, the value of weaponizing space should, in part, be judged by its effect on those relations. It is quite possible that weaponizing space may turn out to be unacceptably provocative—particularly in the post-cold-war world—leading to global instability and deteriorating US foreign relations.

Space weapons are provocative because they inherently possess offensive utility. Consider that war in space is much like the infamous shoot-out at the OK Corral. In that gunfight, armed men constituted an enduring offensive threat to all other gunslingers. There were no defensive shots, and at all times anybody was a potential target. Space is similar. The laws of astrodynamics routinely give space weapons (ground-

and space-based) clear line of sight to the satellites or territories of other nations. Such weapons could be fired instantaneously and without warning. Significantly, these circumstances encourage future space combatants to preempt adversaries by shooting first. This destabilizing result is discussed below in more detail.

Even if space weapons could be understood as defensive, the US's current treaty obligations make it likely that steps toward weaponizing space will strain its international relations. The 1972 ABM Treaty, for example, bans development, testing, and deployment of space-based ABM systems or components. The treaty also limits the United States and Russia each to a single ABM site with no more than one hundred missiles.[157] Except for the protection of National Technical Means of Verification granted in Article XII of the same treaty, international law is ambiguous if not silent on the subject of ASATs.[158] The traditional international precedent of "that which is not prohibited is permitted" would seem to remove ASATs from treaty constraints. The difficulty in distinguishing between ASATs and ABMs makes this problematic since a powerful ASAT weapon also threatens ballistic missiles. Therefore, a concerted US effort to develop any weapons that project destructive force into or from space will foster protest from those sensitive to violations of the 1972 ABM Treaty. Objections from the Russians are particularly worrisome since they have clearly linked both Strategic Arms Reduction Talks (START) treaties to continued US compliance with the ABM Treaty. Under these accords, thousands of missiles will be destroyed by the United States and Russia. Clearly, preserving these accords is well within the US's national interest. In the words of one of the ABM Treaty negotiators, "A missile scrapped is a missile that does not have to be shot down."[159]

If space weapons are indeed offensive by nature and if they unavoidably challenge international law, then US actions to weaponize space could easily aggravate the security dilemma that fosters arms races. Nations exist in a setting where no diplomatic sovereign arbitrates international conflicts. Each must ultimately rely on its own strength for protection and constantly look for shifts in relative power.[160] This preoccupation with relative position means that even arms acquisitions intended purely for self-protection are destined to menace one's global neighbors.[161] "What one state views as insurance, the adversary will see as encirclement."[162] In this way, US initiatives to strengthen its relative posture in space could drive other nations to follow suit—even if each is motivated by what it sees as peaceful goals. It is the classic prisoner's dilemma: each state pursuing its own self-interests in space only to find in the end that all are worse off than if they had cooperated.[163] Those familiar with game theory know the opportunity to break this cycle occurs when a principal player risks compromising immediate self-interests for the longer-term good of all. Since the United States undoubtedly leads the world in space weapon technology, the question becomes: Will America lead the world toward cooperation or conflict?

The traditional view of space power as a symbol of international prestige is another force driving nations to keep pace with US technology. In their book *The Prestige Trap,* Roger B. Handberg and Joan Johnson-Freese study what motivated the US, European, and Japanese space programs. They specifically address the question of why these nations made serious resource commitments to exploiting a medium that promised little in the way of immediate return.[164] The answer, in all three cases, was primarily prestige and national pride (with a dash of scientific curiosity).[165] While acknowledging that these early space efforts were often civilian in character, the authors note that "civilian space policy has clear links to the military-industrial policies within most societies. The technologies and technical skills involved in civilian space endeavors in many cases have clear and ready applications to military technology . . . the boundary is thin and easily breached."[166] On either side of this boundary, US strategists should expect their international competitors to keep pace with US developments.

Some strategists might remain relatively unfazed by competition from staunch allies like the Europeans and Japanese. They should pause to reflect, however, because the introduction of space weapons might jeopardize those alliances. From his study of contemporary history, Stephen M. Walt concluded that nations are far more likely to ally against dominant threats than they are to bandwagon with them.[167] This balancing behavior occurs because nations recognize their odds for survival are improved by confronting a rising hegemon before it becomes too strong to resist. Since allying with a hegemon entails the gamble of trusting it, the safer strategy is to join forces with other less threatening nations.[168] The factors that incite this reaction to an emerging hegemon are the hegemon's aggregate power, proximity, offensive capability, and offensive intentions.[169] Nations will be more prone to balance as the threat gets stronger, closer, more offensively capable, and more hostile. This framework poses problems for US strategists planning to weaponize space. Space weapons increase US power with systems already noted as inherently offensive. In his paper on the implications of space weapons, Dr. Karl Mueller postulates that space weapons will also "increase the effective proximity of the United States to previously distant states."[170] The net effect of these changes might well foster an international perception that a new and different US threat is emerging. This perception could lead nations presently friendly or neutral toward the United States to balance against it when US space weapons are deployed. At a minimum, nations may at least become less willing to cooperate with the United States.[171] Such was Germany's fate when Admiral Tirpitz built a formidable battle fleet as a means of coaxing Britain's alliance. Instead, the British redoubled their own shipbuilding and moved diplomatically closer to France and Russia.[172]

In general, the United States tends to underestimate how its actions affect the security dilemma and international balancing. The United States sincerely believes its actions are categorically peaceful and are perceived as such by other nations. However, this is not the way the rest of the world—including allies—always views the United States. In a multipolar world, the United States is the single most powerful competitor. This distinction naturally impels other nations to observe the United States with at least some suspicion. As an illustration, US Space Command acknowledged that it officially "predicts when selected satellites will be in position to perform intelligence collection against US forces and military/military-related installations, and makes these predictions available to installation commanders." Most Americans would clearly cast this statement in a benign light. They would view such a capability as defensive—the inherent right of US forces to remain aware of when they are being observed. There are reportedly some in the international community who have a different interpretation, however. They link this US Space Command mission with US Army statements that justify the KE ASAT program as fulfilling a requirement to deny hostile remote sensing and reconnaissance capabilities. According to *Military Space*, that "potential linkage . . . generated some uneasiness, especially among foreign space officials."[173]

Whatever the reaction of the international community, the introduction of weapons into space would be strategically destabilizing. Robert Jervis postulates that the military stability of the international system resides in two variables: first, whether defensive weapons can be distinguished from offensive ones and second, whether defensive or offensive weapons are superior.[174] Since space weapons were shown earlier to be inherently offensive, the question of international stability ultimately depends on whether one believes space weapons are superior. Certainly, the US Air Force suspects that they are. The new Air Force strategic vision, approved at the 1996 Corona meetings, states, "We are now transitioning from an Air Force into an air and space force, on an evolving path to a space and air force."[175] What Air Force leaders have apparently concluded is that space is becoming a dominant medium of the future. If they are right, Jervis's framework predicts that space weapons will tend to destabilize the international order. Such weapons favor the side that strikes first and penalize the side that hesitates. In warning, Thomas C. Schelling wrote, "The whole idea of accidental or inadvertent war, of a war that is not entirely premeditated, rests in a crucial premise—that there is such an advantage, in the event of war, in being the one to start it."[176] The US Congress Office of Technology Assessment echoed similar thoughts years later: "Pre-emptive attack would be an attractive countermeasure to space-based ASAT weapons. If each side feared that only a pre-emptive attack could counter the risk of being defeated by enemy preemption, then a crisis situation could be extremely unstable."[177] This particular congressional assessment, and that of Jervis and Schelling, invite US caution with space weapons. The United States may weaponize space only to fight a war that otherwise need not have occurred.

If the future does in fact find the United States in a war featuring space combat, advocates of space weapons assume the United States will prevail. They believe that US technological prowess and industrial power will preserve space superiority. There is no guarantee, however, that the United States will indefinitely possess space superiority—a grave reality since pursuing it may mean forfeiture of the US's hard-won and tentative superiority in the air, land, and sea arenas. Consider the implications of space weapons for US defense spending.

From fiscal year 1996 through fiscal year 2002, defense budgets projected by Congress and the president are expected to decline an average of 20 percent from fiscal year 1995 spending. The Congressional Budget Office reports that the administration remains about $101 billion short of the money required for a fully modernized Bottom-Up Review force.[178] Those shortfalls are further exacerbated by the continuing pattern of diverting procurement funds to pay for operations and maintenance (O&M) costs associated with US peace enforcement forces abroad.[179]

In this budget-constrained environment, funding for space weapons could only come at the expense of other US defense forces. These forces are constantly challenged by global competitors for technological and operational superiority. So far, the United States has done well to preserve its advantage through relentless modernization of its systems. Those modernizations are expensive and today are stretched out beyond the life cycle of the systems they replace. While acknowledging that today's force can handle today's threats, the current chief of staff of the Air Force recognizes that resources are not available to modernize everything at once. His acquisition plan, therefore, calls for just-in-time modernization. F-22s are phased in to replace today's fighters just as those fighters are made obsolete by foreign developments. The C-17 is delivered just as C-141s retire. "We are phasing in the capabilities so that they arrive when we need them," he states, but "delays in the modernization will create vulnerabilities very soon."[180] Why start an arms buildup in space when budget limitations already threaten essential programs like the joint strike fighter and the evolved expendable launch vehicle? Funds allocated to space weapons undermine the budget upon which the US services' just-in-time modernization is predicated. It gambles that investing in space superiority is worth the resulting decline in relative advantage in the other mediums.

Just as there is no guarantee that the United States will maintain air, land, and sea superiority if it shifts significant funds to space programs, there is also no guarantee that the United States will emerge the winner in the space weapons race itself. It is entirely possible that another nation could beat the United States or "leapfrog" past US accomplishments late in the race. It is widely recognized that several European and Asian nations are rapidly advancing technologically. In fact, the United States no longer leads the world in some sectors. Twenty years ago, for example, the United States launched 80 to 90 percent of all commercial satellites in the world. Today, that figure stands at 27 percent and continues to drop as the Russians, Chinese, and French make inroads.[181] The French alone own more than 50 percent of the launch market share.[182] These statistics and other examples challenge the assumption that the United States could never be bested in a technology that proves to be crucial to war fighting in space. It might be somebody else who first develops some concept as revolutionary as British radar in the Battle of Britain, the German blitzkrieg in the Battle of France, or the Russian sputnik during the cold war.

Not only is it possible that foreign know-how might overpower the United States in some key technology sector, but US know-how might work against the United States in a race for space superiority. Dr. Mueller cites nuclear history as an example of this. Today, an early US nuclear monopoly continues to erode with every additional nation that acquires nuclear weapons. It cannot be ignored that the growing US vulnerability to such weapons is in part compliments of the United States. It was the United States that demonstrated the feasibility of nuclear weapons and paid the tremendous nonrecurring development costs to do so. It was from the United States that atomic secrets leaked to its chief adversary. In general, the growing fraternity of nuclear powers benefited from US hindsight and experience. It ought to be expected that the same thing could be repeated should the United States accelerate development of advanced space weapons.[183]

So far, independent arguments for a sanctuary strategy suggest that weaponizing space in no way guarantees the United States is better postured to meet security challenges. In fact, a practical requirement to cut other US defense expenditures to pay for space weapons may actually make the United States less secure. This could happen if the US's military advantages in space weapons were offset by new disadvantages in the air, land, and sea mediums or if potential adversaries won the contest for space superiority. Even if the United States were to successfully establish an enduring superiority in all mediums, it might prove so provocative as to isolate the United States from the international community. This isolation would undercut the US's stated national interests in physical security, economic well-being, and expansion of democratic values. In addition to the potential impacts on these interests, weaponizing space also jeopardizes US interests in the environment and domestic programs.

US policy makers are growing increasingly concerned that space debris will begin to impede peaceful commercial exploitation of space. This concern dates back to 1967 when the United States signed the Treaty on Principles Governing the Activities of States in the Exploration and Use of Outer Space. Article IX of that treaty requires parties to "conduct exploration . . . so as to avoid their [space and celestial bodies] harmful contamination."[184] In 1996 the president of the United States directed that "the United States will seek to minimize the creation of space debris. . . . The design and operation of space tests, experiments, and systems will minimize or reduce accumulation of space debris consistent with mission requirements and cost-effectiveness. It is in the interest of the US Government to ensure that space debris minimization practices are applied by other spacefaring nations and international organizations. The US government will take a leadership role in international fora to adopt policies and practices aimed at debris minimization."[185] This environmental concern is real and must be factored into the decision to weaponize space. Space combat is potentially very messy—recall that a single test of the US's miniature homing vehicle ASAT produced fragments by the hundreds.[186] Combat of this sort could easily come at the expense of commercial exploitation of space. Driving that point home, the French satellite Cerise was crippled in a collision during 1996. It was destroyed by a fragment of an Ariane booster upper stage.[187] Less than a year later, on 15 February 1997, the space shuttle *Discovery* was forced to dodge a Pegasus upper stage fragment.[188]

US space weapons not only jeopardize the environment, they also threaten US budget deficit reduction and domestic spending. It is not unrealistic to expect that weaponizing space, especially if it occurs in the context of an arms race, could be one of the United States's most expensive military undertakings to date.

Since 1984, SDI and BMD researchers have spent $39 billion and the Congressional Budget Office estimates that an effective space-based missile defense, alone, will cost another $60 billion through 2010.[189] Notably, these estimates assume a benign space environment controlled and exploited by the United States. They do not consider foreign challengers in space nor do they consider future military space operations other than ballistic missile defense. Both considerations promise to hike costs further.

These spending estimates come amidst strident calls to reduce the US national debt—calls that political leaders are slowly heeding. Experts project the US's debt at $5,457 trillion after fiscal year 1997. At the end of the same fiscal year, the annual federal deficit, having narrowed roughly $200 billion from 1992 to 1996, is predicted to widen back to $125.7 billion.[190] Remedying these fiscal conditions could well constitute a national interest more compelling than unilateral US action to accelerate the weaponization of space.

Allocating the nation's scarce dollars to important domestic programs may better serve US interests, as well. In 1996 an estimated 555,000 Americans died of cancer—215,000 more than in 1971. Current trends indicate that by the year 2000, cancer will overtake heart disease as the US's number one killer.[191] Researchers studying cancer are funded from a slice of the National Institutes of Health $12-billion annual budget.[192] In 1994 Congress comprehensively reviewed that budget and the fight against cancer in total. The ensuing report concluded that current research funding is inadequate to "capitalize on unprecedented opportunities in basic science research."[193] Future funding, however, stands in direct competition with that for space weapons. It is a compelling assertion, however, that researchers attacking a disease that every year kills 10 times the number of US combatants lost in Vietnam deserve higher priority than insurance against hypothetical space threats. Consider, also, that cancer research is but one of hundreds of domestic programs in similar circumstances.

In summary, developing space weapons may not serve US national interests. Weaponizing space brings opportunity costs that fundamentally challenge US security interests as defined by the national security strategy. These opportunity costs are steep, and while they may be justified in scenarios where the United States is clearly threatened from space, they appear dubious given the superiority the US military enjoys today.

Summarizing the Independent Argument for Space Sanctuary

In 1996 the Joint Warfighting Center (JWFC) conducted a series of war games to simulate the effectiveness of forces proposed for 2010. In two of the games, US and "red team" forces faced each other with highly capable space weapons in their orders of battle. In both cases, the games opened with what one observer referred to as a "space Armageddon." The flag officers, having quickly discovered that space weapons severely curtailed operational freedom of their air, land, and sea forces, were forced to win total space superiority before proceeding with their terrestrial campaigns.[194]

Advocates of space weapons would be quick to point out that the JWFC war games prove their point—the United States must move *now* to control space or risk losing it in future conflicts. This section, however, indicates that space weapon proponents should look deeper into the issues motivating them to support weaponizing space *now*. It asks them to carefully differentiate the question of *if* space should be weaponized from the question of *when* space should be weaponized. Today, the United States may have better alternatives with which to reduce the vulnerability of US space systems, as well as better alternatives with which to reduce the exposure of US terrestrial forces to enemy space ISR. In addition, strategists should continue to debate the proposition that weaponizing the high ground unquestionably optimizes US national interests. US space weapons, even if advertised as defensive systems, may unacceptably undercut broader US interests related to international relations, global arms stability, military superiority, and domestic concerns. Finally, it is possible that other nations currently have neither the inclination nor the resources to start their own weaponization programs in space. They could well discover that inclination, however, if the United States proceeds with a space weapons program of its own.

Conclusions

> *Strategy . . . is concentrated upon achieving victory over a specific enemy under a specific set of political and geographic circumstances. But strategy must also anticipate the trials of war, and by anticipation to seek where possible to increase one's advantage without unduly jeopardizing the maintenance of peace or the pursuit of other values.*
>
> —Bernard Brodie

Four years after World War II, Bernard Brodie called upon military strategists to make their thinking broader and more sophisticated. Brodie believed uniformed officers well versed in the military links to political, social, economic, and international dynamics were essential to formulating the best US security policies.[195] The nuclear age that followed his comments made this requirement more important as well as more challenging. Clemenceau's assertion that war was too important to be left to generals foreshadowed the predominant role civilians would play in formulating US defense policy after the introduction of nuclear weapons. Civilians like Brodie, Herman Kahn, Schelling, and Albert Wohlstetter were responsible for most of the truly groundbreaking work underpinning the United States's fledgling nuclear strategy—a result fostered as much by military disinterest in strategic policy as it was by civilian interest in the same.

While the value of civilian contributions should never go unappreciated, the absence of substantive military nuclear theorists should never pass as acceptable. Surely US nuclear strategy would have been improved had bright military officers asserted themselves in matters other than execution of policy. Such officers, if properly prepared, might have brought the invaluable perspective of military professionals schooled in the complexities of national and international power.

Today, national strategists debate space weapons in a policy climate not unlike the early days of nuclear strategy. The subject of space weapons also attracts strong civilian intervention and has done so since the 1950s. As was the case with nuclear policy immediately after World War II, there is still no comprehensive theory or strategy for space power. In fact, even the most rudimentary ideas about space power remain undeveloped. One thing is certain. The United States will develop a space theory and strategy in the future. The question is who will develop it. Will military strategists distinguish themselves and be included this time around?

Bearing this question in mind, the 1997 USSPACECOM effort to draft a military space theory and doctrine was an encouraging development.[196] That effort will succeed if those involved strive to see space power in the broadest of terms. Theorists and strategists alike must consider far more than weapon technologies, principles of war, and campaign planning. They must consider, from every angle, the contributions of space to a nation's power and the means by which a state's actions in space do or do not influence other nations. Strategists should recommend courses of action in matters like space weapons only after rigorously considering all perspectives.

The previous section examined the issue of weaponizing space from one such perspective—that of a sanctuary advocate arguing the strongest possible case against further weaponization of space at this time. Since a basic purpose of this study is to give military space thinkers something with which to mentally wrestle on their own, the sanctuary argument was offered without criticizing it. That is left for strategists to do within the context of their specific problems. In addition, the logic behind the convictions of weapon advocates was treated only to the point of establishing the framework upon which to build the sanctuary discussion. No doubt the case for space weapons today could have been articulated in more depth and with greater sophistication. That too was beyond the basic purpose and is also left for future strategists.

There are two final points which are important for strategists who are judging the merits and shortcomings of the sanctuary argument. First, the sanctuary position should never be construed as a passive national strategy. Second, strategists who conclude that US national interests are indeed served by introducing space weapons will still find the sanctuary perspective invaluable to their planning.

It is incorrect to see the sanctuary strategy as passive or to believe that it requires policy makers to stand idly by while competitors seize the initiative. Instead, the sanctuary strategy replaces US investments in space weapons with action through other national avenues. Any deliberate decision to pursue a sanctuary space strategy warrants aggressive diplomatic, informational, military, and economic support. As an illustration, US diplomats might seize the initiative by denouncing space weapons in international forums. In turn, international cooperation in space could be fostered through treaties and agreements. Any sanctuary strategy would undoubtedly require strong investments in national and military systems capable of recognizing treaty violations. Economic trade might be conditionally linked to nations demonstrating "good faith" in space treaty matters. Finally, and consistent with their military tradition, the United States would be wise to maintain a technological posture that always protects its ability to accelerate weapons development to meet threats. This posture recognizes that the conditions conducive to a sanctuary strategy can change over time to favor a weapons-oriented strategy instead.

It is equally mistaken to dismiss the sanctuary perspective as irrelevant if the United States does set out on a strategy to weaponize space. Weaponization occurs in degrees, and at any given time the strategist must carefully balance the merits of further weaponization with the value of preserving the sanctuary which still remains. The best strategy will rarely discount one entirely in favor of the other. There will normally be an optimum point somewhere between the extremes of total weaponization and a complete sanctuary.

Indeed, the United States's first steps toward any hypothetical weaponization of space might be heavily influenced by sanctuary thought. Weapon systems might remain ground-based so as to minimize any provocation associated with space-based weapons. Weaponizing covertly could further defuse the risk of provocation, and sharing key technologies with staunch allies might help assuage their suspicions and fears. Mindful of tentative superiority of American air, land, and sea forces, US strategists might opt to field technologies for space control missions but not for force application. This would minimize the risk of potential adversaries hitchhiking on US force application technologies to undermine our advantage in terrestrial military strength. International and national concerns over space debris might lead the United States to field systems that kill without fragmentation. The possible permutations are numerous and strategists must determine which ones best suit their situations.

The sanctuary perspective helps identify the space infrastructure that will support space weapons in the same way it helps the strategist to tailor the specific nature of the space weapons themselves. Consider space launch systems. The requirement for quick, cost-effective, and reliable access to space is well understood by the military space community. It recognizes that without it, satellite forces become more expensive and prone to gaps in coverage. Sanctuary thought, however, leads space strategists and acquisition decision makers to strengthen the justification for responsive launch beyond the force "push" that it provides.

Earlier, the sanctuary perspective proposed that space weapons were inherently offensive and therefore destabilizing in a crisis. Responsive launch systems, however, help reestablish stability. They permit strategists to create a protected second-strike capability by retaining a significant portion of their space weapons on the ground, hence reducing incentives for preemptive attacks against space systems in orbit. In this way, launch reconstitution plays a stabilizing role similar to the submarine leg of the nuclear triad. Here, then, is a patent case where the sanctuary perspective should lead even a weapons proponent to modify strategy for the better. There are certainly more such cases.

In conclusion, the sanctuary argument broadens the understanding of US strategists wrestling with the question of space weapons. The argument exposes domestic and international issues that might otherwise be overlooked. It allows military strategists to more completely weigh alternatives, thereby strengthening the military's contribution to US space defense policy.

Henry IV once remarked, "I never suffer my mind to be so wedded to any opinions as to refuse to listen to better ones when they are suggested to me."[197] The wisdom of the sixteenth-century king's approach is timeless. Contemporary decision makers should approach any decision on space weapons with a good deal of listening. They should understand the sanctuary perspective not because they are comfortable with its conclusions, but because they are uncomfortable if they never hear it. There is, after all, a lot at stake for the United States.

Notes

1. Basil H. Liddell Hart, *Strategy*, 2d ed. rev. (New York: Meridian, 1991), 321.

2. Russell F. Weigley, *Eisenhower's Lieutenants: The Campaign of France and Germany, 1944–1945* (Bloomington, Ind.: Indiana University Press, 1981), 377.

3. Tim Zimmerman, "Chemical Weapons: Senate Skeptics Ratify a Treaty," *U.S. News & World Report*, 5 May 1997, 44.

4. Warren Ferster, "U.S. Military Develops Plan to Protect Satellites," *Space News* 8, no. 7 (17–24 February 1997): 6.

5. Steven Lambakis, "Space Control in Desert Storm and Beyond," *Orbis*, Summer 1995, 428.

6. Ferster, 26.

7. The White House, *A National Security Strategy of Engagement and Enlargement* (Washington, D.C.: Government Printing Office, February 1996), 11.

8. Statistics on the number, type, and national origins of satellites since 1957 are updated annually by *Air Force Magazine*. See also Tamar A. Mehuron, "Space Almanac," *Air Force Magazine*, August 1996, 38–40. For more details on modern international space activities, see USAF Phillips Laboratory, *Europe and Asia in Space: 1993–1994* (Colorado Springs, Colo.: Kalman Sciences Corp., 1994), 347.

9. On US ABM programs, see B. Bruce-Riggs, *The Shield of Faith* (New York: Simon and Schuster, 1988); and Ernest J. Yanarella, *The Missile Defense Controversy* (Lexington: University of Kentucky Press, 1977).

10. Robyn M. Dawes, *Rational Choice in an Uncertain World* (Orlando, Fla.: Harcourt Brace College Publishers, 1988), 103.

11. Curtis Peebles, *Battle for Space* (New York: Beaufort Books, 1983), 51.

12. Curtis Peebles, *High Frontier: The U.S. Air Force and the Military Space Program*, Air Force History and Museums Program (Washington, D.C.: Government Printing Office, 1997), 4.

13. Paul B. Stares, *The Militarization of Space: U.S. Policy, 1945–1984* (Ithaca, N.Y.: Cornell University Press, 1985), 30.

14. Ibid., 30.

15. Ibid., 48.

16. Ibid., 47–48.

17. Ibid., 49.

18. Ibid.

19. Ibid., 51.

20. Ibid., 55.

21. Ibid., 75.

22. Peebles, *Battle for Space,* 83–85.

23. Stares, 80–81.

24. Peebles, *Battle for Space*, 89–90.

25. Stares, 119; and Peebles, *Battle for Space,* 90.

26. Peebles, 85.

27. Ibid., 92.

28. Stares, 81. The Partial Test Ban Treaty of 1963 prohibited nuclear test explosions in all mediums including space.

29. Ibid., 127.

30. Ibid., 128–29.

31. Ibid., 93.

32. Ibid., 94.

33. Ibid.

34. Ibid., 97.

35. Ibid., 93.

36. Ibid., 103. The Outer Space Treaty also reserved the moon and other celestial bodies exclusively for peaceful purposes, and forbids the testing of any type of weapon, the establishment of military bases, and the conduct of military maneuvers on the moon or other celestial bodies. This is another illustration of the sanctuary paradigm in action.

37. Peebles, *Battle for Space,* 94.

38. Ibid., 103.

39. Ibid., 105.

40. Ibid., 103.

41. Stares, 162–63.

42. Ibid., 165.

43. Ibid., 164.

44. Ibid., 169.

45. Ibid., 170.

46. Ibid.

47. Ibid., 146.

48. Ibid., 145.

49. Peebles, *Battle for Space,* 112.

50. Stares, 174.

51. Ibid., 171.

52. Ibid., 185.

53. Peebles, *Battle for Space,* 111.

54. Ibid., 113–14.

55. Stares, 116.

56. Peebles, *Battle for Space,* 122.

57. Stares, 218.

58. Ibid., 231.

59. Ibid., 232.

60. Peebles, *High Frontier,* 67.

61. AU-18, *Space Handbook: A Warfighter's Guide to Space*, vol. 1 (Maxwell Air Force Base [AFB], Ala.: Air University Press, December 1993), 43.

62. Peebles, *High Frontier,* 67.

63. Edward Reiss, *The Strategic Defense Initiative* (Cambridge, Mass.: Cambridge University Press, 1992), 145.

64. *Star Wars Quotes* (Washington, D.C.: Arms Control Association, July 1986), 115.

65. Ibid., 36.

66. Charles A. Monfort, "ASATs: Star Wars on the Cheap," *Bulletin of Atomic Scientists*, April 1989, 10.

67. Stares, 49.

68. Ibid., 50.

69. Ibid., 109 and 112.

70. Ibid., 76.

71. Peebles, *Battle for Space,* 53–54.

72. Stares, 98.

73. Ibid., 164–65.
74. Monfort, 10.
75. Reiss, 51.
76. Peebles, *High Frontier*, 67–68.
77. Reiss, 91.
78. Ibid., 186–88.
79. General Accounting Office, *Report to the Chairman on Armed Services, U.S. Senate. Strategic Defense Initiative: Estimates of Brilliant Pebbles' Effectiveness Are Based on Many Unproven Assumptions* (Washington, D.C.: Government Accounting Office (GAO), March 1992), 2.
80. Reiss, 56, 88.
81. Ibid., 88.
82. Ibid.
83. Patricia A. Gilmartin, "Successful Neutral Particle Beam Firing Paves Way for More Ambitious SDI Test," *Aviation Week & Space Technology*, 24 July 1989, 31–32.
84. Michael A. Dornheim, "Alpha Chemical Laser Tests Affirm Design of Space-Based Weapon," *Aviation Week & Space Technology*, 1 July 1991, 26.
85. White House, *Fact Sheet: National Space Policy*, 19 September 1996, 5.
86. Ibid., 6.
87. David Mosher and Raymond Hall, "The Clinton Plan for Theater Missile Defenses: Costs and Alternatives," *Arms Control Today*, September 1994, 15.
88. Elizabeth A. Palmer, "Clinton Hews to Narrow View on ABM Treaty," *Congressional Quarterly*, 17 July 1993, 1894.
89. Mosher and Hall, 15–16.
90. Pat Cooper, "ASAT Funds Boosted in Senate," *Space News*, 27 May–2 June 1996, 6.
91. Pat Cooper, "U.S. Political Battles Threaten Antisatellite Project," *Space News*, 24–30 June 1996, 7.
92. Pat Cooper, "ASAT Funds," 6.
93. Pat Towell, "Clinton Signs Republicans' Fortified Defense Bill," *Congressional Quarterly*, 12 October 1996, 2931.
94. Cooper, "U.S. Air Force Considers Antisatellite Weapons," *Space News*, 26 February–3 March 1996, 4.
95. Gen Howell M. Estes III, CINC US Space Command, speech to the Air Force Association Annual Symposium, Beverly Hills Hilton, Los Angeles, Calif., 18 October 1996.
96. Ferster, 26.
97. Ibid., 6.
98. Jennifer Heronema, "A.F. Space Chief Calls War in Space Inevitable," *Space News*, 12–18 August 1996, 4.
99. William B. Scott, "USSC Prepares for Future Combat Missions in Space," *Aviation Week & Space Technology*, 5 August 1996, 51.
100. Steve Weber, "ASAT Proponents Fail to Reverse White House Policy," *Space News*, 19–25 September 1994, 7.

101. Mehuron, 40.
102. "Space Control Study Looks at Shielding Assets," *Military Space,* 30 September 1996, 7.
103. William B. Scott, "New Milspace Doctrine 'Vital,'" *Aviation Week & Space Technology,* 22 April 1996, 26.
104. Lt Gen Thomas S. Moorman Jr., "Space: A New Strategic Frontier," *Airpower Journal,* Spring 1992, 22.
105. Cooper, "ASAT Funds," 6.
106. "SSDC: ASAT Needed for Denial of Sat Recon," *Military Space,* 3 March 1997, 1.
107. "Space Control Study Looks at Shielding Assets," 7.
108. Steven Lambakis, "The United States in Lilliput: The Tragedy of Fleeting Space Power," *Strategic Review,* Winter 1996, 35–36.
109. "Second DOD Forum: 'Guide Stars' and Hail, Farewell," *Military Space,* 17 February 1997, 3.
110. "National Security Space Master Plan Finished," *Military Space,* 17 February 1997, 5.
111. Joseph S. Toma, "Desert Storm Communications," in *The First Information War,* ed. Alan D. Campen (Fairfax, Va.: AFCEA International Press, October 1992), 3.
112. Alan D. Campen, "Iraqi Command and Control: The Information Differential," in *The First Information War,* 175.
113. Larry K. Wentz, "Communications Support for the High Technology Battlefield," in *The First Information War,* 21.
114. Gordon R. W. MacLean, "Will Fiber Optics Threaten Satellite Communications?" *Space Policy,* (May 1995): 95–99.
115. Dr. Joseph N. Pelton, "Why Nicholas Negroponte Is Wrong about the Future of Telecommunications," *Telecommunications,* January 1993, 38.
116. "First Space Architecture Is Released, Others Delayed," *Military Space,* 16 September 1996, 1–3.
117. Maj Gen Kenneth Israel, "High Altitude Endurance Unmanned Aerial Vehicle," DARPA Tactical Technology Office, 13 January 1997, n.p. Online, Internet, 13 January 1997. Available from http://www.arpa.mil/asto/hae.html.
118. Capt Mike Evans, Headquarters C4A, Scott AFB, Ill., telephone interview with author, 10 December 1996.
119. Colin Clark, "Global Hawk Rolls Out; First Flight By Fall," *Defense Week,* 24 February 1997, 7.
120. "EELV, SBIRS Tops Space," *Military Spaces 17 February 1997, 8.*
121. "Global Hawk: Tier II Plus High Altitude Endurance Unmanned Aerial Reconnaissance System," commercial brochure from Teledyne Ryan Aeronautical.
122. Clark, 7.
123. Israel.
124. Gen Thomas S. Moorman Jr., "The Future of United States Air Force Space Operations: The National Security Dimension," address to the

National Security Section, Commonwealth Club, San Francisco, Calif., 1 December 1993, *Vital Speeches*, 60 (15 March 1994): 326.

125. Anne Eisele, "Phillips Moves toward Light, Tiny Satellites," *Space News*, 13–19 January 1997, 17.

126. Christian C. Daehnick, "Blueprints for the Future: Comparing National Security Space Architectures" (master's thesis, School of Advanced Airpower Studies [SAAS], June 1995), 3.

127. "NRO Satellites to Shrink in Size, Technology Director Says," *Space Business News*, 19 February 1997, 8; and "NRO Plans for Smaller Satellites," *Space News*, 17–23 February 1997, 23.

128. "LM Eyes A2100 for SBIRS High," *Military Space*, 20 January 1997, 7.

129. Anne Eisele, "Lower Costs Drive Development in Europe, Japan, and the United States," *Space News*, 17–23 February 1997, 17.

130. Ibid., 8.

131. Robert F. Futrell, *The United States Air Force in Korea, 1950–1953* (Washington, D. C.: Office of Air Force History, 1983), 272–73. On Vietnam and Afghanistan, see Mark Clodfelter, *The Limits of Airpower: The American Bombing of North Vietnam* (New York: Free Press, 1989); and Edward B. Westermann, "The Limits of Soviet Airpower: The Bear versus the Mujahideen in Afghanistan, 1979–1989" (master's thesis, SAAS, June 1997).

132. Sir Peter Anson and Dennis Cummings, "The First Space War: The Contribution of Satellites to the Gulf War," in *The First Information War*, 122.

133. Alasdair McLean, *Western European Military Space Policy* (Aldershot, England: Dartmouth Publishing Co., 1992), 101.

134. Ibid.

135. Lambakis, 421.

136. For a precise accounting of what nations use Intelsat, Inmarsat, and Arabsat, see Andrew Wilson, ed., *Jane's Space Directory: 1996–97*, 12th ed. (Alexandria, Va.: Jane's Information Group, 1996), 289, 297. The situation with Arabsat in the Gulf War is particularly interesting. Arabsat is headquartered in Saudi Arabia and plays a vital role in Middle East communications. Its 21 members include Egypt, Iraq, Kuwait, Qatar, Saudi Arabia, and Syria.

137. David E. Lupton, *On Space Warfare: A Space Power Doctrine* (Maxwell AFB, Ala.: Air University Press, 1988), 69.

138. USAF Phillips Laboratory, *Europe and Asia in Space: 1993–1994* (Colorado Springs, Colo.: Kaman Sciences Corp., 1994), 347.

139. Lupton, 68.

140. Ibid., 69.

141. USAF Phillips Laboratory, 348.

142. Nicolay Novichkov, "Russian Space Chief Voices Dire Warnings," *Aviation Week & Space Technology*, 6 January 1997, 26.

143. "Mission Control," *Military Space*, 17 February 1997, 3.

144. USAF Phillips Laboratory, 20.

145. "Mission Control," 3.

146. USAF Phillips Laboratory, 334.

147. Craig Covault, "Advanced KH-11 Broadens United States Recon Capability," *Aviation Week & Space Technology,* 6 January 1997, 24.

148. Ibid., 24.

149. McLean, 127.

150. Ibid., 119.

151. Joseph P. Keddell Jr., *The Politics of Defense in Japan: Managing Internal and External Pressures* (Armonk, N.Y.: M. E. Sharpe, 1993), xiii, 8.

152. "Increases Seen in Space Early Warning," *Military Space*, 2 September 1996, 7.

153. Craig Covault, "China Seeks Cooperation, Airs New Space Strategy," *Aviation Week & Space Technology,* 14 October 1996, 29–32.

154. Simon Fluendy, "Up in Smoke: Latest Launch Failure Could Cost China Dearly," *Far Eastern Economic Review,* 5 September 1996, 69; and Mark Ward, "China's Exploding Space Program," *World Press Review,* June 1996, 36.

155. Craig Covault, "Delta Explosion Halts $1 Billion in Launches," *Aviation Week & Space Technology,* 27 January 1997, 33.

156. The White House, *A National Security Strategy of Engagement and Enlargement* (Washington, D.C.: Government Printing Office, February 1996), 11.

157. McLean, 179.

158. Ibid., 177.

159. Sidney N. Graybeal and Daniel O. Graham, "Should the United States Build a Space-Based Missile Defense?" *Insight,* 11 September 1995, 19.

160. Robert Jervis, *Perception and Misperception in International Politics* (Princeton, N.J.: Princeton University Press, 1976), 62.

161. Ibid., 63.

162. Ibid., 64.

163. Ibid., 67.

164. Roger B. Handberg and Joan Johnson-Freese, *The Prestige Trap: A Comparative Study of the United States, European, and Japanese Space Programs* (Dubuque, Iowa: Kendall/Hunt Publishing Co., 1994), 212.

165. Ibid.

166. Ibid., 3.

167. Stephen M. Walt, "Alliance Formation and the Balance of World Power," *International Security* 9, no. 4 (Spring 1985), as reprinted in *The Perils of Anarchy: Contemporary Realism and International Security,* ed. Michael Brown, Sean M. Lynn-Jones, and Steven E. Miller (Cambridge, Mass.: MIT Press, 1995), 238.

168. Ibid., 210.

169. Ibid., 214.

170. Karl Mueller, "Why Building Space Weapons Would Threaten U.S. Security: The Perils of Occupying the High Frontier" (paper, SAAS, Maxwell AFB, Ala., 1997), 6.

171. Ibid.

172. Walt, 216.

173. "Project SATRAN Warns of Hostile Recon from Space," *Military Space,* 14 April 1997, 7.

174. Robert Jervis, "Cooperation under the Security Dilemma," *World Politics* 30, no. 2, (January 1978): 187–214.

175. "Global Engagement: A Vision for the 21st Century Air Force," Headquarters USAF, 25 November 1996, n.p. On-line, Internet. 25 November 1996. Available from http://www.af-future.hq.af.mil/21/logi/mist.htm.

176. Thomas C. Schelling, *Arms and Influence* (New Haven, Conn.: Yale University Press, 1966), 227.

177. Reiss, 145.

178. "AIA's Fuqua: Aerospace Recovery Under Way," *Military Space,* 8 January 1996, 7.

179. "QDR Sets Pace, Questions," *Military Space,* 17 February 1997, 6.

180. Johan Benson, "Conversations with General Ronald Fogleman," *Aerospace America,* July 1996, 16.

181. Moorman, 328.

182. Pierre Sparaco, "Arianespace Seeks Non-European Allies," *Aviation Week & Space Technology,* 27 January 1997, 62.

183. Mueller, 9.

184. William J. Burke and Rita C. Sagalyn, "Active Space Experiments Affect Treaty Obligations," *Signal,* June 1990, 74.

185. The White House, "National Space Policy Fact Sheet," September 1996, 14.

186. AU-18, *Space Handbook: A Warfighter's Guide to Space,* vol. 1 (Maxwell AFB, Ala.: Air University Press, December 1993), 43.

187. Leonard David, "Severity of Orbital Debris Questionable," *Space News,* 24 February–2 March 1997, 4.

188. "Shuttle Avoids Collision with Pegasus Debris," *Space News,* 24 February–2 March 1997, 2.

189. Stan Crock, "Star Wars Junior: Will It Fly?" *Business Week,* 15 July 1996, 89.

190. Council of Economic Advisers, "Economic Indicators" (Washington, D.C.: Government Printing Office, December 1996), 32.

191. Rita Rubin, "Special Report: The War on Cancer," *U.S. News & World Report,* 5 February 1996, 54.

192. Gary S. Becker, "The Painful Truth about Medical Research," *Business Week,* 29 July 1996, 18.

193. "Better Coordination, More Funds for Cancer Research Urged," *Chemical and Engineering News,* 10 October 1994, 20.

194. Lt Col Ed Felker, Chief, Joint Vision 2010 Concepts Branch, Joint Warfighting Center, interview with author, 13 January 1997.

195. Bernard Brodie, "Strategy as a Science," *World Politics* 1, no. 4 (July 1949): 499.

196. Gen Howell M. Estes III, CINC USSPACECOM, directed his command to complete a space theory and doctrine by May 1998. In an interview

dated 24 March 1997, the general highlighted the lack of such a work as the single largest obstacle to astute space policy making in the future.

197. Peter G. Tsouras, *Warriors' Words: A Quotation Book* (London, England: Arms and Armour Press, 1992), 289.

PART III

Space Control Perspectives

Chapter 5

Counterspace Operations for Information Dominance

James G. Lee

The launch of the Soviet "sputnik" satellite in October 1957 shocked the world and propelled the rhetoric and the realities of the cold war into the space age. At the same time, the Soviet feat raised the threat of mass destruction from space and served as the basis for strategists to argue for a means to shoot down enemy satellites. Although the arguments used to justify the need for an antisatellite (ASAT) weapon have changed in the years since sputnik, the policy and strategy for its employment have always focused on the need to destroy, or threaten to destroy, Soviet satellites on orbit.

The Need for a Change

Since the mid-1960s, US military strategy has focused on deterrence based on flexible response. US deterrent power is based on a balanced mix of nuclear and conventional forces, augmented by strong alliances, forward basing, and power projection. Likewise, US military space systems were initially developed in a cold war context and viewed as primarily strategic systems—supporting the Strategic Air Command, the intelligence community, and the National Command Authorities. Timely, accurate, and unambiguous strategic and tactical warning information from reconnaissance, surveillance, and communication satellites provided situational awareness of our perceived enemy and became integral to the deterrent power of the triad.

This work was accomplished in partial fulfillment of the master's degree requirements of the School of Advanced Airpower Studies, Air University, Maxwell AFB, Ala., 1996.
Advisor: Col James K. Feldman, PhD
Reader: Lt Col Gary P. Cox, PhD

In essence US military space systems became a de facto hidden leg of the strategic nuclear triad. The stability of US and Soviet nuclear deterrence rested on the ability of space systems to collect, process, and disseminate information. The balance of information provided by space systems resulted in each side having a sufficient degree of timely warning of the other side's actions. Maintaining the balance in warning information prevented one side from achieving surprise and rendering the other side incapable of a nuclear retaliatory strike. In fact, the value of the information from space systems was viewed as essential for cold war stability, and many argued that space must remain a sanctuary to preserve stability. Gen Charles Gabriel, Air Force chief of staff, subscribed to this position when he argued that the value of an ASAT weapon was not as an offensive device intended for creating an imbalance by conducting a first strike attack against the Soviet satellite system, but rather as a weapon deployed to deter attacks on US space systems.[1] If deterrence of Soviet attacks upon US space systems failed, the ASAT was to be employed to restore the balance of information by counterattacking Soviet satellites.

A recent, and perhaps the most compelling, argument for an ASAT was articulated in 1987 by Gen John Piotrowski while serving as commander in chief, United States Space Command. General Piotrowski argued that, while space systems remain integral to the deterrent power of our nuclear triad, space systems have also become critical to the successful conduct of conventional war. General Piotrowski believed the ability to negate enemy satellites would enhance the war-fighting capabilities of US terrestrial forces. Therefore, he concluded the true value of an ASAT rested with its contribution to deterring conventional war with the Soviet Union, and if deterrence failed, its ability to deny the Soviets use of their critical space systems.[2] Piotrowski's cold war argument for an ASAT suggests that a counterspace capability may also be needed in an evolving world to increase deterrence of conventional conflicts, and if deterrence fails, to deny information to the enemy.

The cold war appears to be over, but the world is, in many ways, much more complex. Gone is the relatively simple arrangement of bipolar alliances and loyalties that have characterized the four decades since World War II ended. In one sense the cold war made the US national security strategy and foreign policy straightforward; to a large degree nations were considered either pro-Soviet or anti-Soviet. Today, the traditional and historical ethnic and religious animosities, once held in check by the fear of a common enemy, have reemerged and, in some cases, erupted in civil war. The future may likely be characterized by an increase in regional political instabilities, economic and social dislocation, and a widespread diffusion of conventional military power, coupled with the proliferation of the capability to create and deliver chemical, biological, or nuclear devices.

The thawing of the cold war has also brought changes in US military force structure. The dismantlement of the Warsaw Pact and the Soviet Union has left US political leadership with the perception of a reduced external national security threat. This perception, coupled with what seems to be an out-of-control US national debt, has resulted in a willingness to reduce US strategic and conventional military forces and their forward-based presence overseas.

Although US forward presence is shrinking, the US will remain committed to the North Atlantic Treaty Organization (NATO) and the collective defense of such other nations as Japan, Korea, and some of the nations of Southwest Asia. To project power rapidly and respond effectively to crisis situations worldwide, US conventional forces are becoming lighter, more rapidly deployable, and more expeditionary.

In the future the United States may not have the same opportunity for extended mobilization in preparation for war as was afforded in Operation Desert Shield. Regional crises and conflicts probably will be "come as you are," and the necessity to collect, process, and disseminate strategic and tactical information on the enemy's forces and terrain may become increasingly important to expeditionary forces that must fight effectively in potentially unfamiliar terrain against an unfamiliar enemy. Likewise, allowing an enemy access to information on US force deployments, order of battle, movements, and logistics could jeopardize US ability to stage and deploy forces, and successfully execute US military strategy. Therefore, it would seem that the ability to control information may become increasingly important, and possibly decisive, in future military operations.

Since the ability to collect, process, and disseminate information to field commanders may become a decisive contributor to victory in future conflicts, information warfare actions may emerge as an essential function in crisis response and war. At the operational level, information-warfare denies the enemy the capability to collect, process, and disseminate information with the objective of creating a positive information gap between friendly and enemy forces. This positive information gap has been referred to as information dominance.

Information Dominance

The concept of information dominance first emerged in the writings of Soviet military theorists in the late 1970s as part of a discussion of the concept of military technical revolutions. The Soviets coined the phrase, "military technical revolution," to describe past and future eras in which extreme transformations in warfare occurred or may occur as a result of the exploitation of technology. The Soviets, however, did not see technology in and of itself defining the revolution as the phrase might suggest. Rather, they saw the operational and organizational innovations resulting from the exploitation of the technology as defining a military technical revolution.[3]

The Soviets predicted that the technological advances occurring in US information collection, processing, and dissemination, coupled with the increasing range and accuracy of precision-guided munitions, would lead to the next military technical revolution. They believed, if fully exploited, these technologies could become the basis for logically integrated, yet geographically distributed, weapon systems whose elements perform reconnaissance, surveillance, target acquisition, and target engagement. The increased emphasis of modern weapon systems on the reliance and the ability to collect, process, and disseminate information seems to suggest that the ability to establish information dominance over an adversary could be increasingly important to the conduct of military operations.[4]

Information dominance can be described as a condition in which a nation possesses a greater understanding of the strengths, weaknesses, interdependencies, and centers of gravity of an adversary's military, political, social, and economic infrastructure than the enemy has on friendly sources of national power.[5] Attaining information dominance could mean the difference between success and failure of diplomatic initiatives, successful resolutions of crises, or war, or forfeiture of the element of surprise to the enemy in military operations. Therefore, the ability to attain information dominance can widen the gap between friendly actions and enemy reactions, and allow friendly commanders to manage the enemy's decision cycle by controlling and manipulating the information available to them.[6] On the other hand, failure to achieve information dominance at the onset of hostilities could lead to the inability of friendly forces to conduct military operations successfully.

Today more than ever, information is power. Consequently, military operations to attain information dominance should probably be initiated at the onset of a crisis to facilitate rapid mobilization and power projection sustained through the crisis and, if necessary, through war.[7] Information dominance can be obtained by conducting offensive and/or defensive military operations. Offensively, information dominance can be attained by collapsing an adversary's command and control infrastructure through such offensive operations as the disruption of critical communication links; or by denying access to reconnaissance and surveillance information, such as blinding optical sensors with ground-based lasers. Defensively, measures such as hardening, frequency hopping, and encryption further ensure information dominance by helping to ensure friendly forces have uninhibited access to communications, surveillance, and reconnaissance information provided by space systems.[8] Therefore, delaying and denying a potential adversary information, while providing similar information to friendly forces, can indeed be a valuable mechanism for balancing power during peacetime and a decisive terrestrial force enhancer/multiplier during war.

Role of Space Systems

Just as there is a synergism among air, land, and sea forces, there appears to be an emerging synergism between space systems and terrestrial forces, suggesting that space systems are becoming inseparable to land, sea, and air warfare. Existing military space systems have demonstrated an ability to provide near-real-time command and control, weather, surveillance and reconnaissance, and navigation information to air, land, and sea forces. In Operation Desert Storm, for example, US Air Force space systems provided near-real-time surveillance data of Iraqi Scud missile launches directly to the US Central Command (CENTCOM) command center in Saudi Arabia. This warning data was then used to alert coalition forces and direct Patriot air defense artillery fire against the Scud missile and direct air strikes in counterbattery operations against the Scud launchers. The integration of information from space systems with modern weapon delivery systems and precision munitions during Desert Storm would seem to validate the Soviet vision of the next military technical revolution and the importance of space systems to the concept of information dominance.

As space systems become more valuable to attaining national security and to our ability to support allies and promote international stability, their value to information dominance increases as well. Given the increasing importance of information from space systems to terrestrial military operations, attaining information dominance appears to require the capability to conduct counterspace operations.

However, the ability of the United States to conduct counterspace operations may become increasingly difficult as space systems and technologies proliferate among nations. Indeed, the majority of the world space programs and systems are considered civilian systems and were not initially developed or intended for dedicated military purposes. It may be prudent to assume that nations subsidizing civilian space activities are also exploiting these "nonmilitary" satellites for military and national security information.[9] For example, the French commercial space system *Systeme Probatoire pour l'Observation de la Terre* (SPOT) has demonstrated an intelligence capability by providing commercial photographs of Soviet laser facilities at Sary Shagan.[10] The inherent military capabilities of civilian space systems suggest the proliferation of space systems and technologies could have serious military implications with respect to our ability to establish information dominance.

In the past, the United States and Russia could exercise a degree of control and leverage over the information other nations received from space systems through our collective monopoly on the ability to build and launch satellites.[11] However, France, Japan, China, India, and Israel have all launched and orbited civilian satellites with imaging capabilities. Furthermore, nations such as Brazil, Canada, and Great Britain are also developing satellite systems capable of providing imagery with potential military utility. Indeed, nations do not need to own space systems to have access to information from space. Numerous space-faring nations, such as France, Russia, and Japan offset the cost of developing and deploying space systems by marketing their information.[12] In light of the increasing global instabilities and uncertainties, some nations may find it advantageous to make militarily useful information from civilian satellites available to countries hostile to the United States—Brazil to Libya or China to Iran—for example.[13] It is not unreasonable to speculate that in the future the United States could find itself in a crisis situation, or war, with an adversary either operating its own space system, or relying on information from another nation's space system. In this situation the United States is usually portrayed as having only two options: do nothing, or destroy the enemy's satellite with an ASAT. Under international law it is generally accepted that the destruction of a nation's space system as an act of self-defense is justified.[14] However, in situations where the enemy is acquiring information from a space system owned by a neutral third party, the unilateral destruction of that satellite with an ASAT is considered an act of aggression and a violation of that nation's sovereignty.[15] This suggests that there may be situations in which employing an ASAT to destroy a satellite may simply not be an acceptable alternative.

The apparent trend for global proliferation of space systems and marketing of space information seems to raise doubts regarding the flexibility and responsiveness of our current space control strategy and our ability to achieve information dominance. This work evaluates current space control strategy in terms of the ability to ensure information dominance in the evolving national security environment characterized by the increasing proliferation of space systems. A discussion of the phenomenon of global proliferation of space systems and the military utility of civilian imagery systems is the focus of the next section. The section immediately following it entails an assessment of current space control strategy and policy with respect to the emerging threat from proliferated space capabilities. The last two sections offer both an alternative space control strategy to deny the enemy the use of information from space systems and a means to implement that alternative space control strategy.

Proliferating Space Technology

Nations possessing space capabilities can be divided into three tiers. First-tier space-capable nations possess dedicated military and civilian space capabilities on the cutting edge of technology. Second-tier nations develop and use dual-purpose space systems for both military and civilian purposes. Third-tier nations lease or purchase space capabilities or products for military and civilian purposes from first- and second-tier nations.[16] Table 24 gives examples of nations in each of the three tiers.

Proliferation of Civilian Space Capabilities

Nations within the first tier, the United States and Russia, have disseminated surveillance and reconnaissance products from dedicated military satellite systems to alliance partners for many years. There are also several civilian corporations selling such space products as communication channels, weather information, and earth imagery on the international market to almost any nation able to pay the price. In fact, one of the major sources of earth imagery available on the commercial market is from the US civilian satellite system, Landsat.

Table 24

Space-Capable Nations by Tier Groups

Tier	Nations
First Tier	United States Russia
Second Tier	France Great Britain China Japan India Israel
Third Tier*	Brazil Italy Australia Thailand South Africa Canada Iran Iraq Pakistan

*Not all inclusive, only major nations in this category are listed.

Landsat is an earth-remote sensing satellite system. There are currently two operational Landsat satellites each capable of providing imagery in seven spectral (color) bands, and one black and white panchromatic band. The most recent Landsat launched, Landsat 6 in 1992, is capable of producing black and white images with a ground resolution of 15 meters.

Initially owned and operated by the National Oceanic and Atmospheric Administration (NOAA), the Landsat system was privatized in 1979 and is now operated by a private company, EOSAT, for NOAA. Under the provisions of the Remote Sensing Act, Landsat data must be made available for sale to any individual or nation on a nondiscriminatory basis. The secretary of defense, however, does have the authority to determine customers or circumstances for which the sale of Landsat data can be denied for national security reasons. Presently, the Department of Defense (DOD) has not established any criteria or specific provisions for restricting the sale and distribution of Landsat imagery.

In addition to selling processed Landsat imagery products, NOAA/EOSAT also oversees the establishment and licensing of Landsat ground stations in foreign countries. In addition to the Landsat ground station in the United States, there are currently 13 licensed stations with plans to build another two outside the United States. These Landsat ground stations can receive and process Landsat data directly from the satellites. Table 25 shows the locations of current and projected licensed Landsat ground stations.

The technology and facilities required to build and operate a Landsat ground station are simple and relatively cheap when compared to the cost of developing, launching, and operating a comparable satellite system. Costs to construct a Landsat ground station are about $20 million, plus an additional $3 million a year in operational costs. The NOAA/EOSAT licensing fee is a flat $600 thousand a year.[17] Once licensed, ground stations are permitted to receive, process, and sell Landsat information in accordance with the US policy on nondiscrimination.

Although the technology and equipment to build and operate a Landsat ground station is straightforward and inexpensive, it is also subject to US export controls. The US government uses export controls and its final approval authority for

Table 25

Existing and Projected Landsat Ground Stations

Existing	Projected
United States	Ecuador
Brazil	New Zealand
Argentina	
Spain	
Italy	
South Africa	
Saudi Arabia	
Thailand	
Indonesia	
Australia	
China	
Japan	
Sweden	
Pakistan	

foreign ground station construction as a means to control the proliferation of space technology.

Consequently, no member of the former Soviet bloc has yet received approval to establish a Landsat ground station.[18] Controlling the information from Landsat is, however, a different matter. Presently, the only way to restrict the foreign ground stations from directly receiving and processing downlinked Landsat data would be for EOSAT to command the satellite sensor not to image the area in which data is to be denied.[19] Commanding the sensor "off," however, would also deny imagery data from the specific area to other licensed ground stations and the United States because the current Landsat satellites have no onboard data storage capability.[20] In addition, since most foreign ground stations do not have the capability to command the Landsat, controlling unauthorized direct access to Landsat data appears fairly reliable.

Russia, the other first-tier space nation, also sells photographic imagery of the earth's surface from satellites. This information, however, is derived from their KFA 1,000 camera carried on board the Resurs series military satellites. In 1987 the Russians began to sell, through the Soyuzkharta company, black and white photographic images with five-meter ground resolution of any site/area located in nonsocialist countries. Even though the Russians seem to be in need of hard currency and concerned with the survival of their space program, they have not yet licensed, nor do they appear interested in commercially licensing foreign satellite ground stations.

The Resurs satellite represents older technology and uses a recoverable film canister from the satellite to produce earth imagery rather than processing downlinked digital imagery data like Landsat. Although technologically obsolete compared to Landsat, the five-meter ground resolution of Resurs imagery is one of the best available on the commercial market.

Second-tier space nations are growing in both numbers and capability. France was the first nation to challenge American and Russian dominance in space with its commercial space launcher, Ariane, and is now a third major competitor in the commercial remote sensing market.

The French SPOT can provide multispectral remote sensing data in four spectral bands with ground resolutions of 10 meters in black and white panchromatic imagery, and 20-meter resolution for imagery in other spectral bands. SPOT Imaging Corporation describes the current capabilities of its satellite as having sufficient resolution to allow detection of objects 10 to 30 meters in size, recognition of objects 20 to 60 meters in size, and description of objects 60 meters or larger.[21] In addition, the imaging sensor onboard SPOT satellites has the ability to look 27 degrees to the right or left of the satellite track. This off-nadir imaging capability allows the same area of the Earth to be imaged on successive orbits from different viewing angles. Fusing multiple images of the same area from different viewing angles results in a capability to produce stereo images.[22]

Imagery data from SPOT satellites can be transmitted directly to ground stations or archived on tape recorders on board the satellite for later transmission.[23] Regardless of the source, all imagery data is downlinked to either the SPOT primary control center near Toulouse, France, or the SPOT control center near Kiruna, Sweden.[24] These two ground stations are primarily responsible for processing the imagery data stored on the onboard tape recorders and data collected over the north polar region, Europe, and North Africa.[25]

SPOT Image has also established a global network of receiving stations to receive, process, and disseminate satellite imagery on a similar nondiscriminatory basis as NOAA/EOSAT for the Landsat system. Table 26 shows the location of current and planned SPOT ground stations worldwide. French export controls governing the transfer of technology to establish and operate a SPOT ground station are similar to those employed by the United States. SPOT, however, also restricts the area in which each ground station is authorized to receive and process data.[26] India, for example, is authorized to receive imagery data directly from the SPOT satellite only while the satellite is within a 2,500 kilometer (km) radius of the Indian ground station.[27] Thus the Indian ground station can only receive and process images of its own territory even though it is capable of receiving and processing data encompassing a much greater area. SPOT accomplishes these restrictions by withholding

Table 26

Existing and Projected SPOT Ground Stations

Existing	Projected
France	Ecuador
Sweden	China
Canada	South Africa
India	Taiwan
Canary Islands	Indonesia
Brazil	Saudi Arabia
Pakistan	
Thailand	
Japan	
Israel	
Australia	

certain bits of information regarding the satellite's mode of operation and orbit needed to process data from the satellite.

Through a combination of the receiving restrictions and the onboard tape recorders, SPOT was able to deny Iraq images of the Persian Gulf region during operations Desert Shield and Desert Storm while providing these images to the Coalition forces.[28] SPOT does, however, acknowledge that a ground station could break out the information needed to circumvent the restrictions and gain access to the data from unauthorized zones.[29] Although this ground station would not be able to sell these images overtly, it could provide them to the host country's government for intelligence purposes or sell then clandestinely.

In addition to its civilian space systems, France is also expanding its space program into the military arena by spinning off the civilian SPOT satellite technology to develop a dedicated military reconnaissance satellite called Helios.[30] Helios, a joint development project with Italy and Spain, is reported to have ground resolutions approaching 0.3 meters using both multispectral imagery and a synthetic aperture radar. Although Helios imagery will most likely not be available for purchase on the commercial market, the similarities between

SPOT and Helios technology could result in significant improvements for the SPOT system.

Peter Zimmerman, a physicist at the Carnegie Endowment for International Peace, speculates that with minor improvements in optics SPOT imagery resolution could be improved to 2.5 meters.[31] In fact, the next generation SPOT satellite, SPOT 5, is reported to be capable of providing earth imagery at resolutions less than five meters. Richard Del Bello of the Office of Technology assessment believes the blurring of military and civilian technology will result in one-meter ground resolution becoming a commercial imagery standard by the year 2000.[32] This seems entirely likely and achievable considering the projected resolution capabilities of SPOT 5 and its expected competition with the Russians who are already beginning to market imagery with a 2.5-meter resolution.

Some other second-tier space nations include China, Israel, Japan, and India. China, in addition to operating a licensed Landsat ground station, launched its first photo intelligence satellite in 1975 and has since orbited at least 12 imaging satellites.[33] The Chinese FSW-1 series imaging satellites use a recoverable film canister retrieval method for returning images to Earth after an average mission duration of two weeks.[34] The imaging products derived from the FSW-1 satellites are believed to be capable of less than 80-meter resolutions and clearly support civilian resources and military reconnaissance activities. China is also engaged in a joint program with Brazil to produce and launch the China/Brazil Earth Remote Sensing satellite (CBERS).[36] Projected for a late 1993 launch, CBERS will provide multispectral imagery, similar to SPOT and Landsat, with an expected ground resolution of 20 meters.[37] In addition to developing a remote sensing capability, the Chinese also have an expanding launch capability with the Long March series of boosters. The most recent Chinese booster, Long March 2E, is considered a heavy-lift vehicle with performance between the US Atlas II and Titan IV boosters. The Long March 2E is capable of boosting 9,200 kilograms into low Earth orbit or 3,370 kg into a geosynchronous transfer orbit.[38]

Another second-tier space nation, Israel, started its space program in 1988 as a response to Israeli discontent with having to rely on the United States to provide satellite imagery.[39] Several high-ranking Israeli cabinet officials suspected that the United States withheld satellite imagery prior to the 1973 Yom Kippur War. Therefore, with the assistance of South Africa, Israel built and launched OFFEQ-1 in 1988, and OFFEQ-2 in 1990.[40] Although the Israelis deny the OFFEQ satellites carry a photo-reconnaissance payload, the nature of the orbit, 200 km at the lowest point and 1,500 km at the apogee, is a good indication that they have some intelligence gathering utility.[41]

Japan is another second-tier space nation with a rapidly developing civilian space capability. The Japanese Earth Remote Sensing Satellite (JERS-1), launched in 1992, possesses seven spectral bands capable of producing images with 18-meter ground resolution and a synthetic aperture radar capable of 25-meter ground resolution.[42] Data from the JERS-1 satellite is not available commercially, although Japan's National Space Development Agency (NASDA) may authorize sales of data in the future.[43]

Japan is also actively developing a commercial space launch capability. NASDA has been pursuing a space-launch program since 1969; however, in exchange for US rocket technology, Japan agreed to launch only Japanese payloads.[44] NASDA's newest space launcher, the M-II, is entirely a Japanese design and will allow Japan to enter the commercial launch market. Scheduled for an initial launch in 1993, the H-II is reported to have the ability to place 9,080 kg into low Earth orbit and 3,600 kg into a geosynchronous transfer orbit.[45]

India is another nation actively pursuing self sufficiency in space. The Indian Resources Satellite series (IRS1A-1988, 1B-1991, and 1C-projected for a 1993 launch) has two sets of imaging sensors with ground resolutions of 72 meters and 36 meters respectively.[46] The next generation of Indian remote sensing satellites is projected to have improved sensors giving it a multispectral resolution of 20 meters and a panchromatic imaging resolution of 10 meters.[47]

Third-tier space nations such as Pakistan, Indonesia, and Luxembourg have chosen, for political or economic reasons, not to develop or operate their own satellites. Tier-three nations acquire space information products through direct purchase or through licensing agreements to build ground stations. Although these nations depend on foreign sources for their space needs, this dependence is mitigated to some degree by building their own ground stations and obtaining licensing agreements to receive and process foreign satellite data, as in the case of Landsat and SPOT.

Military Utility

As increasing sophistication of civilian space technology blurs the distinction between military and civilian space capabilities, the probability civilian satellites will be used for military and national security purposes also increases. SPOT Image Corporation, for example, openly advertises the intelligence gathering and military utility of SPOT imagery.[48] Marketed as "The New Way to Win!" SPOT illustrates the potential for nations to exploit the inherent military capabilities of civilian systems for military and national security purposes. As the number of nations developing their own satellites or establishing satellite ground stations to process satellite imagery increases, the proliferation and exploitation of civilian imagery data for military purposes could impact the ability of the United States to prepare for and conduct military operations.

Assessing the military utility of civilian systems requires an understanding of some of the qualitative measures used to evaluate the capabilities and utility of remote sensing/imaging satellites. Spatial resolution, spectral resolution, and revisit time are the most common attributes used to compare and assess the capabilities of imaging satellites. Table 27 shows the spatial and spectral resolution and the revisit frequency of several civilian imaging/remote sensing satellites with commercially available products.[49]

Spatial resolution refers to the size of an object on the ground a sensor can distinguish. For optical sensors, spatial resolution is typically the area on the ground that is observable by a single light-sensitive-sensor element, or pixel. A pixel for an infrared sensor, for example, is a single infrared cell. The area observable by the single sensor pixel is called a sensor's instantaneous field of view (IFOV). A sensor cannot

Table 27

Qualitative Measures of Various Civilian Satellite Systems

Country	Resolution Meters (m)	Spectral Channels	Revisit Cycles
France/SPOT	10–20 m	4	2.5 days
Japan (JERS-1)	25 m	7	44 days
Russia* (Resurs/KFA 1,000 camera)	5 m	2	14 days
USA (Landsat 6)	15 m	8	16 days

*The Russian Resurs satellite was initially developed for military purposes; however, imagery is now marketed for commercial purposes.

detect any object on the ground smaller than its IFOV. Normally it takes at least two pixels to distinguish what a detected object actually is. Therefore, although a satellite with a 10-meter IFOV can detect a 10-meter object on the ground, under normal circumstances it can only distinguish objects 20 meters or larger in size.

For military purposes spatial resolution characterizes the satellite's ability to perform such delineation tasks as detection, general identification, precise identification, description, and technical analysis. *Detection* refers to locating a class of objects or an activity, such as a naval vessel or a rail switching yard. *General identification* is the ability to determine a general target group, while *precise identification* is the ability to discriminate within a target group. General identification of missiles, for example, would distinguish between ballistic missiles and surface to air missiles. Precise identification of missiles, on the other hand would distinguish between Hawk or Patriot surface-to-air missiles. *Description* refers to determining the size/dimension, configuration/layout, component construction, or equipment count of the target group, such as the difference between an F-15E or an F-15C. *Technical analysis* is the detailed analysis of specific equipment within the target group. Imagery supporting technical analysis allows the capability or limitations of a piece of equipment to be evaluated. Table 28 shows the

Table 28

Ground Resolution Requirements for Object Identification (in meters)

Target[a]	Detection[b]	General ID[c]	Precise ID[d]	Description[e]	Technical Analysis[f]
Bridges	6	4.5	1.5	1	0.3
Communications					
Radar	3	1	0.3	0.15	0.015
Radio	3	1.5	0.3	0.15	0.015
Supply Dumps	1.5	0.6	0.3	0.03	0.03
Troop Units (in Bivouac or on Road	6	2	1.2	0.3	0.15
Airfield Facilities	6	4.5	3	0.3	0.15
Rockets/Artillery	1	0.6	0.15	0.05	0.045
Aircraft	4.5	1.5	1	0.15	0.045
C^2 Headquarters	3	1.5	1	0.15	0.09
SSM[g]/SAM[h] Sites	3	1.5	0.6	0.3	0.045
Surface Ships	7.5	4.5	0.6	0.3	0.045
Nuclear Weapons Components	2.5	1.5	0.3	0.03	0.015
Vehicles	1.5	0.6	0.3	0.06	0.045
Land Mines	9	6	1	0.03	0.09
Ports and Harbors	30	15	6	3	0.03
Coasts/Beaches	30	4.5	3	1.5	0.15
Rail Yards and Shops	30	15	6	1.5	0.4
Roads	6–9	6	1.8	0.6	0.4
Urban Areas	60	30	3	3	0.75
Terrain		90	4.5	1.5	0.75
Surfaced Submarines	30	6	1.5	1	0.03

[a]Chart indicates minumum resolution in meters at which target can be detected, identified, described, or analyzed. No source specified which definition of resolution (pixel-size or white-dot) is used but the chart is internally consistent.

[b]Detection: location of a class of units, object, or activity of military unit

[c]General Identification: determination of general target type

[d]Precise Identification: discrimination within a target group

[e]Description: size/dimension, configuration/layout, component construction, equipment count, etc.

[f]Technical Analysis: detailed analysis of specific equipment

[g]Surface-to-surface missile

[h]Surface-to-air missile

ground resolution needed to perform the various delineation tasks for various objects of interest to military planners.[50]

Historically, analysts generally believed that to be useful for military purposes, imagery and remote sensing satellites would need ground resolutions less than 10 meters.[51] Typically satellites with ground resolutions greater than 20 meters were not considered militarily significant, being viewed as useful primarily for terrain analysis and economic purposes.[52] There is, however, growing evidence that satellites with ground resolutions between 10 and 20 meters, such as Landsat and SPOT, can have significant military utility. The United States Defense Mapping Agency, for example, is one of the largest users of SPOT and Landsat imagery. Commercial imagery from Landsat and SPOT have been instrumental in the generation of three-dimensional targeting information for cruise missiles and other precision-guided munitions.[53]

In addition to the potential tactical applications of civilian imagery systems like Landsat and SPOT, there are also possible significant strategic applications. Coupled with a priori knowledge from other sources of intelligence that can identify a general area to be imaged, Landsat and SPOT have also demonstrated some military utility by providing useful strategic intelligence information. Tables 25–28 show how the 10- to 20-meter ground resolution of Landsat and SPOT imagery appears to have more than adequate resolution capabilities to detect and provide general identification of major port and rail facilities, urban areas, and surfaced submarines. The satellite photographs used by the US government in public international forums to substantiate US accusations that the Soviet radar at Krasnoyarsk constituted a violation of the Antiballistic Missile (ABM) Treaty were SPOT images.[54]

Other nations in addition to the United States use commercially available imagery from civilian satellites to augment their military strategic intelligence efforts. West Germany, for example, acknowledged using SPOT images to gather intelligence and confirm the existence of the disputed chemical warfare plant in Libya.[55] Another example is the Japanese, who purchased Landsat photos in 1985 to identify and assess airfield improvements for TU-22 Backfire bombers at Zavitinsk.[56]

Spectral resolution is the second qualitative measurement pertinent to imaging systems. Spectral resolution refers to the various light frequencies, such as infrared, ultraviolet, visible light, X-ray, and so forth, that sensors are designed to detect. Using several spectral bands to observe the same patch of earth simultaneously can provide information that allows the discrimination between vegetation and soil, identification of thermal gradients in the ocean, measurement of surface moisture, and a variety of other analyses. Current civilian technology, however, restricts the data capacity of satellite downlinks; therefore, there are tradeoffs between the number of spectral bands and the spatial resolution of sensors. Typically, the more spectral bands a satellite sensor has the larger the spatial resolution. Conversely, the fewer spectral bands, the smaller the spatial resolution. The total amount of raw data for each image is increased in proportion to the number of spectral bands. Likewise, the amount of raw data for each image is also increased as the spatial resolution decreases. For example, the amount of raw data per image for a sensor with one spectral band is about half as much as a sensor with two spectral bands.

Collecting imagery of the same area in different spectral bands can often provide more information than a high-quality black and white image with ground resolutions of less than 10 meters. This is because various soils and plants have different chemical characteristics and, therefore, reflect light in different frequencies. The variations in the way light is reflected cause soil, plants, and man-made objects to look different in various spectral bands. Table 29 shows spectral bands of the Landsat and SPOT satellites and the capabilities associated with each of the different spectral bands. Imaging an area with a sensor in the green light spectral band, for instance, could not distinguish between real vegetation and green camouflage, but imagery in any of the near- or mid-infrared band could. The use of Landsat and SPOT imagery during Desert Storm provides a good example of the military utility of imagery in different spectral bands. Whenever a vehicle traversed over the ground, sand, or grass, the ground was disturbed. This disruption caused chemical changes in the ter-

Table 29

Landsat and SPOT Spectral Band Applications (in microns)

Landsat	SPOT	Application
.45–.52 (Blue light)		Coastal water mapping soil/vegetation differentiation deciduous/coniferous differentiation
.52–.60 (Green light)	.50–.59	Green reflectance from healthy vegetation iron content in rocks and soil
.63–.69 (Red light)	.61–.68	Chlorophyll absorption for plant differentiation
.76–.90 (Near-Infrared)	.79–.89	Biomass survey water body delineation
.80–1.1 (Mid-Infrared)		Crop vigor
1.55–1.75 (Mid-Infrared)	1.58–1.75	Plant moisture content cloud/snow differentiation
2.08–2.35 (Mid-Infrared)		Soil analysis
10.4–12.5 (Thermal Infrared)		Thermal mapping soil moisture

rain that could be identified using multispectral imagery from Landsat and SPOT and provided US war fighters with useful insights into Iraqi operations.[57] Likewise, imagery from Landsat and SPOT, if made available to the media, could have revealed US plans for the left hook at the start of the ground war.[58] In addition, fusing the data from different spectral bands of the same area on Earth can reveal various surface features undetected by imagery in a single spectral band. Table 30 shows a comparison between the civil applications for multispectral imagery and some of the related military applications of multispectral imagery from satellites such as Landsat and SPOT.[59]

The last qualitative measure for assessing the utility of imaging and remote sensing satellites is timeliness. There are three variables affecting the timeliness of remote sensing im-

Table 30

Civil/Military Uses of Multispectral Imagery

Civil Application	Military Application
Soil features	Terrain delineation Attack planning Trafficability
Surface temperature	ASW support Trafficability Airfield analysis
Vegetation analysis	Terrain delineation Camouflage detection
Clouds	Weather Attack planning
Snow analysis	Area delineation Attack planning
Surface elevation	Mapping, Tercom
Ice analysis	Navigation ASW support
Water analysis	Amphibious assault planning
Cultural features	Targeting, BDA

agery: satellite revisit time, image processing time, and image delivery time. Timeliness, therefore, refers to the "throughput" time—the time it takes from tasking the sensor to delivery and exploitation of the product.

One variable in timeliness is revisit frequency. Revisit frequency is the time, usually in number of days, it takes the satellite to fly over the same point on the Earth twice. For example, a typical orbit for a remote sensing satellite has an altitude of 800 km and an inclination of approximately 98 degrees. Satellites in this type of orbit have a frequent revisit time at high latitudes and an infrequent revisit time at low latitudes. Measured at the equator, the more frequent the revisit time the greater the opportunity to image the area of interest on the ground and the quicker an image can be provided to the war fighter.

Some military planners have suggested that to be useful for weapon system targeting and keying a throughput time of less than two or three days is needed, while throughput times less than 30 days could be useful for ocean surveillance and battle-damage assessment.[60] Throughput times greater than a month, however, would only be considered useful for fixed target surveillance, verification, and terrain analysis.[61]

During Desert Storm, Landsat images were routinely delivered to the theater commander anywhere between five and 12 days after the request.[62] If the area to be imaged was already in EOSAT's database, the delivery time would be less. Given the Landsat revisit time of 16 days, it could take the two Landsat satellites between one and eight days before one of them would image the desired area and another three to four days for EOSAT/NOAA to provide the imagery to the DMA.[63] After DMA had received the imagery, it normally took only one day to forward it to the theater commander.[64] Given the timeliness criteria suggested by military planners, Landsat's throughput range between five and 13 days substantiates its capability to provide targeting, damage assessment, surveillance, and terrain analysis information.

The throughput time for the SPOT system is estimated to be between four and 14 days. Although the revisit time on the SPOT satellite is 26 days, the satellite's capability to view areas up to 27 degrees off centerline enables SPOT to image a given area between three and six days after initial tasking. Image processing normally takes about one day and, depending whether or not the requester has direct access to SPOT data, delivery times can range from zero to seven days. In the final analysis the timeliness of SPOT imagery, between four and 14 days, also appears to have significant military utility for targeting, damage assessment, surveillance, and terrain analysis.

Military Utility

The end of the cold war and the disbanding of the Warsaw Pact, coupled with decreasing US military presence overseas, has motivated US allies in Europe, Asia, and the Pacific to reexamine their security needs. An increasing number of nations is choosing not to remain dependent on the United

States to provide critical space services and products. As a result, they have commenced to develop or purchase commercially available space products.

Proliferating space technologies and products could have significant implications for US national security. First, proliferating space capabilities could provide regional military powers with an advantage over US forces in any future regional conflict. Advantage could be gained by eliminating the US ability to achieve strategic and tactical surprise. The inability of US forces to achieve surprise could lead to protracted engagements.[65] Second, modern warfare is becoming highly dependent on space systems for communication, intelligence gathering, and environmental monitoring. Operation Desert Storm provides a good example of how the control of space may be a decisive factor in dominating the battlefield and the successful execution of a nation's military strategy. Just as air was the "high ground" during World War II, Korea, and Vietnam, space is emerging as today's "new high ground."[66] As the capabilities and military utility of civilian space platforms increase, so does the probability that these systems will be integrated with ballistic missiles and deep strike weapons.[67]

In sum, a new type of space threat seems to be emerging. Although future conflicts for the United States will probably be confined to militarily inferior regional powers, the increasing availability of space technologies and products could offset US military advantages. The United States, therefore, must ensure that its space control policy and strategy is flexible and responsive to deal with the changing world space order.

Traditional Space Control Methods and Strategy

For most of the last 40 years, US national security strategy has focused on the containment of the Soviet Union and the spread of the communist ideology.[68] Consequently, the threat of Soviet military power became institutionalized. The need to counter the threat presented by the Soviets' antisatellite system was the principle rationale for the US antisatellite program.[69] US space control policy and strategy was derived from the threat. The threat from space, however, is changing.

Although Russia remains the only nation capable of challenging US access to space, the proliferation of space technologies and capabilities suggests a potential threat emerging from space against US terrestrial military operations. Having characterized and discussed the proliferating threat, this work now assesses the effectiveness and credibility of current space US control policy and strategy against the threats posed by tier-one, -two, and -three space-capable nations.

Before the effectiveness and credibility of our space control policy and strategy can be assessed, a brief explanation of Air Force framework is necessary. Air Force Manual (AFM) 1-1, *Basic Aerospace Doctrine of the United States Air Force*, March 1992, lays out the framework in which Air Force space control planning and operations are performed and serves as the source of contextual definitions for the roles and missions of space control.

AFM 1-1 integrates space control into the basic role of aerospace control. According to AFM 1-1, the ideal aim of aerospace control is the absolute control of the air and space environment. All military activities having the objective of gaining and maintaining control of the air and space environment fall into two broad mission categories: counterair and counterspace. The purpose of counterspace mission is to gain and maintain control of space through offensive and defensive counterspace operations. According to AFM 1-1, the objective of offensive counterspace operations is to "seek out and neutralize or destroy enemy space forces in orbit or on the ground at a time and place of our choosing."[70] The objective of defensive counterspace operations, on the other hand, can be viewed from the perspective of active and passive counterspace defense. The aim of active counterspace defense is to detect, identify, intercept, and destroy enemy forces in space or passing through space attempting to attack friendly forces, or to penetrate the aerospace environment above friendly surface forces.[71] The objectives of passive counterspace defense are to reduce the vulnerabilities and increase the survivability of friendly satellites and include measures such as frequency hopping, nuclear hardening, and maneuverability. Although the survivability and protection of friendly space assets is essential if the enemy threat against our space forces is significant, typically the most efficient method for achieving control of space is to attack the enemy's assets close to their source.[72] With respect to space systems, this infers attacking satellites in orbit.

Space Policy

The *National Space Policy,* published 2 November 1989, acknowledges the vital role space systems play in achieving national security objectives. This policy states the national security objective of space control is to ensure freedom of action in space.[73]

The Department of Defense (DOD) also recognizes that space control includes both freedom of access to space and the ability to deny this access to a potential enemy. Unlike the balanced approach of the National Space Policy, DOD policy appears to be oriented towards offensive counterspace operations, emphasizing the need for a flexible and responsive mix of antisatellite weapons to degrade the effectiveness of an enemy's ground, air, and sea forces by denying them support from space-based systems.[74] Furthermore, DOD envisioned the ASAT fulfilling a response-in-kind role, acting to deter attacks against US satellites by the Soviet ASAT system.[75]

Gen John Piotrowski, the former commander in chief of United States Space Command, not only reaffirmed the offensive orientation of our current space control policy, but established the strategic objectives of offensive counterspace operations. According to General Piotrowski, an ASAT weapon is needed, not only to deter attacks against US space assets, but as a deterrent against a Soviet decision to go to war and, if deterrence fails, as a needed war-fighting capability.[76]

Traditionally, military planners have envisioned the ASAT war-fighting capability as a hard kill (i.e., physical destruction) weapon system, such as a satellite interceptor missile (kinetic energy) or a ground-based laser (directed energy), engaged in offensive counterspace operations to destroy orbiting enemy satellites. DOD's most recent ASAT project was seeking to develop a ground-based kinetic energy interceptor with provisions in the long term for the development of a directed-energyASAT.[77]

Strategy Implications

As outlined in AFM1-1, our current space control strategy can be summed up as a strategy aimed at achieving space supremacy.[78] In this context, space supremacy means absolute control of the space environment.[79] The ability to achieve space supremacy is presumed, as articulated by General Piotrowski, to deter attacks against US space assets, deter against a Soviet decision to go to war, and, if deterrence fails, serve as a critical war-fighting capability. Any assessment, therefore, of the flexibility and credibility of our strategy for relying on an ASAT weapon for offensive counterspace operations must be made in the context of the condition desired: deterrence and war fighting against the emerging spectrum of potential threats from tier-one, -two, and -three space-capable nations. Before assessing current space control strategy against the emerging threat, one inconsistency regarding our current ASAT policy must be addressed. General Piotrowski stated an ASAT was needed to deter attacks on US space assets. The belief that an ASAT can deter attacks on US satellites did not originate with General Piotrowski; rather it has its basis in the initial argument used by the Air Force to justify an ASAT. According to this argument, the United States is more dependent on space systems than the Soviets and the ASAT will be a strong deterrent against Soviet attacks on US space systems. The inconsistency of this argument lies in the fact that if space systems are actually more important to the United States than to the Soviets, how can threatening Soviet space systems deter an attack on US space systems? This would seem to be analogous to threatening a chess opponent's knight in hopes of deterring him from taking your queen. Rather, the perceived asymmetry between the importance of US and Soviet space systems to their overall war-fighting capability suggests that the threat from the Soviet ASAT could be used to limit US ability to respond in a crisis situation.

Because space systems are becoming increasingly important for successful conventional military operations, the capability to deny critical information and functions from space systems contributes to conventional deterrence and is militarily useful if deterrence fails for other reasons. Of course, the extent to which an ASAT contributes to deterrence depends on the opponent's perception of the importance of his space systems to his ultimate success and the extent to which he believes you have the will to deny him the use of these space systems. It would seem logical to assume that as the space capabilities of nations decrease from tier-one through tier-three, so too does the importance of space to their overall military strategy. Furthermore, as the importance of space systems to a nation's war-fighting capability decreases from tier-one through tier-three, so too does our incentive to use an ASAT weapon. Therefore, it appears that as the space capabilities of a nation decrease across the tiers, the contribution of an ASAT to deterrence also decreases.

The war-fighting utility of an ASAT against the emerging space threat resulting from the proliferation of space technology and products is assessed in the three scenarios that follow. The first scenario looks at a conventional conflict between the United States and another tier-one nation while the second scenario deals with conflict with a tier-three nation. Lastly, the third scenario discusses the utility of the ASAT in conflicts between the US and a tier-two nation.

The first scenario is conventional conflict between the United States and a tier-one space-capable nation. As discussed, the nations currently comprising tier-one are Russia and the United States. In a wartime environment, US and Russian space systems will provide reconnaissance, surveillance, weather, navigation, and mapping/geodesy information as well as provide communication functions essential for combat operations. However, enhancing our terrestrial forces' war-fighting operations is not just a function of how much information can be provided, but also a function of how much information can be denied by the enemy.[80] Consequently, in addition to their extensive dedicated military space systems, Russia also has an operational ASAT weapon that would likely be used to deny critical war-fighting information and functions from our space systems to our national command authorities and theater commanders. It is precisely this scenario that has served as the motivating threat for US space control policy, strategy, and force structure. Clearly, using an ASAT in a conventional war with the Russians to destroy their satellites appears to provide the most reliable means of denying critical military information and functions from space systems.

The second scenario, conflicts with a tier-three space-capable nation, represents the most likely type of conflict we may face in the future. Tier-three space-capable nations are those nations that do not actually possess a space capability but receive satellite information from tier-one or tier-two nations either by direct purchase or by operating licensed satellite ground stations. Regardless of how tier-three nations receive their space information, third-party satellite imagery and surveillance can affect US national security.[81] The war-fighting utility of an ASAT in a conflict with a tier-three nation may be limited because of the political consequences of using an ASAT. These consequences can be illustrated by considering the situation where the United States is engaged in a limited war with a tier-three nation licensed to operate a SPOT ground station. In this situation it is extremely difficult to envision the United States using an ASAT to destroy a French SPOT satellite. First, in accordance with the outer space treaty, attacking a nation's satellites is an act of war. It is unlikely that the United States would commit a unilateral act of war against France over SPOT imagery. Second, an attack on a SPOT satellite would likely result in some sort of retaliation. Retaliation could range from political and economic sanctions involving France and other European countries to some sort of military retaliation. Politically the European community could deny port call privileges, deny overflight, or cancel status-of-forces agreements for forward-based US forces in Europe. Militarily, France could choose to broaden its support or even enter the conflict against the United States. France could also consider executing a response-in-kind option by exploiting the inherent ASAT capability of their strategic ballistic missiles. Any military benefit of attacking a SPOT satellite, therefore, would seem to be overshadowed by the associated risk of conflict escalation.[82]

The third scenario involves the use of an ASAT against a tier-two nation. Second-tier space-capable nations have little or no dedicated military space systems and rely primarily on their civilian space systems for war-fighting information and functions.[83] In addition, most tier-two nations currently do not have a dedicated ASAT capability and do not present a significant threat to orbiting US space assets.[84] The use of an ASAT to destroy a second-tier space-nation's satellite in a conflict situation falls in a gray area. On one hand, similar to the first scenario, destroying a satellite providing information and services to an enemy during war would seem justified with the ASAT being the most reliable means of ensuring the denial of information and those services. On the other hand, most tier-two nations typically sell the data from their satellites on the commercial market to other nations. Therefore, in this scenario, destruction of the satellite not only denies the enemy information and services, but also denies all the licensed operators of foreign ground stations and their customers. The time and cost to reconstitute this capability may result in long-term economic retardation, not only for the tier-two nation, but also the users of the satellite data as well. Economic hardships, coupled with some preexisting political instability, could lead to increased regional instabilities and potential hostilities directed against the United States. This would seem to imply that although the destruction of a civilian satellite may be militarily prudent, the long- and short-term impacts on nonbelligerent countries could result in intolerable political consequences.

Traditional Space Control Methods and Strategy Summary

On the surface, the current space control strategy emphasizing the employment of ASAT weapons might seem viable. However, after assessing this strategy in the context of the existing space threat and the emerging space threat from the proliferation of space technologies and capabilities, there appear to be some weaknesses.

First, although ASATs contribute to our overall conventional deterrent capabilities, the extent they contribute seems to diminish across the threat spectrum. As the space threat decreases from a tier-one to a tier-three nation, the contribution of an ASAT to conventional deterrence also decreases.

Second, regardless of the inherent military utility a civilian satellite may possess, the military benefits of destroying a civilian satellite must be weighed against the potential political backlash created by intentionally targeting and destroying a nonmilitary system.

As Gen Donald J. Kutyna, another former commander in chief of US Space Command, inferred, enhancing terrestrial force operations through offensive counterspace operations is a function of how much information can be denied the enemy.[85] This reinforces the notion that the actual threat from space systems is the information they provide and not the space systems themselves. However, in accordance with our policy, doctrine, and strategy, the stated goal of offensive counterspace operations is to achieve supremacy over the environment (space) to deny the enemy the use of space through the destruction of his space-based assets. This appears to shift the focus away from the information and functions space systems provide, and leads one to focus only on the destruction of the orbiting asset.

The military utility of an ASAT appears to depend on political and military factors limiting the feasibility of destroying satellites. The current focus of offensive counterspace operations on space supremacy through an ASAT seems to lack the flexibility and responsiveness needed to deny potential enemies information across the spectrum of conflict scenarios. This would suggest we refocus our space strategy away from space supremacy and the denial of space for enemy use to a strategy based on the denial of information.

Counterspace Operations for Information Dominance

Before discussing offensive counterspace operations in support of information dominance, an understanding of the strategic objectives of an information dominance strategy is in order.

As presented previously, information dominance should be thought of as a state in which a nation possesses a higher degree of understanding of an adversary's military, political, social, and economic strengths, weaknesses, interdependencies, and centers of gravity, while denying the same information on friendly sources of national power to the adversary.[86] Military actions directed against the enemy should be undertaken with the strategic objective of delaying, disrupting, and denying information used by the enemy leadership for the effective execution of military strategy. The objective is to convince the enemy of his inability to execute his military strategy successfully. Therefore, in an information dominance strategy, the strategic center of gravity is the enemy leadership, both military and civilian, that relies on information to execute the national military strategy. In essence, the end game is to coerce the enemy by increasing his uncertainty regarding his ability to successfully execute his military strategy.

In modern warfare, space systems will be the strategic and tactical eyes and ears of a nation's national security establishment. Therefore, controlling space is essential to achieving information dominance. In an information dominance strategy, however, the objectives of space control must be viewed in a different context. Currently, as outlined in AFM 1-1, the objective of space control is to gain space supremacy or control over the environment of space. The nature of this objective has, historically, tended to focus offensive counterspace operations on the destruction of the satellite in space. Space control under an information dominance strategy, on the other hand, seeks control over the information or products space systems provide. An objective of this nature recognizes that space systems are distributed weapon systems, consisting of three segments: an orbital segment, a ground segment, and a link segment, connecting the orbital and ground segment together and disseminating the information to military and civilian leadership.[87] Controlling the information from space systems can be accomplished by attacking any of these segments and does not necessarily involve the physical destruction of equipment or facilities. The operational objective of offensive counterspace operations for information dominance, therefore, is to delay or deny an enemy's capability to collect, process, and disseminate information by disrupting or destroying, as required, the enemy's space systems.

Operational Concept

Since information dominance can create uncertainty regarding the focus and thrust of the theater campaign, offensive counterspace operations should normally precede other theater operations. To attain information dominance, offensive counterspace operations should use a combination of lethal and nonlethal weapon systems to attack the operational center of gravity of a space system. Depending on the space system, enemy, and level of conflict, the center of gravity can be located in any of the three segments of an enemy's space system.

Operational centers of gravity in the orbital segment of an enemy's space system can be the entire satellite or the satellite subsystems critical for mission performance. This implies a satellite does not have to be destroyed to prevent it from accomplishing its mission. Rather, permanently or temporarily damaging or disrupting vital satellite subsystems can prevent satellites from effectively accomplishing their mission. Examples of vital subsystems include satellite attitude control sensors, mission sensors, uplink/downlink antennas, and power generation systems.

The center of gravity in the link segment is the communications link, the radio frequency used to pass information to and from the satellite. Since most satellites rely on uplinked command and control information from the ground for station keeping, payload management, and satellite health and status functions, attacking a satellite's uplink during critical commanding periods could seriously degrade mission performance. The effectiveness of electronic jamming, however, is limited because of line-of-sight restrictions and increased satellite autonomy; therefore, attacking the downlink, rather than the uplink, is usually easier and more reliable at disrupting a space system. Since the satellite downlink telemetry contains the mission information and health and status information on the spacecraft and the satellite's sensor, successfully attacking the downlink directly attacks information flow and, therefore, has a more immediate effect on achieving information dominance.

The centers of gravity in the ground segment include satellite launch facilities, command and control facilities, and processing stations (airborne, sea-based, and fixed or mobile land-based). All parts of the ground segment are vulnerable to attack from various means such as clandestine operations, air attack, and direct ground attack.

Weapons for Offensive Counterspace Operations

What type of technology is needed to conduct offensive counterspace operations for information dominance? Historically, doctrine and policy addressing space control has focused primarily on the hard-kill technologies to destroy orbiting satellites. Other technologies, however, can be used to achieve offensive counterspace objectives without physical destruction of the orbiting satellite. Nondestructive soft-kill (e.g., mission-kill) technologies can permanently disable the satellite without destruction while nonlethal technologies can achieve nonpermanent space-system mission degradation and disruption. The specific technologies used for offensive counterspace operations can be grouped according to the segment they are targeted against: orbital, link, or ground.

Offensive counterspace weapons used to attack the orbital segment of a space system usually fall into two technology categories: kinetic energy and directed energy. Kinetic energy is a hard-kill technology causing physical destruction of the orbiting satellite. Weapons based on kinetic energy employ projectiles that can be launched into space to destroy orbiting satellites through the shock of impact. There are various types of kinetic energy ASAT weapons: exploding fragmentary warheads, guided nonexplosive warheads that collide with satellites, and space mines. The benefit of using a kinetic energy ASAT weapon is the high probability or certainty of denying the information from the attacked satellite. The disadvantages, on the other hand, include a lack of plausible deniability regarding the reason the satellite failed and the originator of the attack.

Perhaps the most flexible of the technologies used for offensive counterspace weapons is directed energy. Directed energy weapons can be employed to achieve a destructive hard kill, a nondestructive soft-kill, or a nonlethal temporary disruption or degradation. Examples of directed-energy weapons are lasers and high-power microwave weapons. Lasers use electromagnetic radiation (light) for either lethal or nonlethal attacks on satellites.[88] Depending on their power, lasers can damage, disrupt, or destroy a satellite by overheating its surface, puncturing the outer surface of the spacecraft to expose internal equipment, or by blinding critical onboard mission or control sensors.[89] Ground-based lasers, such as the Russian laser at Sary Shagan, are estimated to have a satellite hard-kill capability up to 400 km and a soft-kill capability up to 1,200 km.[90] Another directed-energy technology that can be used for offensive counterspace operations is high-power microwave. High-power microwave weapons employ radio frequencies to damage satellite electronics. Unlike kinetic energy and some types of laser attacks, high-power microwave weapons achieve satellite subsystem failure rather than vehicle failure.[91] Intelligence estimates suggest it is possible to construct a microwave radiation weapon today with a satellite soft-kill capability of about 500 km. In addition, microwave radiation at lower power levels can be effectively used for satellite jamming.[92] There are several advantages of using directed-energy weapons against the orbital segment in offensive counterspace operations. First, directed energy attacks take place at the speed of light, therefore, the result of the attack is near instantaneous, thereby minimizing the effectiveness of enemy defenses. Second, there is plausible deniability associated with soft-kill and nonlethal satellite attacks. Potential adversaries may not have the capability to detect the nature, nor the source, nor whether a hostile action actually occurred. Hence, plausible deniability can be useful in politically sensitive situations. Third, the desired results can be tailored from nonpermanent disruption and degradation to permanent degradation and destruction.

The link segment, as mentioned earlier, consists of the electromagnetic energy used for space system uplink, downlink, and in some cases a crosslink. Given that the link segment is made up of electromagnetic energy, the primary technology used to attack the link segment is electronic warfare. There are two ways of using electronic warfare to attack the link segment: jamming and spoofing. Jamming is essentially transmitting a high-power, bogus electronic signal that causes the bit error rate in the satellite's uplink or downlink signals to increase, resulting in the satellite or ground station receiver's losing lock.[93]

Attacking the link segment by spoofing involves taking over the space system by appearing as an authorized user, such as establishing a command link with an enemy satellite and sending anomalous commands to degrade its performance.[94] Spoofing is one of the most discrete and deniable nonlethal methods available for offensive counterspace operations.[95]

Offensive counterspace operations directed against the ground segment include all offensive actions directed against a satellite launch complex, satellite command and control facilities, and satellite ground processing stations. The ground segment is vulnerable to all types of terrestrial attacks from special operations to strategic attack with gravity bombs. While the ground segment is the most vulnerable segment in a space system, it may also represent the higher political and military risk. Typically, ground segments for space systems are distributed within the enemy's homeland to reduce single point failures and to reduce their vulnerability to attack. In addition, high development costs associated with dedicated military space systems and rapidly advancing commercial technology possessing inherent military utility has resulted in an increase of dual use (military/civilian) space systems. Therefore, in many tier-two and tier-three space-capable nations, ground segment targets are usually located near urban areas susceptible to collateral damage and civilian casualties. Although susceptible to all forms of direct attack, it may be more politically acceptable and less risky militarily to attack ground segment targets with highly accurate precision munitions in discriminating attacks.

In the final analysis the available technologies for conducting offensive counterspace operations appear flexible and responsive; however, the employment options are situation dependent.

Offensive Counterspace Options

As discussed, the biggest drawback of our current offensive counterspace strategy is that there are some conflict situations in which destroying an enemy's satellite with an ASAT is not an attractive or realistic option. However, an information-dominance strategy has as a primary objective the delay or denial of information; therefore, employment options for offensive counterspace operations can exist for all threat nations, at all conflict levels, against all segments of a space system.

Employment options for conducting offensive counterspace operations in an information-dominance strategy are influenced by three major variables: the threat (e.g., tier-one, -two, or -three), the level of conflict (e.g., peace, crisis, or war), and the segment of the space systems to be attacked (orbital, link, or ground). Figure 3 illustrates how options for offensive counterspace operations can be viewed discretely depending on the combination of variables the situation represents. Depending on the threat and the level of conflict, employment options for offensive counterspace operations applicable to the three segments of a space system can range from "no option" at the low end of the spectrum, to ASAT attacks against the satellite or

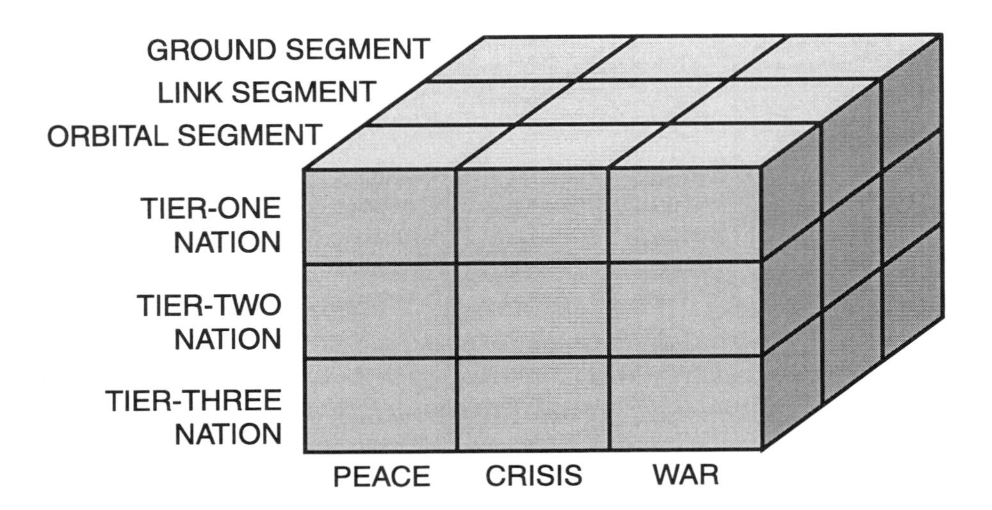

Figure 3. Offensive Counterspace Options

strategic attack against the ground station at the high end of the spectrum.

Examples of suggested offensive counterspace employment options for tier-one, -two, and -three space-capable nations are shown in figures 4, 5, and 6. Although an information-dominance strategy provides our military planners with greater flexibility for conducting counterspace operations, examination of figures 4, 5 and 6 reveals two trends shaping offensive counterspace operations. First, as the level of conflict moves from peace to war within a tier group, the different segments of a space system subject to attack increases and the level of acceptable violence of the attack also increases. For example, figure 5 shows that during a crisis the orbital segment of a second-tier nation could be attacked with nonlethal disruption weapons whereas during war, the orbital segment could be attacked by either hard- or soft-kill mechanisms.

Second, as the threat from space decreases across the tier groups from tier one to tier three, the conflict threshold for attacking space systems segments increases while the level of acceptable violence of the attack decreases. This is illustrated by comparing the available options for attacking the orbital

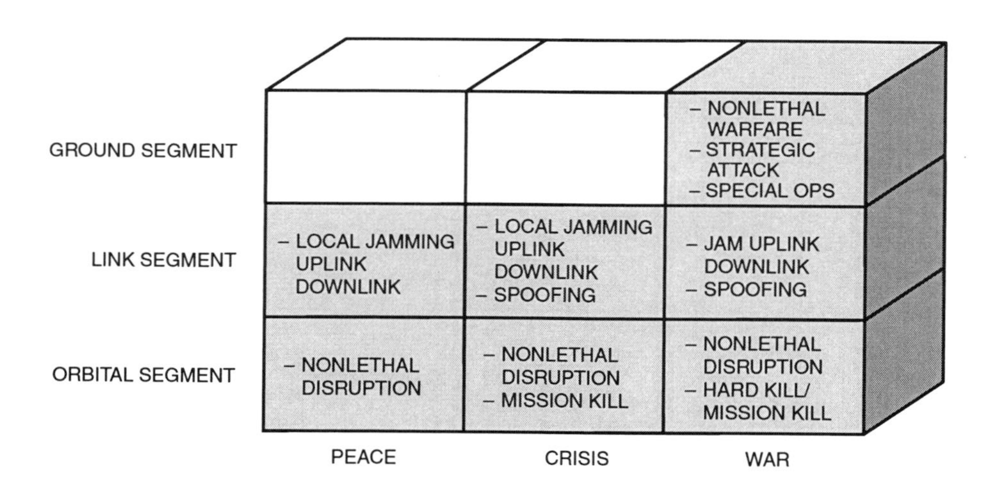

Figure 4. Offensive Counterspace Options for Tier-One Nations

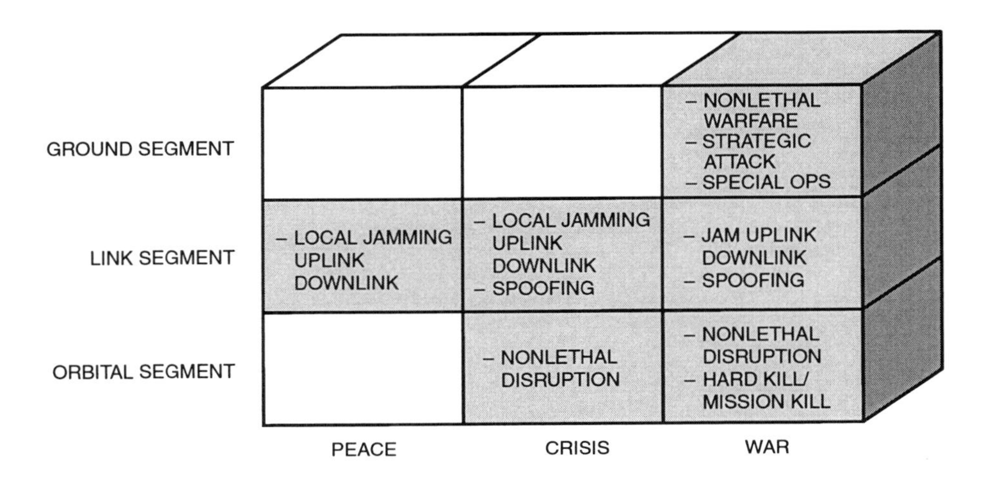

Figure 5. Offensive Counterspace Options for Tier-Two Nations

segment during war on both figures 4 and 6. In no case is attacking an orbital segment of a tier-three nation with hard- and soft-kill mechanisms viewed as being politically acceptable, whereas it would be against a tier-one nation.

The ability to delay and/or deny information from space systems, at all levels of conflict, permits the establishment of information dominance during peacetime and its sustainment through crisis and war. Determining options for offensive counterspace operations for information dominance can be illustrated in the following scenario. The potential for a crisis exists between the United States and a tier-three space nation with a licensed SPOT ground station. If a crisis erupts, the US wants to be prepared with a rapid show of force in the theater of operations and has, therefore, issued a warning order to preposition forces. To ensure secrecy, the theater commander has requested offensive counterspace operations be conducted to deny the enemy nation information from the SPOT system that could reveal the force mobilization. As shown in figure 6, the only available option for offensive counterspace operations during peacetime is electronic warfare against the link seg-

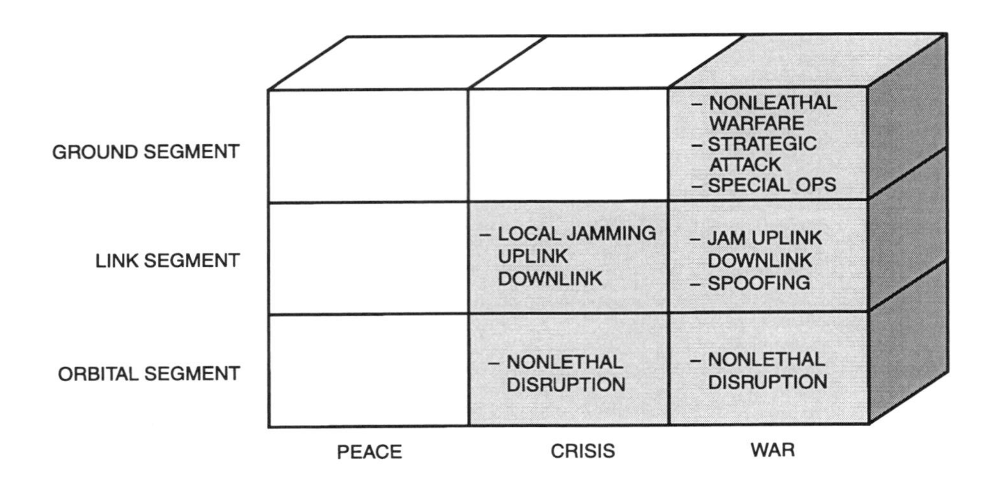

Figure 6. Offensive Counterspace Options for Tier-Three Nations

ment. If the situation escalated to a crisis, or to war, the options for counterspace operations would expand and eventually span all segments of the space and cut across the spectrum of violence from nonlethal to lethal soft-kill, to lethal hard-kill.

Summary

Information dominance strategy as an alternative to the current space control strategy has several advantages. First, because the strategy focuses on the denial of information rather than the denial of the environment, the link and ground segments of the space system correctly reemerge with an increased relevance to offensive counterspace operations. This total systems approach has essentially increased operational flexibility of offensive counterspace operations by increasing the operational centers of gravity that can be targeted. Second, the total systems approach, coupled with a philosophy that satellite destruction is no longer essential, has resulted in an increase of available technologies for offensive counterspace operations. Options for employing existing capabilities such as nonlethal directed energy, electronic countermeasures (ECM), and precision-guided munitions seem more politically viable than the destructive ASAT, which in the past has been questioned by many within Congress. Finally, the increased number of space system targets subject to attack, coupled with the ability to employ a broader assortment of lethal and nonlethal technologies, creates options for employing offensive counterspace operations across the spectrum of conflict.

Offensive Counterspace Operations in Support of the Theater Campaign

Traditionally, offensive counterspace operations have been synonymous with an ASAT capability employed to deny the medium of space through the destruction of the enemy's orbiting satellites. However, the United States does not currently have a capability to destroy satellites on orbit, and judging from the political opposition to such weapons, is not likely to get one. Indeed, even if the United States had an operational ASAT capability, situations exist in which the attack and physical destruction of an adversary's satellite is not politically desirable. If the United States is to deny the enemy critical war-fighting information from satellites, it must adopt an offensive counterspace strategy capable of defeating the enemy's space order of battle within existing political constraints.

Conducting offensive counterspace operations with an objective of attaining information dominance does, however, offer an alternative strategy for controlling space information to the operationally limited strategy of space supremacy. Offensive counterspace operations under an information-dominance strategy center on delaying and denying the information and support, space systems provide by disrupting or destroying, as required, targets within the orbital, link, and ground segments of the enemy's space system. Consequently, a total systems approach to targeting, encompassing the link and ground segments, in addition to the orbital segment, is employed.

Implementing an offensive counterspace strategy based on information dominance in support of a theater campaign requires the resolution of two major issues: organizational responsibility for implementing an information-dominance strategy and the need for a comprehensive space order of battle for the emerging threat from space. The first issue to be resolved is that of organizational responsibilities, or more specifically, who is responsible for developing and implementing the offensive counterspace strategy.

The Unified Command Plan assigns the responsibility of the space control mission to the commander in chief United States Space Command (USCINCSPACE); however, the issue of who is responsible for developing and implementing the strategy for offensive counterspace operations in support of a theater campaign would seem to be driven by the responsibilities of the supported commander vis-à-vis the supporting commander. According to Joint Pub 5-02.1, *Joint Operations Planning System* (JOPS), vol. 1, the supported commander is responsible for coordinating and synchronizing war-fighting activities of the supporting commander's military forces in conjunction with his own forces.[96] In addition, the supported commander normally has the authority to designate the objectives and the timing and duration of the supporting commander's actions within the theater.[97] The supporting commander, on the other hand, is responsible for determining the needs of the supported force and taking actions to fulfill them by providing forces and/or developing a plan for supporting the supported commander.[98] Although the supported commander has the authority to determine objectives for the supporting commander, assigning the plan development function to the supporting commander would suggest the responsibility for strategy development and implementation also rests with the supporting commander. According to Joint Pub 0-2, *Unified Action Armed Forces,* the supporting force gives support or operates in support of another force—the supported force. Because of their war-fighting role in theater campaigns, space forces are normally designated supporting forces. Consequently, USCINCSPACE, as the supporting commander, would be responsible for developing the theater plan for counterspace operations in support of the supported CINC's objectives.

Although USCINCSPACE is responsible for implementing offensive counterspace strategy within the theater, there remains no one within the theater specifically identified for integrating counterspace operations into the theater campaign plan. The emergence of space power as a potentially decisive war-fighting capability in the aftermath of Desert Storm provides some incentive to identify an individual or organization responsible for integrating counterspace operations into the theater campaign plan.

One alternative would be to create a Joint Forces Space Component Commander (JFSCC) responsible to the Joint Force Commander (JFC). Since Congress chose not to assign the space warfare mission to any single service, but rather to the unified command US Space Command (USSPACECOM), the organizational relationship of the JFSCC to the other services and the unified command for space is not clear. Realistically the JFSCC should be some sort of USSPACECOM element reporting directly to the JFC, and conceptually be similar to a subunified command. One major problem with the JFSCC concept, however, is that as a component command, the forces assigned to the JFSCC would normally be under the operational command of the JFC. However, the operational command of USSPACECOM space forces will not chop to the JFC.[99] Therefore, the JFSCC would essentially be a facilitator or coordinator with USSPACECOM for the surveillance, reconnaissance, communications, and weather support requirements of the theater component forces. Although facilitating and coordinating the space requirements into the theater campaign is an important function, creating a new component command, led presumably by a general officer, to perform coordination activities that could be performed by existing staff elements, seems to be a misappropriation of resources.

With respect to counterspace operations, the JFSCC would coordinate between USCINCSPACE and the JFC and component commanders to ensure the space control strategy is consistent with the overall theater strategy and the counterspace operations are integrated into the theater campaign plan. In this capacity the JFSCC would have a role similar to that of the US Transportation Command liaison, who, also has no forces assigned.

Another, and perhaps more dependable, alternative for offensive counterspace operations in support of a theater campaign would be to establish a space planning and operations cell under the JFACC. One potential organization capable of assuming the planning function of offensive counterspace operations for information dominance would be Air Force Space Command's (AFSPACECOM) Forward Space Support in Theater (FSST) team.

The objective of the FSST team is to provide regional CINCs space expertise to facilitate the near-term theater-level integration of air and space.[100] FSST teams are currently assigned to Air Force component commands to assist in developing operations plans (OPLAN), training, and ensuring integrated space support.[101] While the primary focus of the FSST teams currently centers around the force-enhancing attributes of space forces, adding counterspace operations responsibilities appears feasible. Given the propensity for offensive counterspace operations to be conducted by air and electronic warfare forces, subordinating the space planning and operations cell to the JFACC would appear to facilitate the integration of the enemy space order of battle into the overall air operations planning effort and the resulting air tasking order (ATO).

The second issue to resolve for offensive counterspace operations in support of the theater campaign is the requirement for a comprehensive space order of battle for potential enemies. The space order of battle required to support offensive counterspace operations for information dominance must have the same total systems approach as the targeting philosophy. The enemy's order of battle for the orbital segment of a space system includes information such as ephemeris, subsystem vulnerabilities, maneuverability, sensor configuration, and periods of natural disruption such as solar interference, satellite eclipse, and proximity operations. Ground segment order of battle information would include information such as the locations of ground stations and control facilities, the existence of mobile ground stations, and ground station vulnerabilities (such as electrical power). Likewise, order of battle information on the link segment would include information on the number of up/downlinks, frequencies, and antijam/encryption capabilities. In addition to the information relating to the physical attributes of a space system, space order of battle should also include such operational information as how the system is used, an assessment of its potential contribution to the enemy's overall military strategy, system reconstitution capabilities, and periods of critical commanding. The existence of a comprehensive space order of battle will facilitate the integration of offensive counterspace operations into the theater operations plan and inclusion of space order of battle targets into the ATO and electronic warfare plan.

Integral to USCINCSPACE's responsibility for planning counterspace operations within the theater is the task of developing and maintaining the space order of battle for the threats from space. Currently, USSPACECOM's Space Defense Operations Center (SPADOC) is responsible for developing and maintaining the space order of battle with data provided from the Joint Space Intelligence Center and the Space Surveillance Center. Because of the cold war legacy imprinted on our space control strategy, space order of battle is oriented on the Soviet space threat and focuses primarily on the orbiting satellites and includes information such as the satellite function, configuration, orbital parameters, and overflight predictions. However, since an information-dominance strategy focuses on attacking the entire space system, the level of effort needed to develop and maintain a space order of battle for counterspace operations appears to exceed the current capabilities of the SPADOC.

As space technology proliferates, the need for a US strategy to exercise control over potentially threatening space systems increases. Basing our offensive counterspace operations on a strategy of information dominance seems to be a logical approach for determining the focus of a space control campaign. Even though the United States has no dedicated operational ASAT capability to provide the lethal, hard-kill options, there are many operational weapon systems that possess inherent capabilities for lethal soft-kill, or nonlethal counterspace applications.

It is increasingly clear that space capabilities are becoming more decisive in the outcome of war. In the current political environment, there is a need to be more creative and innovative in approaches to solving national security problems. Information dominance represents a different approach for confronting the threat from multilateral space capabilities and for viewing the objectives of the space control mission.

Notes

1. House, Committee on Appropriations, *Department of Defense Appropriations for 1985,* 98th Cong., 2d sess., pt. 2, 192.

2. Gen John L. Piotrowski, to Congressman William Dickinson, letter, 7 July 1989.

3. Lt Col Andrew D. Krepenivich Jr., *The Military Technical Revolution, A Preliminary Assessment* (Washington, D.C.: Office of the Secretary of Defense, Office of Net Assessment, July 1992), 3.

4. Ibid., 14.

5. Ibid., 22.

6. Col Gordon Middleton, USAF, "Information-based Warfare of the 21st Century," Air Force Blue Ribbon Panel on Space, November 1992, 2.

7. Krepenivich, 22.

8. "Building Space Power for the 21st Century," Air Force Blue Ribbon Panel on Space, unpublished white paper, 4 November 1992.

9. "The New Hierarchy in Space," in *Commercial Observation Satellites and International Security*, ed. Michael Krepon, et al. (New York: St. Martin's Press, 1990), 19.

10. James T. Hackett and Robin Ranger, "Proliferating Satellites Drive US ASAT Need," *Signal*, May 1990, 156.

11. Thomas Mahnken, "Why Third World Space Systems Matter," *Orbis*, Fall 1991, 577.

12. Col Richard Szafranski, USAF, *GEO, LEO, and the Future* (Maxwell Air Force Base [AFB], Ala.: Air University Press, August 1991), 2.

13. Hackett and Ranger, 155.

14. Sylvia Maureen Williams, "International Law and the Military Uses of Outer Space," *International Relations*, May 1989, 413.

15. Ibid., 413.

16. Mahnken, 564.

17. Mary Umberger, "Commercial Observation Satellite Capabilities," in *Commercial Observation Satellites and International Security*, ed. Michael Krepon, et al. (New York: St. Martin's Press, 1990), 12.

18. Leonard S. Spector, "Not So Open Skies," *Space Policy*, February 1990, 16.

19. Judy Collins, Earth Observation Satellite Corp., telephone interview with author, 26 May 1993.

20. Ibid.

21. Jeffrey T. Richelson, "Implications for Nations Without Space-based Intelligence Collection Capabilities," in *Commercial Observation Satellites and International Security*, ed. Michael Krepon, et al. (New York: St. Martin's Press, 1990), 59.

22. Hugh DeSantis, "Commercial Observation Satellites and their Military Implications: A Speculative Assessment," *Washington Quarterly*, Summer 1989, 186.

23. Forcast International/DMS Market Intelligence Report, *Forecast International*, March 1993.

24. Ibid.

25. Ibid.

26. Spector, 14.

27. Forecast International/DMS Market Intelligence Report, *Forecast International*, March 1993.

28. Marcia S. Smith, "Military and Civilian Satellites in Support of Allied Forces in the Persian Gulf War," *CRS Report for Congress*, 27 February 1991, 7.

29. Spector, 14.

30. Krepon, 19.

31. DeSantis, 189.

32. Ibid., 189.

33. Jeffrey T. Richelson, "The Future of Space Reconnaissance," *Scientific American*, January 1991, 42.

34. "Chinese Space Program Sets Aggressive Pace," *Aviation Week & Space Technology*, 5 October 1992, 48.

35. Lt Col Bret Watterson, USAF, "Proliferation of Remote Sensing Systems," Office of the Assistant Secretary of the Air Force, briefing, 24 December 1991.

36. Ibid.

37. Ibid.

38. "Chinese Space Program Sets Aggressive Pace," 48.

39. Richelson, "The Future," 42.

40. Ibid.

41. Ibid.

42. Ann M. Florini, "The Opening Skies: Third Party Imaging Satellites and US Security," *International Security*, Fall 1988, 107.

43. Robin Riccitiello, "Radar Imagery Sales Grow at Slow Pace," *Space News*, 30 November–6 December, 1992, 6.

44. Elizabeth Corcoran and Tim Beardsley, "The New Space Race," *Scientific American*, July 1990, 76.

45. Ibid., 75.

46. Watterson.

47. Mahnken, 573.

48. "Spot a New Way to Win," advertisement, *Defense Electronics*, November 1988, 68.

49. The Russian Resurs satellite was initially developed as a dedicated military system. Umberger, 11.
50. Florini, 98.
51. Ibid., 101.
52. Ibid., 95.
53. Krepon, 18.
54. Richelson, "Implications for Nations," 55.
55. William J. Broad, "Non Super Powers Are Developing Their Own Spy Satellite Systems," *New York Times*, 3 September 1989.
56. Richelson, "Implications for Nations," 55.
57. US Space Command, *United States Space Command Operations Desert Shield and Desert Storm Assessment* (U) (Secret), 43. Information extracted is unclassified.
58. Watterson.
59. Ibid.
60. Ibid.
61. Ibid.
62. US Space Command, 45.
63. Ibid.
64. Ibid., 46.
65. Vincent Kiernan, "Gulf War Led to Appreciation of Military Space," *Space News*, 25-31 January 1993, 3.
66. Earl D. Cooper and Steven M. Shaker, "The Commercial Use of Space," *Defense and Diplomacy*, July/August 1989, 35.
67. DeSantis, 189.
68. *National Security Strategy of the United States*, August 1991, 1.
69. Paul B. Stares, *Space and National Security* (Washington, D.C.: Brookings Institution, 1987), 73.
70. Air Force Manual (AFM) 1-1, *Basic Aerospace Doctrine of the United States Air Force*, vol. 1, March 1992, 6.
71. Ibid.
72. Ibid., 11.
73. *US National Space Policy,* 16 November 1989, 10.
74. Department of Defense, *Report to the Congress on Space Control* (U), July 1989, iv (Secret). Information extracted is unclassified.
75. Ibid.
76. Gen John L. Piotrowski, letter to Congressman William Dickinson, 7 July 1989.
77. Maj Steven R. Petersen, USAF, *Space Control and the Role of Antisatellite Weapons* (Maxwell AFB, Ala.: Air University Press, May 1991), 69.
78. AFM 1-1, vol. 1, 10.
79. Ibid.
80. Gen Donald J. Kutyna, CINC US Space Command, testimony to Senate Armed Service Committee, 23 April 1991.
81. Ann M. Florini, "The Opening Skies: Third Party Imaging Satellites and US Security," *International Security*, Fall 1988, 122.

82. Stares, 122.
83. France is perhaps the current exception with the Helios satellite.
84. This is recognition of the fact that nuclear-capable nations with a ballistic missile or space launch program possess an inherent ASAT capability. Nuclear ASATs, however, have limited operational utility due to the persistence of residual electromagnetic pulse known as the Argus effect.
85. Kutyna, "Spy Satellite Systems," *New York Times*, 3 September 1989.
86. Krepinevich, 8.
87. Stares, 74.
88. Ibid., 75.
89. Ibid., 76.
90. LCDR James S. Green, USN, *US Plan for Space Control* (Maxwell AFB, Ala.: Air War College, 3 March 1989), 7.
91. Ralph Zirkland, *Evaluation of Soviet Exploitation of CM/CCM in Space Warfare*, Defense Technical Information Center, Defense Logistics Agency (U), November 1983 (Secret). Information extracted is unclassified.
92. Maj William O'Dell, *Application of Electronic Countermeasures in the Defense of US Space Objects* (U) (Maxwell AFB, Ala.: Air Command and Staff College, May 1973) (Secret). Information extracted is unclassified.
93. *Electronic Warfare Threat to US Communication Links* (U), Defense Intelligence Agency (Secret). Information extracted is unclassified.
94. Ibid.
95. Stares, 82.
96. Joint Pub 3-14, "Joint Doctrine: Tactics, Techniques, and Procedures (TTP) for Space Operations," final draft, 15 April 1992, I–17.
97. Joint Pub 3-14, II–15.
98. Joint Pub 3-14, I–17.
99. AFSC Pub 1, *The Joint Staff Officer's Guide*, 1991,1–34.
100. AFM 2-25, "Air Force Operational Doctrine: Space Operations," initial draft, April 1993, 20.
101. AFM 2-25, 20.

Bibliography

Air Force Blue Ribbon Panel on Space, "Building Space Power for the 21st Century." Unpublished white paper, November 1992.

Air Force Manual (AFM) 1-1, vol. I. *Basic Aerospace Doctrine of the United States Air Force*, March 1992.

AFM 2-25. "Air Force Operational Doctrine: Space Operations." Initial draft, April 1993.

Air Force Specialty Code Pub 1. *The Joint Staff Officer's Guide 1991*. Norfolk, Va.: Armed Forces Staff College, n.d.

Augustine, Norman. "The American Military Space Program in Changing Times." *Space Times*, September–October 1992.

Berkowitz, Mark J. "Future U.S. Security Hinges on Dominant Role in Space." *Signal*, May 1992.

Broad, William J. "Russia is Now Selling Spy Photos from Space." *New York Times*, 4 October 1990.

———. "Non Super Powers are Developing their own Spy Satellite Systems." *New York Times*, 3 September 1989.

Brown, George E., Congressman. "Future of the Landsat Program." *Space Times*, March–April 1992.

Chandrashekar, S. "Missile Technology Control and the Third World: Are There Alternatives." *Space Policy*, November 1990.

"Chinese Space Program Sets Aggressive Pace." *Aviation Week & Space Technology*, 5 October 1992.

"Chinese Fen Yun 1B Satellites Transmit High Resolution, True-Color Images of Earth." *Aviation Week & Space Technology*, 19 November 1990.

Cooper, Earl D., and Steven M. Shaker, "The Commercial Use of Space." *Defense and Diplomacy*, July/August 1989.

Corcoran, Elizabeth, and Tim Beardsley. "The New Space Race." *Scientific American*, July 1990.

DeSantis, Hugh. "Commercial Observation Satellites and Their Military Implications: A Speculative Assessment." *Washington Quarterly*, Summer 1989.

Edwards, David T. "Commercial Remote Sensing." *Space Times*, March–April 1992.

Florini, Ann M. "The Opening Skies: Third Party Imaging Satellites and U.S. Security." *International Security*, Fall 1988.

"Global Reach-Global Power." Department of the Air Force, White Paper, June 1990.

Green, James S., LCDR, USN. *A U.S. Plan for Space Control.* Newport, R.I.: Naval War College, 3 March 1989.

Hackett, James T., and Dr. Robin Ranger. "Proliferating Satellites Drive U.S. ASAT Need." *Signal*, May 1990.

Hough, Harold. "Eyes In the Sky: Satellite Surveillance for the Masses." *Soldier of Fortune*, May 1991.

Hunter, Roger C., Maj, USAF. "USAF Policy for a Multi-Polar World." Maxwell AFB, Ala.: School of Advanced Airpower Studies, May 1992.

Johnson, Nicholas, L. "Space Control and Soviet Military Strategy." *Defense Electronics*, May 1988.

Kiernan, Vincent. "Gulf War led to Appreciation of Military Space." *Space News*, 25-31 January 1993.

Kingwell, Jeff. "The Militarization of Space: A Policy Out of Step with World Events." *Space Policy*, May 1990.

Krepon, Michael, et al. *Commercial Observation Satellites and International Security.* New York: St. Martin's Press, Carnegie Endowment for International Peace, 1990.

Lay, Christopher D. "Space Control Predominates as Multi-Polar Access Grows." *Signal*, June 1990.

Levy, Louis J. and Susan B. Chodakewitz. "The Commercialization of Satellite Imagery." *Space Policy*, August 1990.

Mahnken, Thomas G. "Why Third World Space Systems Matter." *Orbis*, Fall 1991.

Middleton, G., Col, USAF. "International Space Issues." Staff study, 29 October 1992.

———. "Information-Based Warfare of the 21st Century." Staff study. 29 October 1992.

"Military Technical Revolution," Staff study. Office of the Secretary of Defense.

O'Dell, William, Maj, USAF. "Application of Electronic Countermeasures in the Defense of US Space Objects." Research paper. Maxwell AFB, Ala.: Air Command and Staff College, May 1973.

Pearson-Mackie, Nancy. "The Need to Know: The Proliferation of Space-based Surveillance." *Arms Control*, May 1991.

Petersen, Steven, R., Maj, USAF. *Space Control and the Role of Space Weapons*. Maxwell AFB, Ala.: Air University Press, 1991.

Piotrowski, John, Gen, USAF, to Congressman William Dickinson. Letter. 7 July 1989.

Pollack, Herman. "International Relations in Space: A U.S. View." *Space Policy*, February 1988.

Power, John W., Capt, USAF, "Space Control in the Post Cold War Era." *Airpower Journal*, Winter 1990.

"Report to the Congress on Space Control." Department of Defense, July 1989.

Riccitiello, Robin. "Radar Imagery Sales Grow at Slow Pace." *Space News*, 30 November 1992.

Richelson, Jeffrey T. "The Future of Space Reconnaissance." *Scientific American*, January 1991.

Smith, Marcia S. "Military and Civilian Satellites in Support of Allied Forces in the Persian Gulf War." *CRS Report for Congress*, 27 February 1991.

Spector, Leonard S. "Not So Open Skies." *Space Policy*, February 1990.

"SPOT A New way to Win." Advertisement, *Defense Electronics*, November 1988.

Stares, Paul. *Space and National Security*. Washington, D.C.: Brookings Institution, 1987.

Szafranski, Richard, Col, USAF. "GEO, LEO, and the Future." Maxwell AFB, Ala.: Air University Press, August 1991.

Watterson, Brett, Lt Col, USAF. "Proliferation of Remote Sensing Systems: Considerations and Implications." Briefing, Office of the Assistant Secretary of the Air Force (Space), 24 December 1991.

Williams, Sylvia Maureen. "International Law and the Military Uses of Outer Space." *International Relations*, May 1989.

Zimmerman, Peter. "Remote Sensing Satellites, Superpower Relations and Public Diplomacy." *Space Policy*, February 1990.

Zirkland, Ralph. *Evaluation Of Soviet Exploitation of CM? CCM in Space Warfare.* Defense Technical Information Center, Defense Logistics Agency, November 1983.

Chapter 6

When the Enemy Has Our Eyes

Cynthia A. S. McKinley

On 17 January 1991, the United States entered a war that turned the military space community upside down. Until then the military space community's focus was locked on the strategic concepts that were developed and refined throughout the cold war. The Gulf War expanded that focus to include the operational and tactical levels of warfare. This change is causing space strategists to consider a broader spectrum of space functions for enhancement, and perhaps most importantly a broader spectrum of measures for space control.

In addition to this expanded focus, the reconnaissance satellite playing field continues to undergo significant changes. During all but a few years of the cold war, there were only two players in the spy satellite game. This was slowly changing toward the end of the cold war. At the time of its invasion of Kuwait, the Iraqi military was receiving support from the Soviets and purchasing satellite imagery from the French. Soon after the invasion, the Soviets joined many other nations in their condemnation of the Iraqi government's behavior and the French refused to sell imagery products. This left the United States in possession of a temporary monopoly on the ability to routinely and unobtrusively probe the enemy's battlefield with highly accurate reconnaissance satellites. Those space assets revealed volumes about the Iraqi capabilities and intentions for battle. The United States assured its Gulf War victory through the combined strengths of its overwhelming offensive power and its unprecedented knowledge of the battlefield. As the world watched this display, it quickly

This work was accomplished in partial fulfillment of the master's degree requirements of the School of Advanced Airpower Studies, Air University, Maxwell AFB, Ala., 1996.
Advisor: Maj Bruce M. DeBlois, PhD
Reader: Dr Karl Mueller, PhD

learned that future warfare success may require a similar illumination of the battlefield.

As the lessons of the Gulf War are being internalized, national and international actors are endeavoring to participate on the high ground of space reconnaissance. The movement to gain access to high-quality satellite photoreconnaissance data has turned into a stampede in only four years. For a nation such as France that has been in the photoreconnaissance business for nearly a decade, this stampede is enabling it to move a rung higher on the international competitive ladder. For Russia it represents an opportunity to regain stature and much-needed wealth. It also shows the world that Russia remains a superpower in the space business, one of the most prestigious of all fields for national pride.

Combining the modified space operations focus and the multipolar space systems playing field, the next war is likely to differ from the Gulf War. Indeed, in the next war, it is likely that the enemy will have our eyes. The United States must be prepared to pursue active space control measures to deny the enemy's access to critical reconnaissance information. However, this problem cannot be solved out of context; the space control mission does not stand alone. It is shrouded in nearly 40 years of history. Furthermore, space control must be achievable within the constraints of current and future international environments. Space control's history and environments need to be unraveled to reach an understanding of how the United States can execute space control in the contemporary world. This monograph provides information that may be helpful to future space strategists and decision makers in determining how to accomplish this mission.

This work integrates research, analysis, and synthesis to take the reader through the study's three subdivisions of the past, the future, and the challenge. Each subdivision offers unique information to help the reader understand the space community's focus during the cold war and how that is changing, and to place the space control mission in its context before attempting to offer space control methods.

Part 1, "The Past," recounts the rise of strategic space intelligence, explains the revolution brought about by digital image processing technology, and elaborates on the changes resulting from the employment of space's strategic assets in modern theater warfare. Part 2, "The Future," speculates on the forms of modern warfare and imagery's potential role in them. Part 3, "Meeting the Challenge," discusses the space control mission and various denial methods that will be considered for employment against the commercial reconnaissance system.

Part One: The Past

The Rise of Strategic Space Intelligence

Strategic space intelligence is one of the first products of the cold war. Today, it remains one of the United States military's most important assets. Its formative years were molded by three themes: competition to lead the nation's space program, the strategic nuclear threat posed by the Soviet Union, and the technological challenges of the new frontier.

Planting the Seed. The evolution of America's space-based reconnaissance systems traces to the conceptual seed planted by Wernher von Braun in May 1945. Von Braun, developer of the V-1 and V-2 rockets for Nazi Germany, is credited with reuniting Adolph Hitler's Peenemuende rocket team to form the nucleus of America's civilian and military space programs. Using the knowledge he gained from his rocketry work, Von Braun provided a report to the United States Army that examined German views on the potential of rocket-launched satellites.[1] This seed quickly grew into an inter- and intraservice rivalry that drove the Army, Navy, and Army Air Forces into a competition to become the agency responsible for future military satellite vehicles. By October 1945, the Navy had published its views on the use of satellites. Already behind the power curve, Maj Gen Curtis E. LeMay, director of research and development for the Army Air Forces, commissioned the RAND Corporation to conduct a three-week crash study on the feasibility of space satellites.[2] General LeMay and Gen Carl A. Spaatz, commanding general of the Army Air Forces, quickly realized that this new frontier was another mission area that could help justify the formation of an independent air force.[3]

Thus, Army Air Forces involvement, along with the intense interservice rivalries, encouraged this little-understood domain to become a fertile arena for the competitive exchange of ideas.

During the ensuing years, the scientific and military communities studied the feasibility and operability of potential satellite systems. With both strong proponents and opponents arguing the potentials and limitations of such technological challenges, the research and development path was by no means smooth. Despite these difficulties, by 1951, the Air Force was able to define its requirements for an operational satellite system. There were three primary requirements for an Air Force satellite system: (1) an ability to produce photography of sufficient quality to enable trained interpreters to identify objects such as harbors, airfields, oil storage areas, large residential areas, and industrial areas; (2) a capability to provide continuous daytime observation of the Soviet Union, cover its land mass in a matter of weeks, and record the data collected; and (3) an ability to produce a quality photographic product suitable for the revision of aeronautical charts and maps.[4]

During these early days of concept exploration and requirements definition, many agencies worked independently without the benefit of oversight. This changed in December 1953 when the Air Research and Development Center gathered many of the proliferating aspects of the research and development groups into a single project entitled Project 409-40. Project 409-40's mandate was to provide the first operational imagery satellite system. The prospective satellite system for this project was given the weapons system designation of WS-117L.[5] The satellite was to be based on state-of-the-art television and videotape recorder technology. However, its engineers soon realized that the 144-foot resolution that this system could provide was inadequate for the task. This technological problem fueled the skepticism and hostility of many Department of Defense personnel who doubted that such systems could ever be of value. But the believers persisted, due in part to President Dwight Eisenhower's vision and his determination to gain information on the Soviet Union's nuclear weapons delivery vehicle capabilities.

The Technological Capabilities Panel formed by President Eisenhower in 1954 provided a briefing in February 1955 on the options for obtaining intelligence data about the Soviet Union. The panel included such notables as Massachusetts Institute of Technology president James R. Killian Jr., Polaroid founder Edwin H. Land, Harvard astronomer James G. Baker, and Washington University's Joseph W. Kennedy.[6] These academic and industry leaders advised President Eisenhower that there were three options for gaining photoreconnaissance data on the Soviet Union: build strategic reconnaissance aircraft, attempt balloon reconnaissance, or develop a satellite reconnaissance system.[7] Supporters of satellite systems hoped the committee would recommend the satellite solution as the top priority, but the committee's official recommendation was to build strategic reconnaissance aircraft.

Not swayed by the committee's focus on near-term solutions, the Air Force quickly issued General Operational Requirement Number 80. Issued less than a month after the committee's report to the president, this document established an official requirement for an advanced reconnaissance satellite.[8] By November 1955, the basic technical tasks were defined and approved and the project was given the code name Pied Piper. Pied Piper's goals were to provide a complete satellite reconnaissance system, including ground facilities for analyzing and disseminating imagery, and to be fully operational by the third quarter of 1963. Three corporations competed for the rights to build this visionary project: Radio Corporation of America, Glenn L. Martin, and Lockheed Aircraft.[9] By October 1956, the Air Force had made the contract award decision. Lockheed was notified to proceed with its development of an advanced reconnaissance satellite as well as the upper stage Agena vehicle that would propel the satellite into low Earth orbit.[10]

The Threat: Soviet Strategic Nuclear Attack. Work on the WS-117L project progressed at a steady pace until the Soviets shocked America with their launch of sputnik on 4 October 1957. This unsettling event shook the foundations of the military and scientific communities, the government, the population of the United States, and helped consolidate the communities' work toward meeting the challenge and threat posed by the Soviets. On 22 January 1958, the National Security Council issued directive number 1846, assigning the highest priority status to the development of an operational reconnaissance satellite.

By February 1958, space experts were briefing President Eisenhower on the two potential imagery acquisition methods using space platforms. One was the original method proposed in Project 409-40, that is, using a film-scanning technique, and the other used a film and satellite recovery method. President Eisenhower decided that the film and satellite recovery system offered hope of immediate payoffs and decided to assign program development responsibilities to the Central Intelligence Agency (CIA). Several factors led to these decisions. President Eisenhower was concerned that the Pied Piper nonrecoverable technology would not yield an operable satellite as quickly as needed, was not enthusiastic about an Army role in space, was concerned about security failings, and had confidence in the CIA's ability to lead the program because of its experience with the secret development of the U-2 airborne imagery collection system. Thus, at the February 1958 meeting, President Eisenhower approved the infamous Corona project. The Corona system was designed to quickly provide an operational spy satellite through development of a recoverable capsule system. The CIA's marching orders were to have the system ready for use by the spring of 1959.[11]

The cover for the Corona program was the Discoverer satellite program. Additionally, the government established a military research and development agency, the Advanced Research Projects Agency, to handle the public aspects of the project. The portions of the WS-117L project that pertained to reconnaissance satellites were canceled and restarted in the highly secretive world of the CIA under the Corona cover. The Air Force was tasked only with the responsibility of testing techniques for recovery of a capsule ejected from an orbiting satellite. After a February 1959 launch failure and the Soviet recovery of a capsule launched in April 1959, the CIA's Corona project met with success in 1960. It was in that year that American space experts successfully launched and recovered two film capsules. By 1961, the CIA's film recovery program was stable and provided regular imagery of the Soviet Union.

Using state-of-the-art equipment, the CIA secretly acquired imagery of great military significance throughout the 1960s. The imagery met the specifications laid out in 1951 by the Air Force, and more importantly, could identify exactly what the Soviets were accomplishing in the strategic nuclear arena. This program and its follow-ons were deemed highly successful at providing high-quality photographic imagery for the United States until the program was superseded in October 1984.

Film-Based Solutions Today, Electro-Optics Tomorrow. The decision to pursue the film-based recovery system was a prudent decision considering the technological capabilities of electro-optics in the late 1950s. Eventually however, electro-optical technology would evolve to the point where its product would match that of the film-based systems and surpass the latter's ability to provide near-real-time intelligence data. Believing this to be true, the many proponents of electro-optical systems continued to develop and refine this emerging technology.

Although it may not have been viewed this way in the 1960s, what appears to have emerged is a dual-track technology progression. One track was the logical short-term solution and the other was the long-term method for providing cold war strategic intelligence. Figure 7 provides an analysis and synopsis of this dual-track technology progression.

Despite official cancellation of all Air Force satellite activities except recovery techniques, work on the onboard film scanning system continued to challenge its proponents. The earliest available evidence that anyone was pursuing digital processing technologies for military application appears in a 1957 report. In its report to the Air Force, Radio Corporation of America recommended—and the Air Force accepted—the idea of a combined film and digital-based system.[12] This plan called for using a film scanning technique in which a conventional camera photographed the target and the film was developed on board. Once developed, the film was scanned with a fine light beam and the resulting signal was sent to a ground receiving station. The ground station translated the signal

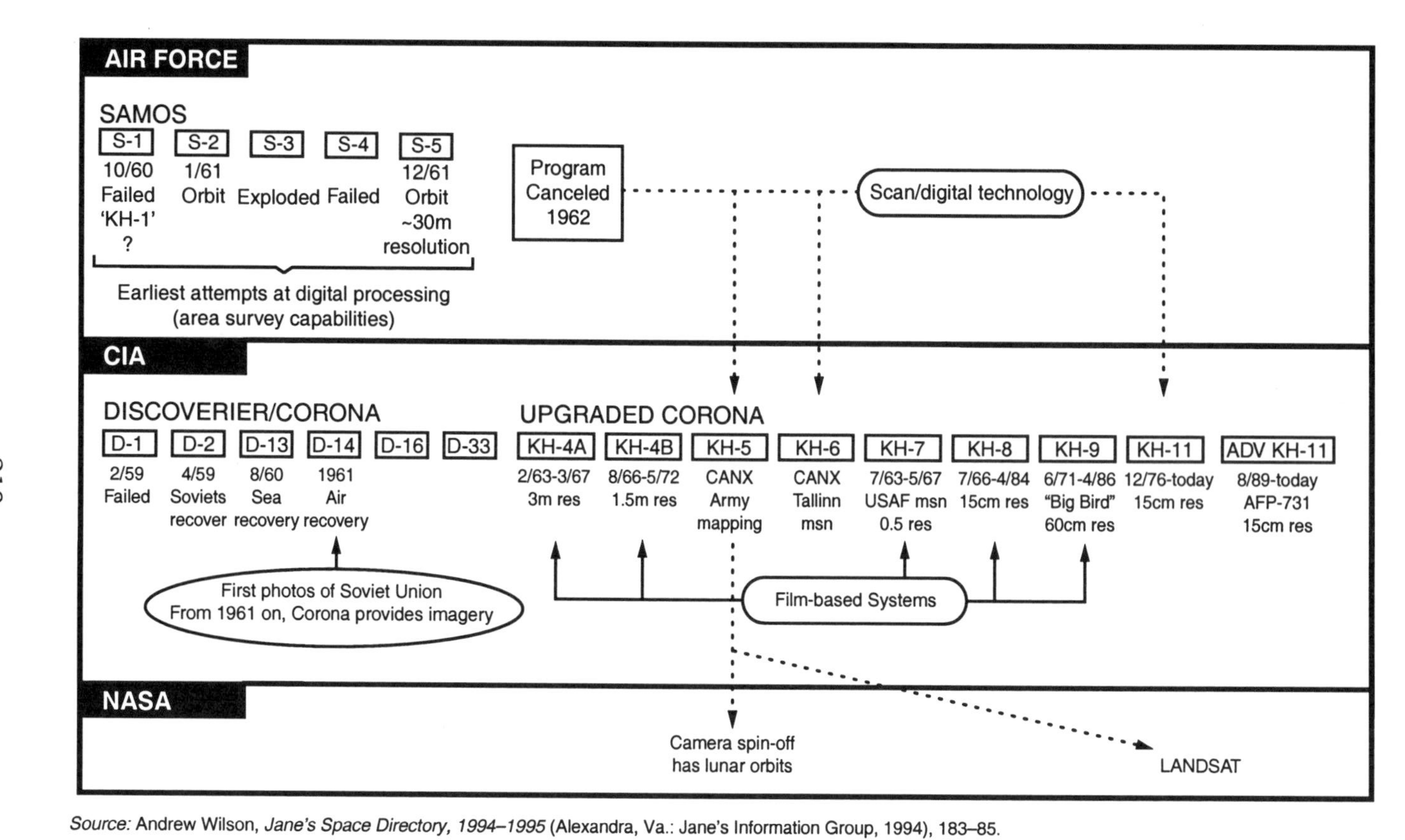

Source: Andrew Wilson, *Jane's Space Directory, 1994–1995* (Alexandra, Va.: Jane's Information Group, 1994), 183–85.

Figure 7. The Dual Track Technology Progression (Film-Based and Digital Processing)

back into an image. The Air Force program that used this technology was the Samos program.[13]

The Air Force's Samos program launched its first satellite in October 1960. After achieving only two successful launches in its five attempts between 1960 and 1962, this program was deemed a failure and officially canceled in 1962. However, this cancellation may have been an attempt to divert the notice of the Soviets and others. Samos-5, the last of the Samos launches, was the most successful and provided imagery resolution in the 30-meter range, not much worse than that provided by the multispectral imagers of today's *Systeme Probatoire pour l'Observation de la Terre* (SPOT) systems.[14] Although officially canceled, its technological advancements reappeared that same year in the CIA's second generation program code named Keyhole. While some of the Keyhole satellites continued to exploit the successful film-based system, others were designed to use the film scanning technique. [15] Although never launched, some development work occurred through the KH-5 Army mapping and the KH-6 Tallinn mission satellites.[16]

It appears that the first successful use of low-quality digital image processing technology occurred at the National Aeronautics and Space Administration (NASA). NASA used the KH-5 film scanning camera on its lunar orbiters, *Ranger* and *Surveyor,* in the mid-1960s and later in *Mariner 4, Mariner 9,* and *LANDSAT.*[17] When Mariner 4 was launched in 1964, it was advertised as using the first all-digital imaging system. Seven years later, Mars became the first planet to be mapped entirely from digital remotely sensed data. The use of digital image processing technology for lunar and planetary exploration continued throughout the 1970s with the launches of *Pioneer, Viking,* and *Voyager* series spacecraft. By 1972 NASA was ready to apply the technology to earth remote sensing and on 23 July launched the first *LANDSAT* satellite. *LANDSAT* was the first American spacecraft to provide multispectral imagery.[18]

By 1976, with 15 years of space reconnaissance work behind them, America's imagery network was established and performing well. It was about to become even better. America's space strategists and scientists were about to elevate the satellite imagery program to an advanced technological plateau.

While they comfortably sat on the successes of the film-based imagery systems, analysts realized that NASA's digital imaging systems were small indicators of the CIA's spy satellite work. In the bicentennial year America succeeded in turning the tables on the Soviet Union. In 1976 America also rattled the bear's cage by launching the first KH-11 reconnaissance satellite into its near-polar orbit.

The Digital Imagery Revolution

By 1976 the United States and the Soviet Union were experts at the orbital cat-and-mouse game of satellite reconnaissance. Both nations used film and satellite recovery systems and routinized their operational procedures. Despite eloquent protests about spy satellites' violation of national sovereignty, both nations acquiesced to the Open Skies policy.[19] Both nations realized that these systems provided insights into each other's strategic nuclear activities and thus provided some stability in a dangerous world.[20]

Soon after celebrating its two hundredth anniversary, the United States launched a satellite that revolutionized the photoreconnaissance business and set the stage for its later use in operational- and tactical-level warfare. For a few short years, the United States operated alone on this plateau of technological achievement. However, achievement breeds imitation. The Soviet Union and France soon developed similar systems.

Charge Coupled Devices and Computers: Keys to the KH-11. The KH-11 satellite launched on 19 December 1976 was the first photoreconnaissance satellite to provide high-quality non-film-based imagery. The KH-11's real-time sensing systems and high-resolution charge coupled device (CCD) cameras enabled it to distinguish military from civilian personnel.[21] The infrared and multispectral sensing devices of the later models can locate missiles, trains, and missile launchers by day or night, and can distinguish camouflage and artificial vegetation from living plants and trees. Space analyst Jeffrey Richelson claims the KH-11 is capable of 15-centimeters (approximately six inches) resolution using a mirror at least two to three meters diameter (similar to the Hubble Space Telescope).[22] The launch of the

KH-11 was a significant milestone in the achievement of space-based imagery products and represented a personal triumph for Leslie Dirks, the CIA's deputy director of science and technology.

The KH-11's roots reach back to RAND's 1945 concept of a television-type imagery return system.[23] Realizing that the technology of the 1950s and 1960s was inadequate to provide the near-real-time data that the national reconnaissance community wanted, Dirks continued to believe it would be available in the future. The breakthrough technology by which the KH-11 became capable of collecting and transmitting imagery in real time lay in its use of CCD. The CCD originated at Bell Telephone Laboratories in the late 1960s, when two researchers, William S. Boyle and George E. Smith, sought to invent a type of memory circuit.[24]

For those in the government who had access to the revolutionary digital imagery provided by the KH-11, they realized the significance was obvious and immediate. Although initially limited to data collection for only a few hours each day, a system that could provide near-real-time images of Earth gave decision makers a near instantaneous ability to see exactly what the adversary was doing.[25] For the analysts, this new system released them from the light table and stereoscope.[26] With digital image processing technology, the analysts began using the much more flexible and dynamic medium of computers.

Using computers, the analysts recalled imagery from the database and manipulated it through a variety of viewing options. For example, the analysts changed the contrast to increase the visibility of objects that were in shadows, obscured by haze or thin cloud cover, or photographed with too much or too little exposure.[27] Computers began performing the task of object detection. Changes in a particular target area were determined using a technique known as electronic optical subtraction. Among the other computer advantages were the ability to improve the image resolution and the ability to delete distortions inherent in photographic systems.[28]

The American Monopoly. From 1976 to 1982, the United States was the only nation utilizing digital image processing technology in its reconnaissance satellites. Combining this technology with its older film-recovery systems and airborne platforms such as the U-2 and SR-71, America's ability to acquire strategic intelligence surpassed that of any other nation.[29] A few of the important bits of strategic intelligence data that these systems provided were nuclear weapons developments and tests, adherence to arms control agreements, locations of strategic and tactical aircraft, troop deployments, and military construction.

The United States' monopoly on digital image processing technology crumbled in 1982, when the Soviets launched their fifth-generation reconnaissance system. With this system, the Soviets followed the Americans in liberating themselves from reliance upon the film-recovery system. Their fifth-generation satellite offers 20-centimeter resolution, nearly the same as the KH-11.[30] The speed with which the Soviets were able to bridge the technology gap with the Americans is probably explained by the several thefts of KH-11 documents that occurred shortly after the first KH-11 was launched.[31]

Several copies of the specifications for the KH-11 system appeared in the Soviet Union in the late 1970s. The first arrived through William Kampiles, a Greek-American who began working for the CIA in 1977.[32] Unhappy with his pay, tedious work, and unglamorous watch tours, Kampiles resigned from the CIA after less than a year and journeyed to Greece in 1978. Packed in his suitcase was a copy of number 155 of the *KH-11 System Technical Manual.*[33] Once in Greece, Kampiles approached a Soviet Embassy official and offered to provide American intelligence documents. Although he requested $10,000 for the KH-11 document, Kampiles received a mere $3,000 for the technical manual that opened the door to one of America's greatest technological achievements.[34]

Aided by America's technology secrets, the Soviets were ready to launch their first digital imagery satellite system on 28 December 1982.[35] Analysts know little about this first all-digital Soviet satellite. In fact, some analysts still question whether it actually represents the Soviets' first attempt to use digital processing. Their first undisputed use of digital technology occurred with the launch of Cosmos 1552 on 14 May

1984. Collection systems have not detected signals from this or subsequent fifth-generation satellites, so analysts believe that the data is retrieved by way of Molniya or geosynchronous communications satellite links.[36] Russia continues to use its Generation 5 satellites today and has apparently developed a Generation 6 follow-on to this initial successful use of digital processing technology.

The French Go Commercial. France, one of the five acknowledged nuclear powers, joined the digital image processing world only four years after the Soviets. Unlike its American and Soviet predecessors, French entry into this domain occurred in the commercial marketplace. The French government began the SPOT program in 1978 and first exploited digital image processing technology satellites with its launch of SPOT-1 in 1986. The SPOT system does not offer the high resolution of its military counterparts; it provides 20-meter multispectral and 10-meter panchromatic resolution imagery. Also unlike its American and Soviet counterparts, the French government did not attempt to underwrite all of SPOT's developmental costs. From its inception, national and international government and private firms have participated in the program. Over a dozen French, Belgian, and Swedish agencies had a stake in the success of SPOT-1. Today, the expansive SPOT Imagery Corporation provides imagery to customers on every continent.

The French entry into the commercial exploitation of digital image processing technology could have signaled the beginning of the transformation of photoreconnaissance imagery to operational and tactical use, but it wasn't until the United States needed such data in a regional war that the military space community began to realize that a fundamental transformation was under way.

Transformation: Strategic Intelligence in Theater Warfare

Throughout the cold war, space-based strategic intelligence enhanced global stability by enabling governments to monitor crises and watch for remote nuclear weapons tests.[37] Observation satellites monitored possible threats to the regimes established by the 1963 Limited Test Ban Treaty, the 1970 Nuclear Non-Proliferation Treaty, and the 1972 Strategic Arms Limitation Treaty. These treaties played an active role in monitoring the 1971 Indian-Pakistani war, the 1973 Arab-Israeli war, and the Iran-Iraq war of the 1980s. Utilizing their strategic eyes, the superpowers kept watch over turmoil in many theaters.

However, it wasn't until the Gulf War of 1991 that America's strategic eyes were actively integrated into every phase of theater warfare.[38] This integration was, and is necessary for the modern battlefield commander to monitor today's expanded theater of operations. Just as telescopes once provided extended vision to the horse-mounted commander, reconnaissance satellites help modern commanders control, manage, and coordinate simultaneous operations over thousands of square miles. Because of the immense complexities of modern warfare, the orbiting remote sensing systems provide critical information that helps the commander achieve success. Aware of these modern warfare demands, it is now easy to see that in the fall of 1990, the most secretive strategic intelligence program in America's unknowingly sat on the doorstep of radical change.

We Have No Maps! When the coalition forces were deployed to the Persian Gulf region, the maps of Kuwait, Iraq, and Saudi Arabia were old and out of date.[39] To correct this deficiency, multispectral imagery satellite systems were used to prepare precise maps of the Gulf area. Multispectral images were used to show features of Earth that exceed human visual detection. With the ability to provide seasonally adjusted battlefield maps, the multispectral imagery analysis identified land cover, healthy and stressed vegetation, soil boundaries, soil moisture content, fording locations, and potential landing or drop zones. These images also allowed analysts to identify shallow water areas near the coastline and earth surface areas in which spectral changes had occurred. With this information, military operational personnel gleaned data that would help achieve military victory. Desert Shield and Desert Storm engineers had valuable data that enabled plans for military airfield construction; Marines knew which areas were best for amphibious assault; land forces could monitor enemy operations; and air attackers could examine attack routes, verify target coordinates, and identify potential landing zones.

One of the great values of Desert Storm's multispectral imagery was its use for aerial combat mission planning and operations. It was combined with other Defense Mapping Agency (DMA) databases and used by pilots to display attack routes and targets as they should appear at flight and attack altitudes. Prior to the air campaign, the military electronically overlaid SPOT images of Iraq on digital terrain maps for mission rehearsals. Additionally, these images were displayed in the Mission Support Systems (MSS) vans deployed in the theater. The MSS heralded the first in-theater use of mobile downlink stations.[40] These units permitted processing and analysis of data by battlefield intelligence units. For combat operations, imagery was a standard part of target folders, and aircrews expected its uninterrupted availability. When reviewing their tasking orders, aircrews wanted and expected to see a picture of every target.[41]

Examples of the use of SPOT imagery in the air campaign include both destructive and constructive applications. The imagery was a key element in the rapid planning and launch of a successful F-111 attack on a single building in Kuwait City to eliminate key elements of the Iraqi military leadership.[42] The SPOT panchromatic imagery closely resembles the resolution and visual appearance of infrared targeting displays.[43] Thus, the images were helpful during flight operations. F-117A stealth aircraft pilots carried the imagery from the onset of hostilities. The SPOT pictures helped them attack such targets as the Iraqi air defense operations center, ministry of defense, intelligence center, and other high priority targets.[44] To assist in the Scud hunts, SPOT imagery was used to identify terrain or man-made features where Iraqi missile launchers might hide.[45]

Equally important, the SPOT imagery helped avoid the loss of civilian lives by identifying the locations of mosques, hospitals, schools, and residential areas. Attack angles for specific weapons were calculated so that bombs or missiles with long- or short-range impact had the least chance of causing collateral damage.[46] On at least one occasion, SPOT imagery assisted in the rescue of a downed F-16 pilot.[47] Rescue mission planners used the images to examine the topography of the area where the pilot ejected. They made judgments about where he would likely go based on seeing the same topography from ground level. During the rescue operation, the imagery was used to guide forces to the area.

With its low-resolution quality, SPOT's main contributions came from its ability to provide bathymetric, hydrographic, and terrain categorization in support of air, naval, and ground combat operations. In short, this exceptional view of the territory and composition of the land and waterways gave coalition forces an unprecedented insight into the environment.[48]

The Soviets' Views. The Soviets were impressed by America's space abilities in theater warfare.[49] As a provider of much of Iraq's war equipment, the soviets were dismayed that space-based reconnaissance, systems detected—and smart weapons quickly destroyed—much of Iraq's modern equipment. Despite the coalition's success in this area, the Soviets were pleased with the Iraqi *maskirovka* techniques.[50] The effectiveness of Iraqi camouflage techniques drew positive remarks from several Soviet officers.[51] The late Marshal Sergei Akhromeyev commented that Iraqi systems of decoy targets and decoy target groupings caused problems for coalition forces in the first weeks of the air war. General Maltsev speculated that nearly 50 percent of the first Coalition strikes were carried out on false targets because of Iraq's extensive deployment of sophisticated dummy air defense systems.[52] Of even greater significance, Iraq was able to use basic camouflage and dispersal techniques to conceal ballistic missiles, chemical and nuclear weapons equipment, and probably other information as well.[53]

The Uniqueness of the Gulf War. As a commercial resource, SPOT's value in theater warfare has led many to speculate on the threat posed by a future adversary's acquisition of high-quality imagery. One of the unique features of the Gulf War was the broad allied Coalition that included the majority of space-based reconnaissance-capable nations. Although Iraq had procured SPOT imagery prior to its invasion of Kuwait, the French terminated all sales of Gulf-related imagery within days of the invasion.[54] Viewing SPOT as a commercial venture, the board of directors stated their intent to sustain a nonmilitary image. A spokesperson for the corporation stated that the board of directors did not want SPOT to appear to the general public as a company that aggressively follows military developments.[55]

The official statements do not indicate a categorical refusal to allow SPOT to provide imagery during conflict or war situations. SPOT officials have repeatedly reminded the world of the corporation's open access policy and refusal to censor its imagery products.[56] Rather, it was the unique circumstances surrounding the Gulf War that caused the French corporation to temporarily modify its policy. When SPOT has viewed a conflict situation as an opportunity to provide newsworthy imagery, it has readily offered to do so.[57] Thus the unique high level of belligerence and subsequent world condemnation of Iraq's invasion led SPOT officials to refuse to supply imagery and to publicly state that it is not their role to track military forces.[58] Interestingly, their altruism in this situation would have quickly disintegrated if any other imagery agency had decided to provide similar data.[59] At the time, the only other agency that could have made such a decision was the Earth Observation Satellite (EOS) Company that operates *LANDSAT*. According to Phillipe Renault, deputy director-general of SPOT Image, if EOS had sold *LANDSAT* images to Iraq, SPOT Image would have done likewise in the interest of business competition.[60]

As the world approaches the twenty-first century, international economic competition is preparing it for unprecedented access to high-quality imagery data. Thirty years of technological evolution and international competition have significantly altered strategic space intelligence. Its employment has changed and its ownership expanded. Imagery intelligence has emerged from its highly secretive cocoon; it has experienced an enormous technological revolution; and most recently, its value has been applied to the operational and tactical levels of warfare. Having reached the end of this short review of the emergence, development, and transformation of strategic space intelligence and the military space community, this monograph now looks to the future. Part 2 provides a perspective on modern warfare and the context for the use of imagery intelligence data.

Part Two: The Future

The Forms of Modern Warfare

Before attempting to speculate on a future adversary's use of imagery intelligence data in warfare, the strategist needs an understanding of some of the variations of modern warfare. This is critical because strategists must recognize that not all adversaries are the same, nor are many at an evolutionary position similar to that of the United States. Each of the potential adversaries the United States may face occupies its own region on a multidimensional warfare evolutionary scale. Each adversary's technological, organizational, and conceptual capabilities will widely vary. Thus, they cannot be engaged in like manner.

A singular employment strategy will not work against diverse adversaries and should not be blindly pursued. The discussion that follows is a departure from traditional warfare analysis. It is offered as another perspective of the evolution and complexities of modern warfare.

Understanding Warfare. Modern warfare is a multifaceted enterprise, one whose evolutionary complexity has mirrored that of human society. This complexity ensures that humanity's attempts to explain modern warfare are as taxing today as they were for primitive humankind.[61] While primitive humankind grappled with the rudimentary skills that characterized early warfare, humanity must attempt to put its arms around many forms of warfare that include highly technical tools and complex organizational and doctrinal concepts. While no individual can master all of the complexities of modern warfare, those complexities can be described by manageable concepts and frameworks.

Warfare is the human expression of the battle for ascendancy. At its roots lay differences about the desirability of the status quo. Status quo issues may concern territory, power, legitimacy, dominance, ideology, or a host of other topics.

Each entity or actor on the international or national landscape has a variety of tools and methods for preserving or attempting to change the status quo. The international battle for ascendancy remains the purview of a small subset of humanity until one side determines that a core interest, value, or belief is threatened or perhaps that the status quo power is incapable of representing the interests of a subset. While a state of war may be referred to metaphorically very early on (for example, a trade war), the military is accustomed to referring to the existence of a state of war only when it is directed to and becomes engaged in force application against the tools of an opposing force. Once a military force is engaged, there are three possible outcomes: the status quo is changed, the forces languish in stalemate, or there is no change to the status quo. If the group seeking change is victorious, it becomes the guardian of the contemporary status quo. The defeated force then becomes the entity seeking to change the status quo at a later point in time. A diagrammatic interpretation of this concept is offered in figure 8.

The Forms of Modern Warfare. In trying to gain a perspective on this "visible" portion of the warfare spectrum, it becomes apparent that throughout the evolution of civilization, people have improved their war-fighting skills by unlocking technological and cognitive secrets. Using technological advancements as a categorical base, humankind has developed three definable forms of warfare. This categorization is organized by the concept that certain technological advancements have produced significant evolutionary fractures. The fractures have signaled a major change in size of the adversarial group that an entity is able to coerce. The three forms of warfare are primordial, industrialized, and nuclear warfare. Figure 9 provides an overview of two dimensions of this multidimensional framework.[62]

As humanity develops each new form of warfare, it continues to maintain and refine its earlier forms. As the secrets within each form are unlocked, humanity modifies the range of its technical coercive capability. Additionally, each new technical ability challenges humanity to harness that new power, focus it, and exploit it through higher orders of organ-

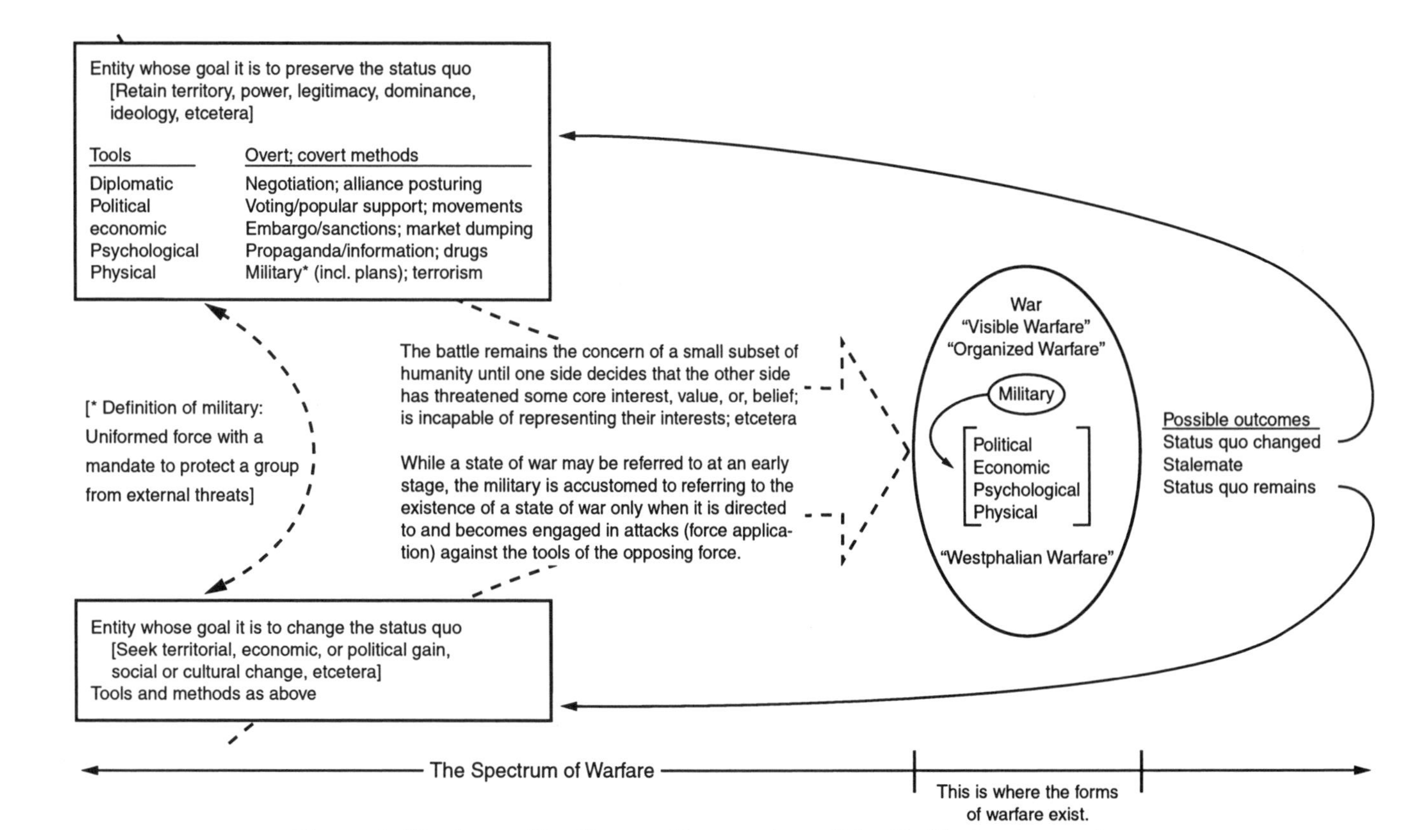

Figure 8. Warfare—The Human Expression of the Battle for Ascendancy

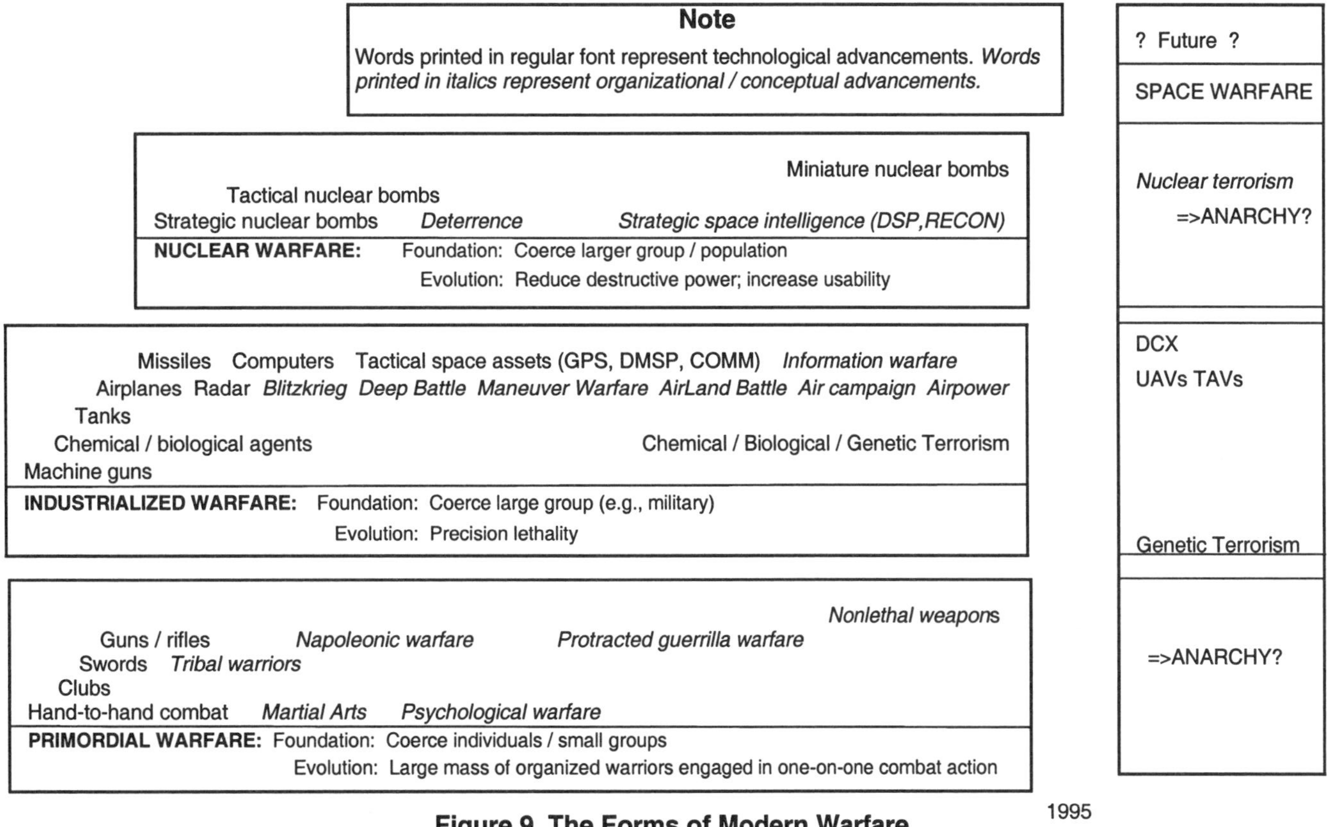

Note

Words printed in regular font represent technological advancements. *Words printed in italics represent organizational / conceptual advancements.*

Miniature nuclear bombs
Tactical nuclear bombs
Strategic nuclear bombs *Deterrence* *Strategic space intelligence (DSP,RECON)*

NUCLEAR WARFARE: Foundation: Coerce larger group / population
Evolution: Reduce destructive power; increase usability

Missiles Computers Tactical space assets (GPS, DMSP, COMM) *Information warfare*
Airplanes Radar *Blitzkrieg* *Deep Battle* *Maneuver Warfare* *AirLand Battle* *Air campaign* *Airpower*
Tanks
Chemical / biological agents Chemical / Biological / Genetic Terrorism
Machine guns

INDUSTRIALIZED WARFARE: Foundation: Coerce large group (e.g., military)
Evolution: Precision lethality

Nonlethal weapons
Guns / rifles *Napoleonic warfare* *Protracted guerrilla warfare*
Swords *Tribal warriors*
Clubs
Hand-to-hand combat *Martial Arts* *Psychological warfare*

PRIMORDIAL WARFARE: Foundation: Coerce individuals / small groups
Evolution: Large mass of organized warriors engaged in one-on-one combat action

? Future ?

SPACE WARFARE

Nuclear terrorism
=>ANARCHY?

DCX
UAVs TAVs
Genetic Terrorism

=>ANARCHY?

1995

Figure 9. The Forms of Modern Warfare
(Categorized by Mankind's Development of Group Lethality Tools)

izational and conceptual abstraction. In some cases, for example, Napoleonic warfare, the warriors' lethality was increased through organizational improvements. For example, in nuclear warfare the owners have attempted to harness the latest destructive tool to make it more useable. The technological, organizational, and conceptual achievements are pursued in the belief that they will elude the adversary and thus provide success in warfare. Without attempting to delve too deeply into the three forms of warfare or reach into other aspects of this multidimensional analysis, a superficial examination is in order.

The foundation of primordial warfare is based on an individual or group's need to coerce individuals or small groups. This form includes hand-to-hand combat and the use of elementary weapons as clubs, swords, and small firearms. Organizationally, its evolution has been expressed through Napoleonic, tribal, and protracted guerrilla warfare. Recent technological developments in this form of warfare seek coercion through the use of nonlethal weapons.

Industrialized warfare is the form of warfare that members of militaries prefer to deal with because it typically concerns forces that resemble themselves and which operate in what is commonly referred to as conventional warfare. The foundation of industrialized warfare is based on an individual or group's need to coerce a larger organized force. It includes all of the nonnuclear tools that industrialized society has created for use in warfare. Examples of such tools include the machine gun, tanks, airplanes, missiles, many of the space assets, and information technologies. The many entities that have gained industrialized warfare capabilities provide extensive variety to this form of warfare. Each has mastered its own unique level of technological, organizational, and conceptual sophistication. Additionally, in this form of warfare, humanity has succeeded in organizing systems of tools into complex and coordinated attack systems. For example, this form includes Germany's concept of blitzkrieg, the United States Army's AirLand Battle doctrine, air campaigns, and emerging concepts of information warfare. In this form of warfare humanity now spends most of its physical, organizational, and conceptual resources in both vertical (advanced technology) and horizontal (advanced organizational and conceptual) development.

The foundation of nuclear warfare is based on an individual or group's need to coerce a very large group or entire population. In this form of warfare, the use of nuclear weapons can quickly transform the objective from coercion to annihilation. This form of warfare so preoccupied politicians and strategists during the cold war that its definitional and organizational complexity quickly overshadowed that of preceding forms of warfare despite its shorter history. Strategic thinkers wrote volumes about deterrence; superpower nations devoted enormous treasures to placing photoreconnaissance and infrared satellites on orbit to locate nuclear weapons and to alert its citizens of their employment. Current developments in this form of warfare may prove to be very difficult to deal with in the near future. Recent efforts in this form of warfare have concentrated on ways to make nuclear weapons less destructive and thus more useful. There are frequent reports that uranium is trickling out of the former Soviet Union. Worse yet, "red mercury," potentially the key ingredient of miniature nuclear bombs, may actually exist despite the skepticism of some experts.[63]

These three forms of warfare (primordial, industrialized, and nuclear) encapsulate modern warfare. Probably the single most important fact to keep in mind is that these warfare forms exist in today's world. Technical, organizational, or conceptual developments within one form of warfare do not negate or supplant the other forms. Humanity merely continues to refine each form to fully exploit the advantages within each. Additionally, an individual or group may combine portions of the forms of warfare or elements within the forms to coerce an adversary. As an example, during the Vietnam War, the North Vietnamese combined portions of industrialized and primordial warfare. Being able to extract from the warfare forms those elements that best fit a country's capabilities and resources provides a great deal of flexibility when seeking to coerce an adversary. Today's modern warfare reservoir offers great variety and complexity depending upon the limitations of the political objectives: the physical, organizational, and conceptual capabilities of the actors and financial resources.

The future warfare bazaar may reveal the transition of current technological breakthroughs as creators of new fractures that enable different forms of coercion. This in turn would modify the framework to include additional forms of warfare. One possibility is that an actor will choose to depart from international agreements and deploy space weapons that are capable of holding the planet hostage. This would definitely cause an evolutionary fracture, adding space warfare as a new form. Another possibility is that the proliferation of emerging chemical, biological, nuclear, or genetic terrorism weapons will reveal themselves as similar transitionary devices. They could become originators of humankind's transition from today's state-ordered system to one dominated by anarchy. On this note, this study turns to look at imagery's role in modern warfare.

Imagery in Future Warfare

How important will imagery be in future warfare? Furthermore, how will the United States respond to an adversary's acquisition of indigenous or commercial space reconnaissance products?

Because modern warfare comes in many variations, the United States must be able to analyze an opponent's physical, organizational, and conceptual capabilities. Among the assets the United States will use to unravel these capabilities are its well-established satellite imagery assets. National and international actors who appreciate imagery's value and see the ease with which it may be acquired will seek to exploit it. With its recent Desert Storm experience and continued technological superiority, the United States military will continue to lead this evolution.

Imagery's Role in Modern Warfare. Imagery's primary value will remain at the strategic level of warfare because of its continuing importance with respect to combating nuclear warfare. In a world where nuclear proliferation is a growing concern, the ability of reconnaissance satellites' to peer into restricted areas will continue to prove that strategic space intelligence is necessary for maintaining peace in a nuclear world.

Imagery's value in industrialized warfare will vary depending upon the adversary's capabilities. For the United States, reconnaissance satellites comprise a portion of the system of systems it uses to gain strategic, operational, and tactical intelligence during warfare.[64] This highly evolved system of systems includes satellites, manned and unmanned aircraft, and surface forces. Potential adversaries' capabilities are less evolved, but could include some combination of these forces being used at all three levels of warfare. The technologically sophisticated actors will have indigenous imagery capabilities and healthy imagery databases; other actors will have purchased imagery and may have similarly healthy databases; some will not have any imagery capability. A few will have achieved capabilities comparable to those of the United States. These nations may own indigenous capabilities or have purchased strategic imagery and be capable of augmenting it with operational and tactical unmanned aerial vehicles.[65] Other adversaries will assume less-evolved stages of development, perhaps they will be able to employ imagery only for general information.

As expected from many industrialized warfare tools, satellite imagery intelligence offers less in primordial warfare than in the other forms. In this form of warfare, it may provide only general strategic intelligence information. More useful media for imagery intelligence in this form of war are airborne strategic and tactical reconnaissance platforms, in particular, unmanned aerial vehicles.

If not critical in the conduct of primordial warfare, imagery intelligence will continue to be one of the most important military uses of satellites in industrialized and nuclear warfare. Its denial to an adversary in those forms of warfare could prove critical for the United States.

Understanding the Challenge. Satellite imagery is no longer the preserve of major powers and specialized units with top secret clearances. Japan, China, India, and Israel have all launched and placed in orbit imaging satellites with varying capabilities. Brazil, Canada, and Great Britain have plans to develop imagery systems. Twenty-meter resolution multispectral and 10-meter resolution panchromatic imagery is commercially available from SPOT Image Corporation; five-meter resolution panchromatic imagery is available from Russia's Soyuzkarta agency. By the year 2000, several corporations will provide imagery of one-meter resolution quality.

SPOT Image Corporation's commercial network extends beyond that of any other supplier and continues to grow.[66] The military value of SPOT imagery during the Gulf War is resulting in millions of dollars in procurements from international military users.[67] As of 1994, SPOT Image Corporation was operating ground receiving stations in 14 countries and selling imagery products on an unrestricted basis.[68] In addition to its current capabilities, SPOT Image plans to upgrade its network by launching SPOT-5 in the year 2000. SPOT-5 will provide five-meter resolution quality data.[69]

Having surprised the world in 1987 when its Soyuzkarta agency announced its intention to begin selling high-quality imagery of the Soviet Union, the Commonwealth of the Independent States (CIS) continues to offer strong competition in the satellite imagery business. Currently, CIS's KFA-100 cameras provide the best commercially available imagery data.[70] Although this imagery is advertised as being of five-meter resolution quality, customers have received imagery assessed at 1.3-meters resolution.[71]

For many years, the United States refused to launch commercial satellites whose resolution was better than 10 meters; however, with the combined pressures imposed by the Soyuzkarta sales and the lack of restrictions on non-US commercial space agencies, it was clear to the US government that the superpower monopoly on high-quality satellite imagery was ending.[72] As a result, early in 1987, the United States announced that it had lifted its 10-meter resolution launch limit.[73] By the end of 1995, a commercial US corporation was providing three-meter resolution imagery. By 2000, two US firms plan to begin offering one-meter resolution quality data. Table 31 provides a summary of the types of imagery data that will be available in the next few years.[74]

Table 31

Planned Imagery Systems, 1995–2000

1995	USA	World View	3m				
	Russia	Almaz-1B (radar)	5m	1996	Japan	ADEOS 1 & 2	8m
	Canada	Radarsat (radar)	10m	1997	USA	Eyeglass	1m
	India	IRS-1C	10m	1999	USA	Space Imaging Inc.	1m
	France	SPOT-4	10m	2000	France	SPOT-5	5m

Source: Berner, Lamphier and Associates, "Many Nations Feed Commercial Imagery Markets," *Space News,* 6–12 March 1995, 9.

With these developments in the remote sensing world, it is likely that future adversaries will own or have access to high-quality imagery data. Iraq's limited access along with SPOT Image Corporation's willingness to restrict its data minimized the risk of exposure of American combat deployments, movements, and battle plans in 1990 and 1991. Additionally, Iraq could not begin to cope with the extent of the coalition's satellite and airborne reconnaissance capabilities. In the Gulf War, these, along with America's other overwhelming capabilities, were the exclusive province of coalition forces.[75] However, with the numerous sources discussed above, the Gulf War may be the last in which the United States holds an overwhelming imagery advantage. It appears certain that in future warfare, the enemy will have our eyes. But what exactly do these eyes give to an adversary?

Militarily Useful Imagery Data. For surveillance data to be useful for military purposes, the resolution quality needs to be 25 meters or better.[76] With 25-meter resolution, an analyst can identify such things as large buildings, road structures, rivers, and lakes. According to Maj Gen William K. James, director of the DMA in 1991, effective military mapping requires a system with a ground resolution ranging from three-to five-meters, five-band spectral resolution, precise metric data, stereoscopic coverage, and broad area collection.[77] This kind of imagery provides the ability to identify, for example, bombers on an airstrip, ingress and egress routes; differentiate between soil types and elevation; and, if provided in digital format, a medium that allows pilots and soldiers access to a volatile display system capable of providing battlefield familiarization.

For terrain analysis or general detection capabilities, low resolution imagery systems work well. For precise equipment identification, the best system is that which provides the highest resolution. If one is viewing, for example, a TU-95 Bear bomber that is 49.5 meters long and has a wing span of 51.2 meters, the aircraft can be detected using the 10-meter resolution provided by SPOT panchromatic imagery. General identification can be attained using the five-meter resolution imagery provided by the Soyuzkarta agency. To begin to see, for example, engine details, the analyst needs the more precise imagery that will be commercially available by the year 2000. If one is viewing a much smaller object, for example, a MiG-29 Fulcrum fighter aircraft, the minimum resolution required for detection is 4.6 meters. For general identification, one needs 1.5-meter resolution. For precise identity, one needs 0.9-meter resolution. For description, one needs 0.15-meter resolution. Table 32 provides a synopsis of the value of imagery of various qualities.

Looking to the future, the US military must assume that its most technologically advanced adversaries will seek to achieve a level of proficiency similar to or better than that achieved by the United States in Desert Storm. The US military continues to analyze its Gulf War successes and failures in the never-ending quest to ensure that US forces are the most capable in the world. Because of the lessons of this war, significant changes are under way in the US military.

The Leading Edge. The unparalleled experience of the United States reconnaissance satellite world and recent Desert Storm experience affords a position at the leading edge of imagery exploitation. The United States has quickly moved to internalize some of the Desert Storm lessons and prepare for future war-

Table 32

Resolution Required for Specific Military Tasks

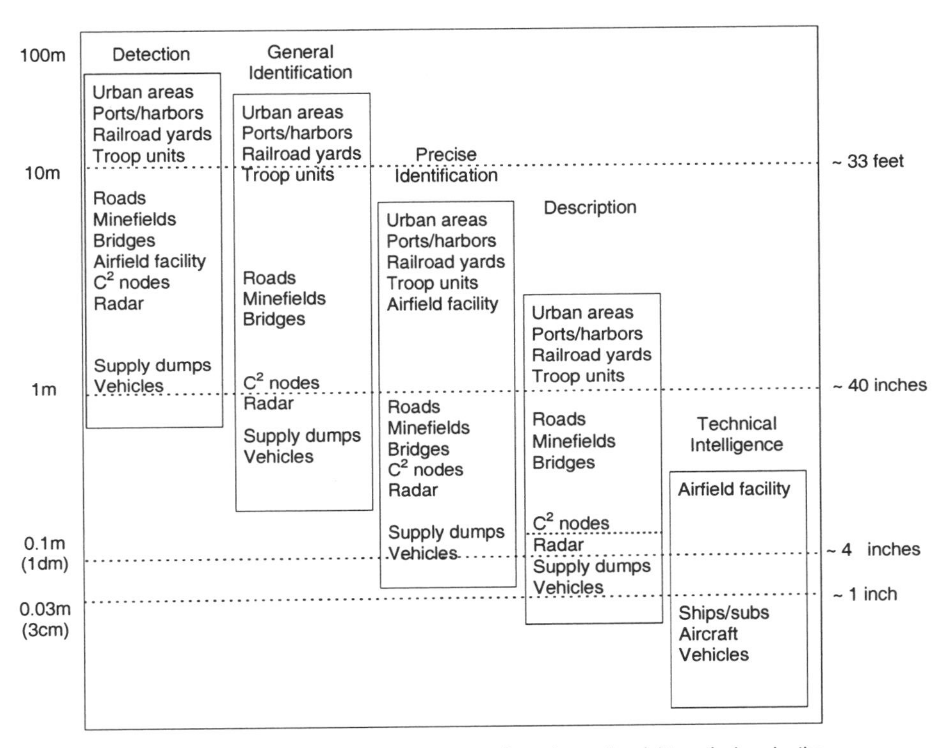

(vertical scale on left lists resolution using the metric scale; on the right vertical scale, the resolution is listed using the English scale)

Source: Capt James R. Wolf, "Implications of Space-Based Observation," *Military Review,* April 1994; and Lyn Dutton et al., *Military Space,* Washington, Brassey's, 1990.

fare. Two exciting developments in this area are the Digital Warrior and Eagle Vision programs.

Digital Warrior allows US combat units to merge intelligence data and computerized mission planning. Using this capability, units can load their mission programs into simulators to practice upcoming missions or into weapons computers to carry out attacks.[78] The Digital Warrior system that enables this uses commercially available desktop personal computers to bring together intelligence data, weapons specifications, and information updates as the mission unfolds.[79]

Both the Air Force and Navy recommended that SPOT imagery become an integral part of a much more operationally oriented space-based imaging reconnaissance capability.[80] The Air Force's Eagle Vision program will allow small mobile ground stations to receive SPOT imagery directly from spacecraft.[81] In many instances, the US military has found that the broad fields of view provided by SPOT imagery were much more useful and available than the narrow fields of view provided by advanced national spacecraft. According to Air Force planners, if the aircrews had been limited to using standard maps, they had approximately a 30 percent chance of destroying a target. Using SPOT data, the first strike success rate jumped to 70 percent.[82]

The significance of these changes is that the Gulf War marked a turning point for the US military's use of space-derived imagery data. The Gulf War proved that the strategic systems developed during the cold war had operational and tactical value.[83] The ongoing acquisition of imagery satellites and their products by potential adversaries alters the future warfare equation and thus raises the priority of the space control mission. Part 3 discusses the space control mission in the context of employment against imagery systems and analyzes a commercial surveillance system to determine how best to deny such information to a future adversary.

Part Three: Meeting the Challenge

Denial

With the preceding knowledge of imagery's history as well as a different perspective on modern warfare and imagery's potential role, the stage is set for the final section of this study. This work now moves forward to meet the challenge of space control in a multipolar, technologically advanced world. For the space strategist, this means understanding the space control mission and having the ability to work through the problem of effective post cold war, post-Gulf War space control measures. To be effective, the strategist needs an awareness of the effects of targeting various portions of the space systems infrastructure. With the international stampede to acquire imagery data, reconnaissance satellites may represent one of the first categories of space systems against which the United States may need to exercise active space control measures.

The Space Control Mission. Space control is an amorphous term whose current definition has lost its connection with United States Air Force doctrine. According to Air Force Manual (AFM) 1-1, *Basic Aerospace Doctrine of the United Air Force*, control has two sides: it permits friendly forces to operate more effectively, and it denies these advantages to the enemy.[84] Joint Doctrine, Tactics, Techniques, and Procedures (JDTTP) 3-14, *Space Operations*, defines space control as combat against enemy forces in space and their infrastructure.[85] This narrow definition lacks the substance and flexibility of the Air Force's use of the term *control* and thus may close a strategist's mind to the broad spectrum through which US instruments of power may successfully deny the use of space assets.

A more encompassing definition of space control should acknowledge both sides of the term. It could describe the denial portion of the term as the diversion, delay, disruption, or destruction of an adversary's space capability. The dual objectives of access and denial require a variety of capabilities, ranging from protective measures for friendly satellite systems to destruction of an adversary's spacecraft. Both lethal and nonlethal means can be employed to limit or deny an adversary's capability to use space systems or to distort the information they provide.

Space-derived intelligence data provide early indications and warnings of crises; ensure dissemination of targeting and planning data; remove uncertainties about the weather and the location and synchronization of forces; and facilitate effective command and control of forces. These data help national leaders exercise the political, economic, and diplomatic instruments of national power. US forces may be directed to deny access to these capabilities through effective space control actions.

To deny the adversary's use of space, US forces may target a wide range of assets including, but not limited to orbiting spacecraft, launch sites, production facilities, research and development laboratories, operations headquarters, fixed- and mobile-command and control ground sites, data reception and analysis sites, power generation facilities, data links, and the many technicians, operators, analysts, and management personnel who create and operate these highly technical systems.

The spectrum of denial ranges from achieving temporary or limited data loss to causing extensive long-term systemic loss. Some examples of denial actions include but are not limited to implementing an international agreement to shut off a satellite's downlink, terminating imagery sales, destroying ground sites, destroying or disrupting system software programs, spoofing or jamming link signals, damaging or disrupting satellite subsystems, and disabling or destroying the satellite. Before choosing any of these measures, the strategist must determine the outcome being sought and what tools can be employed to achieve the intended effect while concurrently minimizing unintended effects.

Before proceeding with the denial analysis, a short discussion of political constraints is in order. Current international law curbs direct attacks against satellites. Because satellites are the sovereign territory of the satellite owner, attacks against them are considered violations of national sovereignty.[86] In spite of this fact, most of the literature on space control is monocular in its discussion of the means of space control. Few nations have emerged from the trappings of cold war concepts to see space control as anything beyond employment of antisatellite weapons.[87] That topic has kept the military space community and many national-level strategists in its grip since the dawn of the space age and hindered analytical thought about how to deny access to space's bounty.

Furthermore, armed forces' personnel at the tip of the warfare spear spend most of their time focusing on the weapons of war. For those at the leading edge of technology, it is easy to understand this focus. The weapons are high technology "toys" that fascinate the imagination. This preoccupation is inadequate for the strategist. The strategist must move beyond this fascination and focus on the outcome that is being sought when the United States uses its military or other instruments of power.

Denying the Advantages. During the cold war, the US military's adversarial problems were in clear focus. There was a much better understanding of who the enemy was and what capabilities he would bring to a conflict. The problem is much more complex today and for future war fighters. Considering only the space control portion of the conflict equation, during the cold war, the United States faced an adversary who owned highly capable reconnaissance satellites and existed in a closed society within a large land mass. With the limited technologies of the day, ASAT attack was perhaps the only viable method of space control. Today, the landscape is much different. Tomorrow's adversary may be receiving imagery data from foreign or domestic commercial vendors. US governmental agencies may be receiving data from some of those very same sources or include their use in contingency plans. This concurrently complicates yet broadens the scope of the denial portion of space control's mission. Now, more than ever before, denial efforts cannot be executed without considering the political, economic, and physical ramifications of those efforts.

The strategist must determine what effect is needed and how best to achieve that effect. Looking at the problem of satellite imagery control, as mentioned earlier, the range of objectives extends from temporary or limited data loss to the long-term future loss of related space systems. For each of these, there are numerous ways to achieve the objective depending upon the circumstances surrounding the actors, the linkages between the actors and those unrelated to the conflict situation, the conflict situation itself, the space systems, the actors' capabilities, and so on. Many such factors will impact the national strategy. Table 33 was developed as an aid for developing space strategy.[88] It does not purport to represent all possible effects, weapons, or means. Rather, it is offered as a tool for the space strategist who is attempting to approach strategy through rational analysis.

As shown in table 33, the strategist will achieve effects for varying lengths of time. The effect may be felt by an isolated

Table 33

Some Means of Denial

Objective	Tool or Weapon	Means of Denial
Immediate temporary or limited data loss	Diplomacy/ownership	Agreement to terminate downlink
		Shutter control
		Terminate imagery sales
	Physical attack	Attack a ground C^2 or receiver site that has a backup system
	Software virus/worm	Destroy software coding to temporarily disrupt spacecraft or site operations
	Electronic warfare	Spoof or jam signal links to disrupt or degrade spacecraft or site operations
	Directed energy	Temporarily disrupt or degrade spacecraft operations
Immediate long term data loss	Physical attack	Attack ground receiver station that does not have any backup system
	Directed energy	Cause repairable damage to spacecraft
Immediate satellite destruction	Physical attack	Destroy all ground C^2 sites to cause spacecraft malfunction / destruction
	Electronic warfare	Spoof/jam C^2 links to cause spacecraft failure
	Directed energy	Cause irreparable damage to spacecraft
	Kinetic kill	Use ASAT to destroy spacecraft
Potential future or long term impact	Physical attack	Headquarters
		Remove replacement capability by destroying spacecraft storage or fabrication facilities
		Remove replacement capability by destroying launch site

group of users or by all agencies who depend upon a launch site to gain access to space. During the heat of battle, one common goal is immediacy. This is especially true for the United States, where there is a desire to terminate conflicts quickly. For an on orbit, operational space system, this raises the importance of the first group of options and lowers the importance of attacking, for example, the spacecraft's headquarters facilities, fabrication facilities, and launch sites.

To gain an immediate effect on the adversary while avoiding costly and unintended effects against friendly users, the strategist must look to the measures shown at the top of table 33. Of those means, some are much easier to implement than others. For example, while an exacting software virus or worm could be employed to achieve temporal and specific results, the war fighter must either have had access or had gain access to the system to employ this tool. This may require access long before the current adversary was considered to be a potential threat. Most space systems employ highly secure cryptology devices to avoid such problems. Thus, while listed as an option, this weapon may not be feasible.

Looking at another of the options in this category, a physical attack against a receiver site that is known to have a backup system may provide the temporal success needed. It may remove the space assets from the adversary's tool box without causing significant long-term effects. It may also maintain the flexibility for more inclusive measures at a later time.

Switching to the middle categories, directly attacking the satellite will have immediate effects, but it also has many unattractive consequences. First of all, in today's interdependent information-based society, destruction of a satellite may effect more actors than desired. It may effect a very large group of users, some of whom may be allies or even the United States government. For those with orbital analysis or astronautical experience, the idea of shattering large in orbit satellites immediately brings to mind two nightmares. The first is the orbital analyst's nightmare of trying to identify and track (perhaps for hundreds of years) the hundreds of resultant debris objects. The second is the orbital analyst's and astronauts' nightmare of determining and reacting to the destructive effects that those pieces of debris may have on friendly satellites or manned spacecraft.[89] Additionally, such an attack minimizes the coercive leverage gained by lesser-destructive measures. Because of the far reaching unintended effects caused by spacecraft destruction, it is one of the least preferred of the attack options. Permanent or temporary disablement through other means, for example, electronic warfare, may achieve the desired effect without risk to US or allied manned or unmanned spacecraft.

Looking briefly at the final category, when dealing with an in orbit system, these targets appear to be the least beneficial of the targeting categories. It is possible that none of them will have an immediate effect upon the conflict.[90] For example, attacking a spacecraft's headquarters will not immediately stop the data flow from the satellite to the command and control station or to the receiver station. At best, such an attack will have an unknown future effect on operations due to the loss of financial and management support. Similarly, attacking the launch facility, or spaceport, may deny the adversary's ability to launch replacement satellites, but it may also remove that spaceport from the small inventory of available launch facilities and cause far-reaching, long-term effects on the entire space industry.[91]

In concluding this overview of selected denial measures, one final item is important to keep in mind. When considering the options, the decision maker must remember that many commanders are involved in all phases of conflict and that they may require different measures to achieve their campaign objectives. Without close coordination, one commander may demand the elimination of a satellite or ground station while another needs to keep that same system operating to permit deception operations. This is where the demands of warfare and the global nature of satellite systems require that a space denial campaign be centralized in the hands of a space systems expert.[92]

The space control mission is becoming increasingly important as the world's powers become proficient in exploiting space resources. Regardless of who becomes an adversary of the United States, the military must be prepared to advise its decision makers about the most effective means for achieving space control. The complex and interconnected contemporary world demands that this advice be given only after completing an analysis similar to that offered above. Simply advocating, for example, spacecraft destruction will not answer the question of how to achieve space control. In what follows, the results of the above analysis will be applied to the world's leading commercial imagery provider to illuminate the implications of attempting to deny its data to a future adversary.

An Analysis of the SPOT System

While the SPOT system may not be the only means by which to procure high-quality imagery intelligence in the near future, it is analyzed here because it has the most extensive network and an aggressive marketing plan to ensure its continued relevance.[93] As with any space system, there are four critical components for SPOT operations: spacecraft, ground stations, communications links, and personnel. To deny SPOT imagery data to an adversary, some or all of these components may need to be targeted. Exact target selection would need to consider the conflict level, constraints, and the desired outcome.

An Overview of the SPOT System. The main contractor for the SPOT program is the *Centre National d'Etudes Spatiales* (CNES) headquartered in Toulouse, France. CNES is responsible for orbit maintenance, payload programming, and data reception and preprocessing. Upon successful launch from the spaceport at Kourou, French Guiana, the SPOT spacecraft travels in a circular sun-synchronous orbit, designed to provide imaging coverage at approximately 10:30 AM local time. Its altitude is 832 kilometers[94] and its orbital inclination is 98.7 degrees.

The SPOT spacecraft includes twin high-resolution sensors called high-resolution visible imagers that acquire either panchromatic imagery in the 0.51-to-0.73 micrometer wavelength range or multispectral imagery at lower spatial resolution.[95] The high-resolution visible instruments measure the reflected solar energy radiated from Earth's surface to create an image. The imagers are comprised of a camera (including the optical system), light-sensitive detectors, and an electrical subsystem for signal processing and camera control.[96] Recorders on SPOT-1, -2, and -3 can hold 22 minutes of data. With SPOT-4 and SPOT-5, this will increase to 40 minutes of data.

Moving to the ground segment, space systems such as SPOT require extensive data processing facilities. The complete remote sensing system must provide capabilities for command and control of the spacecraft, imaging sensor command and control, telemetry data acquisition, telemetry decommutation, extraction of the digital imagery from the telemetry, formatting and display of the imagery, and delivery of data products to the users.[97] Processing operations are categorized into several levels of sophistication.[98] The resultant products are supplied in several formats including standard computer-compatible tapes, MS-DOS diskettes, photographic format, and CD-ROM.

There are two types of fixed-data reception stations are Station Reception des Images Spatiales (SRIS) and SPOT Direct Receiving Stations (SDRS). The primary SRIS receiving stations are located at Aussaguel (SRIS-T), near Toulouse, France, and at Esrange (SRIS-K), near Kiruna, Sweden. These two SRIS receive real-time data as SPOT passes over the north polar region, Europe, and North Africa within a 2,500-kilometer range. Together, the reception capacity of these two stations is 500,000 images per year. The equivalent of seven hundred scenes are archived every 24-hour period at each site.

There are 15 SDRS around the world.[99] These stations only receive real-time imagery and are thus limited to the amount of data stored on board SPOT as it comes within range. The locations of the SDRS are listed in table 34.

Table 34

SPOT Direct Receiving Stations

Prince Albert	Canada	Islamabad	Pakistan
Gatineau	Canada	Hyderabad	India
Cotopaxi	Ecuador	Alice Springs	Australia
Cuiaba	Brazil	Lad Krabang	Thailand
Maspalomas	Spain	Pare-Pare	Indonesia
Riyadh	Saudi Arabia	Taipeh	Taiwan
Tel Aviv	Israel	Hatoyama	Japan
Hartebeesthoek	South Africa		

Source: Andrew Wilson, Jane's Space Directory 1994–1995 (Alexandria, Va.: Jane's Information Group 1994), 393.

In addition to the SRIS and SDRS fixed-data reception sites, the Gulf War coalition forces utilized the first mobile reception systems. Those provided for the conflict were called MSS. Figure 10 provides a summary of the SPOT system network.

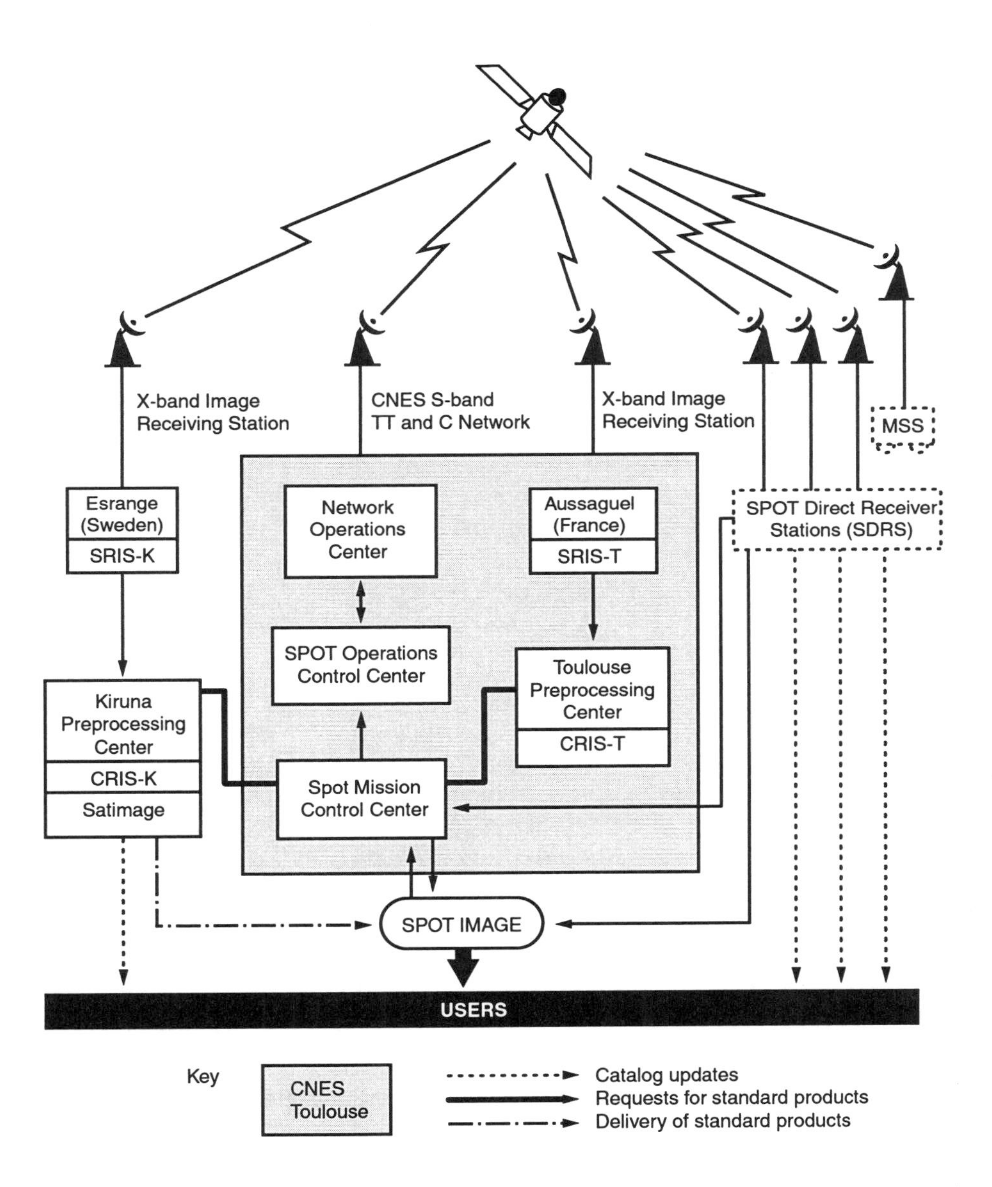

Source: Andrew Wilson, Jane's Space Directory 1994–1995 (Alexandria, Va.: Jane's Information Group 1994), 393.

Figure 10. SPOT Control and Data Reception Network

Denying the SPOT Imagery Advantage

Denying SPOT imagery data to an adversary may be a challenging task. The strategic-level intelligence data it provides can be accumulated over time and kept on file. If the adversary is acquiring timely operational and theater-level data, it may be possible to take measures to deny this information.

To cause the immediate temporary or limited loss of data to an adversary, the United States could enlist the services of its diplomatic personnel to convince the French government that SPOT Image should terminate sales or downlink services to the adversary. The precedent for this would be SPOT Image's willingness to terminate sales of Middle East imagery during the Gulf War.[100] Air, ground, or sea force attacks could terminate operations at a fixed SDRS or mobile MSS. Human intelligence operatives could be employed to eliminate particular data transfers or to provide misinformation. For an internationally intertwined system such as SPOT, these may be the only options by which to attempt space control due to the negative side effects of other means.

The options that seem likely to be politically, economically, technically, or operationally unwise or unfeasible include installing a software virus or worm, spoofing or jamming the spacecraft signal links by way of electronic warfare or directed-energy weapons, attacking the CNES Mission and Operations Control Center and Network Operations Center, the SRIS sites, destroying the SPOT satellite by way of directed or kinetic energy weapons, attacking CNES headquarters, destroying SPOT fabrication facilities and destroying the Kourou spaceport. These actions would affect all SPOT customers, cause multinational discord, disrupt US military use of the data, obligate the US government for expensive replacement costs, and if the spaceport is destroyed, cause enormous financial losses across the entire space industry.

Considering the unique problems that this system provides, it is apparent that the SPOT system's international value minimizes the feasible space control actions that can be accomplished by the military. Of the actions discussed, only two, physical attack against a fixed or mobile regional receiver site and human intelligence (HUMINT) activities, appear as viable means to gain immediate temporary or limited data loss. Assuming these measures can in fact be successfully conducted, one may achieve limited results. If the adversary's database is intact, he may still have access to strategic-level intelligence and perhaps some operational-level data. Perhaps the best that can be expected is the termination of temporal-operational or tactical-level data.

One final note on the above analysis is essential. The risk of applying the results of a conceptual analysis to a particular system is that one then falsely extrapolates the attributes of the part (SPOT) to the whole (all space systems). The author has attempted to avoid this shortcoming but also realizes that SPOT represents today's most sophisticated commercial imagery network. It may thus be among the first to cause headaches for American leaders who are tasked with engaging a technologically adroit adversary. An analysis similar to that accomplished above for the SPOT system may yield similar results for other satellite systems, in particular, many of the communications satellites. For actors with indigenous systems designed to serve a solitary actor, America would be at greater liberty to take aggressive space control actions. It is thus critical that the strategist have the background information that will clarify the constraints under which the space control measures must operate.

Conclusions and Implications

The author has attempted to provide information for space strategists and nonspace personnel alike. The work has acquainted readers with some strategic space intelligence history, described its revolution, touched upon the significance of using strategic space intelligence in theater warfare, offered a unique outlook on modern warfare, and contributed analytical tools to the space control mission. To place the study in perspective, the conclusions reach into the future by considering the implications.

America's first space strategists and scientists were determined to provide the tools necessary to ensure security and stability in a nuclear world. Their work provides an example of how human vision can stimulate the achievement of technological breakthroughs that change history. As is so often true, once technological secrets are unlocked and mastered, it is only a short time before the closest competitor closes the gap. Once the digital imagery genie was out of the bottle, it took only one theater war to enlighten the rest of the United States competitors about the potential of imagery in warfare. For the military space community, the Gulf War represented a fundamental of the United States break with the past. During the cold war, attempts to demonstrate the operational and tactical value of strategic space assets were seldom encouraged. There was little deviation from the strategic missions and only a few personnel were involved in exploring space systems' value for auxiliary missions. It is thus not surprising that when looking at the military space community's Gulf War shortcomings, critics latch on to the lack of space doctrine for theater warfare. They were destined to find little theater-level doctrine because, quite simply, the use of the United States's strategic space assets for theater warfare was not a primary, secondary, or tertiary mission during the 30 years of cold war military space operations. The only group actively seeking use of strategic space assets in theater warfare were those involved in tactical exploitation of national capabilities program (TENCAP). The space forces used for the Gulf War did not provide everything the operations personnel wanted or needed, but they rose to the challenge of turning their world upside down and are credited with significant contributions to the coalition's success. From this, there are two important reminders for the future. First, obtaining such a fundamental change in focus cannot be executed overnight. Cold war mind-sets and procedures still permeate every military space subcommunity. Resistance to change will continue until leaders help those communities grasp the requirements of the post cold war, post-Gulf War world. Secondly, the strategic space missions have not been replaced by the new operational and tactical ones. Instead, the scope of the latter missions has expanded and their significance has been raised. During their enthusiastic exploration of space systems' potentials for theater warfare, space strategists must remember to accept space systems' limitations. Satellite reconnaissance is only one of several methods for procuring timely imagery data. In many cases, the preferred method for acquiring imagery reconnaissance information will still be through the use of airborne strategic or tactical platforms.

As the United States competitors quickly seek space-derived imagery products, the United States faces a future where its adversaries may cloak themselves in different forms of warfare. The United States must understand the unique abilities of the adversary's and combinations of warfare forms and tools before it can successfully engage them in combat. Today, a multitude of actors operate spacecraft and do so in consortia. Although the United States is the world leader in the exploitation of space reconnaissance systems, United States allies and adversaries are closing the gap. They will be skilled at using indigenously produced or commercially procured space imagery. The imagery will assist them in maintaining regional peace as well as in waging war. The responsibility for regional monitoring may become less of a US concern as other nations procure satellite imagery systems and assume monitoring responsibilities. Commercially available imagery data give, at the very minimum, the ability to procure and maintain strategic databases. This allows an adversary to develop attack plans and rehearse missions. The interconnectedness and multiple uses for reconnaissance data suggest that space strategists must analyze the adversary's use of space systems and its international linkages before recommending space control action.

The existence of "many eyes" makes the space control mission more challenging than at any previous time. It does not, as some proclaim, justify procurement and deployment of antisatellite weapons or other space weapons for space control. During the cold war's formative space years, the United States and the Soviet Union could each claim to control space. They owned the space control mission by default: there were no challengers during those early years. Each nation, at various times, owned and demonstrated the antisatellite weapons by which it intended to exercise space control. But those formative years are now part of the military space community's history. The advent of commercially available space reconnaissance data limits the Uunited States ability to control space regardless of the weapons it chooses to develop. If the adversary has developed a strategic database, destruction of portions or all of a space system's infrastructure cannot remove this peacetime endowment. At best, destruction may remove the imagery's operational- or tactical-level application. Knowing this, a worthy adversary will have devised ways to ensure access to the imagery it needs. The adversary may have created a redundant system of systems that includes strategic and tactical airborne reconnaissance platforms. Another option for the adversary might be to attempt to undermine or negate the United States's superior technological capabilities by using unfamiliar or different organizational or doctrinal concepts.

The analysis suggests that both space- and ground-based antisatellite weapons are less viable in today's multipolar world. Because of the interdependence of today's space assets, spacecraft attack will in most cases affect multiple actors. An attack may impose upon United States taxpayers high financial liabilities. The debris cloud caused from satellite destruction may unintentionally damage or destroy friendly manned or unmanned spacecraft. Since the adversary may have already procured the database necessary for military operations, spacecraft attack may not accomplish the original objective of denying data. Thus, spacecraft attack may be an ineffective space control measure in many contemporary warfare scenarios. Their use may escalate the conflict, terminate allied support, and eliminate a resource for US military forces. These facts of space attack are often dismissed or forgotten due to the exotic appeal of space attack weapons. These weapons capture the warriors' imagination because they represent scientific discovery's latest breakthroughs in harnessing man's destructive capabilities. Additionally, they induce warriors to prepare for their employment because they promise to destroy inanimate objects hundreds of miles from the natural human domain. This promise allows their sponsors to peddle them as the necessary and sufficient space control solution. Such trappings do not take into account the realities of spacecraft attack that become apparent through analysis of weapons effects. Antisatellite weapons may have been the only method to achieve space control in the early decades of space exploitation, but they are not as viable in today's information dominated society. Spending vast sums of taxpayer money to procure cold war systems for a twenty-first century world may leave the United States with unusable weapons and ineffective strategy. Space control strategies for the twenty-first century must be based upon more than one option.

The space control mission may be more elusive than in the past, but that does not imply it is beyond the United States's grasp. The best control measures are those which incur the least amount of risk, cost, and unintended consequences. Reasoned space control for the next century may be limited to terrestrial-based activities such as diplomatic bargaining or surgical attacks against certain ground-based operations. Precision surgical attack is a capability that the United States military forces excel at with their air-, land-, and sea-based force application weapons. What is important to remember is that the recommendations the space strategist select should be based on analysis, rather than on reliance upon cold war solutions that are still proposed, yet may not be valid. The United States's achievements in developing the air, land, and sea weapons of industrialized warfare are capable of delivering the temporary space control that it needs for warfare. The United States does not need to step up to the realm of space warfare to achieve space control.

Proceeding along the space warfare path has several unattractive consequences. It affects arms control agreements and raises questions about world dominance and planetary protection. Although the international system is characterized by anarchy, most nations have agreed to live within the limits of international law and to attempt to resolve differences peacefully. In the early days of space achievement, the major powers rejected the calls of military leaders to use space for terrestrial attacks and the moon as a ballistic missile base. They agreed to limit military activities in space. Those who advocate departure from these decisions reopen debates on and concern about two important international arms control issues.

First, US pursuit of space weapons reduces the effectiveness of current arms control agreements. Second, it jeopardizes US ability to gain additional agreements. Ignoring the problems of arms control may propel the world along the path of lawlessness and violence, moving warfare to yet a higher plateau from which humankind is able to destroy itself.

Turning to the world dominance and planetary protection problems, these two concerns stem from the fact that an entity in charge of space weapons is capable of threatening any spot on the planet. If the United States were in charge of space weapons, then it could be in a position to dominate the world or claim that its destiny is to become the protector of the planet and its peoples. In either case, an adversary who disagrees with these roles may attack the United States' homeland or assets abroad. An entity who does not want the United States to dominate the world or act as its police force may be encouraged to execute preemptive strikes, perhaps through the use of nuclear, chemical, biological, or genetic weapons. Assuming that the United States has no designs on planetary domination and that the nations of the world agree that the United States should police the planet, how might this impact the United States? In addition to making the United States a more attractive target for attack, assumption of this role could result in the United States becoming embroiled in every regional conflict. As the United States takes this path, other nations may use their resources to pursue national objectives. To gain insight on how this path may develop, consider some of the CIS activities.

The Commonwealth of Independent States holds nearly all of the space achievement records. CIS names normally precede those of Americans in the record books. Their knowledge of space exceeds that of any other nation. They hold all space endurance records and currently operate the world's only space station. Since the demise of the Soviet Union, CIS has gained access to the world's space markets, including those of the United States. They seek further cooperative space endeavors yet do not protest calls by members of the US military, industry, or Congress to deploy space weapons. If Congress approves the deployment of US space weapons, CIS space experts could use their space resources for interplanetary exploration and exploitation. This once again opens the door for them to place their names before those of Americans; they could become the first nation to establish a space colony charged with extracting another planet's precious materials. If the United States is physically and economically embroiled in solving regional problems, it may miss the opportunity to tag along as a junior partner in this endeavor.

The space weapons path is not an attractive path for the United States. As the current generation of war fighters identifies space threats and industry responds with cold war solutions, both forget that the contemporary world is much different from that in which they spent their formative years. Pursuing space weaponry could place arms control agreements at risk, could lead to perceptions that the United States wants to dominate the planet, and could lead to US assumption of the role of planetary protector. The ability of the United States's scientists to unlock new destructive capabilities does not necessitate the development or use of these capabilities. Before advocating weapons development or procurement, space strategists must understand the past, present, and future environments and analyze how to achieve the desired objective. Furthermore, the strategists must project the consequences of space procurement decisions. The military space community has not yet emerged from its cold war mind-set. The challenge for present-day space strategists is to redefine their raison d'être and the scope of their missions in an intertwined international environment. The twenty-first century will not accept cold war solutions. The space procurement decisions for the next century must provide a force structure that is based on the challenges of future space operations.

Although intended primarily for space strategists, this work provides nonspace operations personnel with some of the complexities and realities of space operations and modern warfare. With a message for present and future strategists, planners, and decision makers, it interjects a measure of reality into warfare plans. That reality demands an analysis of effects rather than blind allegiance to exotic weapons and visionless adherence to predetermined employment concepts.

For those who immediately leap to destructive methods of coercion, it reminds them of the value of other instruments of power. For those who, unversed in space operations, transfer air and terrestrial warfare concepts to space without understanding the medium or the consequences of their proposed actions, it offers a more rational approach to decision making. Those who persist at ignoring the differences between air and space or proselytize about the powers of space exploitation without solid historical, experiential, or analytical foundations may be destined to follow in the footsteps of airpower theorists who have kept many air strategists' ideas imprisoned in binary thought. This work takes a first step toward avoiding that affliction for the United States's future space strategists.

Notes

1. Jeffrey T. Richelson, *America's Secret Eyes in Space: The US Keyhole Spy Satellite Program* (New York: Harper & Row, 1990), 2.
2. Ibid.
3. Ibid.
4. Ibid., 5.
5. Ibid., 9.
6. Ibid., 12.
7. Ibid.
8. Ibid., 12–13.
9. Ibid., 13.
10. Ibid., 17.
11. Ibid., 26–27.
12. Ibid., 19.
13. Ibid., 44. According to Richelson, Samos did not stand for the commonly referred to Satellite and Missile Observation System. Richelson's sources stated that the WS-117L project director selected the name "Samos" in the belief that no one would produce an acronym from it.
14. SPOT is the French civil earth observation program operated by SPOT Image Corporation. SPOT Image is headquartered in Toulouse, France. The *Centre National d'Etudes Spatiales (CNDES)* is the main contractor for the program. See Andrew Wilson, *Jane's Space Directory 1994–1995* (Alexandria, Va.: Jane's Information Group, 1996), 183. SPOT Image Corporation, CNES, *The Catalogue of SPOT Products and Services*, 1989, 3. SPOT's multispectral imagers provide 20-meter resolution data.
15. Wilson, 183-84. The Keyhole satellites that continued to exploit the film recovery system included the KH-4A, KH-4B, KH-7, KH-8, and finally, the KH-9 known as "Big Bird."

16. Richelson, 61. The major objective of the KH-6 Tallinn mission satellite was to obtain close-look photography of the site near the Estonian city of Tallinn, where 1961 Corona photographs showed possible antiballistic missile deployments. The National Photographic Intelligence Center's interpreters believed that the photos showed construction for the deployment of SA-5 Gammon interceptor missiles.

17. William B. Green, *Digital Image Processing: A Systems Approach*, 2d ed. (New York: Van Nostrand Reinhold, 1989), 2.

18. Ibid., 2–3.

19. Alvin and Heidi Toffler, *War and Anti-War: Survival at the Dawn of the 21st Century* (Boston: Little, Brown, 1993), 230–31. The Open Skies policy stems from a proposal made by President Dwight Eisenhower to Soviet Premier Nikita Khrushchev at a summit meeting on 21 July 1955. The policy was formally rejected by Khrushchev but became a de facto agreement as each nation pursued the use of imagery satellites to view the other nation.

20. Others argue that the imagery satellites destabilized the strategic nuclear standoff. Because the satellites provide precise targeting data, the superpowers could develop a counterforce vice countervalue nuclear strategy. This was seen as destabilizing because the counterforce strategy encourages preemptive strategic nuclear attack to ensure national survival.

21. Wilson, 184. The KH-11 provides eight to 12 images per minute. For details on the workings of CCD imaging devices, see Green, 37–38.

22. Wilson, 184.

23. Richelson, 126.

24. Ibid., 128–29. Using this technology, the KH-11 scans its targets in long, narrow strips and focuses the light onto an array of CCDs with several thousand elements. The light falling on each CCD during a short, fixed period of time is then transformed into a proportional amount of electric charge. In turn, the electrical charge is read off and fed into an amplifier, which converts the current into a whole number between zero and 256. The number represents a shade of color ranging from pure black to pure white. Each picture can be transmitted as a string of numbers, one from each element.

25. Ibid., 134. Data collection was limited to only a few hours each day due to power limitations. The power required to transmit the data to the relay satellite was so great that it drained power far faster than it could be replenished by the satellite's solar panels. To increase its effectiveness, the KH-11 operated in concert with the KH-8 and KH-9 until this limitation was solved.

26. Ibid., 135. Stereoscopes allow photo-interpreters to superimpose photos taken from different angles. Such superimposition can yield a three-dimensional effect that makes it easy to determine the height and length of weapons bunkers, space launch vehicles, or other objects of interest.

27. Ibid., 136.

28. Ibid., 137.

29. One could argue that the United States's needs in this area uniquely exceed those of other nations. Being such an open society, it is easy for other nations to acquire data about the United States. Potential adversaries of the United States excel at keeping their societies closed and information about their strategic systems hidden. Thus, one could say that the most sophisticated imagery systems emerged naturally in the United States because "necessity is the mother of invention."

30. Wilson, 158.

31. To learn more about the national betrayals that occurred in the spy satellite world during the 1970s, see Richelson, 157–84; and Robert Lindsay, *The Falcon and the Snowman: A True Story of Friendship and Espionage* (New York: Simon and Schuster, 1979).

32. Richelson, 157–58.

33. Ibid., 158.

34. Ibid., 161.

35. Wilson, 159.

36. Wilson, 154; and Paul B. Stares, *Space and National Security* (Washington, D.C.: Brookings Institution, 1987), 209. Molniya satellites provide government and military communications. They circle Earth in a highly elliptical orbit (commonly referred to as a Molniya orbit) that provides extended dwell time over the northern hemisphere.

37. Daniel Deudney, "Unlocking Space," *Foreign Policy,* Winter 1983–84, 95. One public example of observation satellites' monitoring value was the nuclear test facility construction site spotted in South Africa by a Soviet satellite in 1977. After intense diplomatic pressure, South Africa dismantled the site.

38. Two factors help set the stage for this transformation: the Air Force's Tactical Exploitation of National Capabilities (TENCAP) project and the fall of the former Soviet Union. TENCAP began using the national satellites for tactical operations in the 1980s, but its scope expanded significantly after the Gulf War, when the Air Force created the Space Warfare Center (SWC) to give the war fighter direct access to space's bounty. TENCAP was one of the first projects assigned to the SWC and has undergone significant growth since SWC operations began in November 1993. The fall of the Soviet Union changed the international focus from a battle between the superpowers to regional contingencies. While these two factors set the stage for the shift to operational and tactical warfare, it took the Gulf War to hasten the military space community's metamorphosis. For information on TENCAP, see *The Space Tactics Bulletin* (Falcon AFB, Colo.: Space Warfare Center).

39. Vice Adm William A. Dougherty, US Navy, "Storm From Space," *Proceedings,* August 1992, 51.

40. Guy Durham, "Satellites' Keen Eyes Help Allies See Victory," *Air Force Times,* 3 June 1991, 26.

41. Eliot Cohen, *Gulf War Air Power Survey Summary* (Maxwell AFB, Ala.: Air Command and Staff College, 28 March 1993), chaps. 4 and 14.

42. Craig Covault, "USAF Urges Greater Use of SPOT Based on Gulf War Experience," *Aviation Week & Space Technology,* 13 July 1992, 63.

43. Ibid.

44. Ibid.

45. Ibid., 65.

46. Ibid., 63.

47. Ibid., 65.

48. Gen Colin L. Powell, chairman of the Joint Chiefs of Staff, argued that the intelligence available to the coalition was the best in military history. Gen Colin L. Powell, *Report on the Roles, Missions, and Functions of the Armed Forces of the United States* (Washington, D.C.: CJCS, February 1993), 11–13. See also Cohen, chaps. 4 and 12; and Dougherty, 51.

49. Capt Brian Collins, "Soviet View of the Storm," *Air Force Magazine,* July 1992, 73.

50. *Maskirovia* is a CIS term which, loosely translated, means deception.

51. Collins, 74.

52. Ibid.

53. William Burrows, "Give Space Reconnaissance Systems a B+," *Space News,* 5–18 August 1991, 21; and Cohen, chaps. 2, 4, and 15.

54. Peter B. de Selding and Andrew Lawler, "SPOT Halts Sales of Gulf Area Imagery," *Space News,* 13–19 August 1990, 3.

55. Phillipe Renault, deputy director-general of SPOT Image, as quoted in de Selding and Lawler, 21.

56. Ibid.

57. Ibid. One example of such an occurrence is the controversial images provided on a Libyan chemical weapons plant. The Department of Defense claimed the plant had been destroyed by fire.

58. Ibid.

59. Ibid.

60. Ibid.

61. Had he been asked to explain warfare, the task before primitive man would appear to us to be a rather simple affair. He had few tools with which to work and if warfare involved anyone other than the two opponents, it may have involved only a few others in a loosely organized group. However, this stretched the limits of his cognitive dexterity. Explaining his form of warfare could have appeared as complex a task as that which humanity undertakes today in trying to understand modern warfare.

62. Discussions on other portions of this multidimensional interpretation are beyond the scope of this project. Many people refer to the Gulf War as the first space war. Author disagrees with this. Space systems were critical in the Gulf War, but to call this the first space war implies that space forces played more than a supporting role. To have a space war, there must be at least two space-faring entities battling for control of the planet or other resources through the use of terrestrial weapons to attack space objects or space weapons to attack terrestrial or space assets. As of today, all space systems are passive. Thus far, humanity has been successful in

curtailing the deployment and employment of space weapons through adherence to such treaties as the Treaty on the Principles of the Activity of States in the Exploration and Use of Outer Space Including the Moon and Other Celestial Bodies (Outer Space Treaty).

63. Red mercury, also known as "cherry red" because of its color, is a semiliquid compound of pure mercury and mercury antimony oxide. It could be used to make a baseball-sized neutron bomb capable of killing everyone within 600 meters (approximately 0.37 mile) of the explosion. Experts believe this mercury-based explosive was developed in the former Soviet Union. "Bomb-Making 'Red Mercury' May Exist After All," *Military Newswire*, 28 April 1995. See also Britain's *New Scientist*, April 1995.

64. The "system of systems" can be divided into three general categories of reconnaissance assets: space-based strategic intelligence assets including DSP and the various types of reconnaissance satellites; strategic airborne reconnaissance assets including RC-135 Rivet Joint and U-2/TR-1; and tactical airborne reconnaissance assets including RF-4Cs and unmanned aerial vehicles. Despite the wishes of every warrior, space reconnaissance continues to be limited by weather conditions and orbital mechanics. It is thus unable to provide the continuous imagery coverage required for theater warfare. For the dynamic conditions of theater warfare, strategic intelligence assets will only augment the airborne strategic and tactical reconnaissance platforms that will continue to be the primary providers of timely imagery intelligence.

65. As many as 30 countries are currently producing unmanned aerial vehicles.

66. The SPOT system's technical details are discussed shortly.

67. Covault, 63.

68. Wilson, 393.

69. Ibid.

70. Renee Saunders, "War in Iraq Enhances Value of Commercial Remote Sensing," *Space News*, 21 January-3 February 1991, 16.

71. Lyn Dutton et al., *Military Space* (Washington, D.C.: Brassey's, 1990), 99.

72. Ibid., 98.

73. Ibid, 98–99.

74. Assessments on the number of entities who will possess military or commercial imagery satellites vary. Some naval personnel estimate that 15 nations could belong to this group by 2000. In their count the Navy lists Canada, Germany, Israel, Italy, Pakistan, South Africa, South Korea, Spain, and Taiwan as those expected to have imagery satellites. After the year 2000, Argentina and Brazil may join the group. Offering a significantly larger estimate, an Air Force Space Command planning directorate report estimates that 30 countries will have reconnaissance capabilities by 2000. See Vincent Kiernan, "War Boosts Anti-Satellite Weapons Proponents," *Space News*, 6–12 May 1991, 7; and Neff Hudson, "Air Force Researching Ground-Based Lasers," *Air Force Times*, 3 May 1993, 3.

75. Cohen, chap. 7, 21.

76. Dutton, de Garis, Winterton, and Harding, 98.

77. Patricia A. Gilmartin, "France's SPOT Satellite Images Helped US Air Force Rehearse Gulf War Missions," *Aviation Week & Space Technology*, 1 July 1991, 22.

78. William Matthews, "The Digital Warrior," *Air Force Times*, 6 January 1995, 36.

79. Ibid. The process starts with a computer that downloads intelligence data from a satellite. The data, which is processed intelligence, includes information ranging from target area weather reports and terrain features to the latest information on enemy force locations, enemy radars, antiaircraft missiles, and other threats. Taking this information, the computer produces maps of the intended targets and threats that troops are likely to encounter en route to them. Currently, Digital Warrior works through Windows software running on a 486 system. Ultimately, Digital Warrior is intended to provide pilots with computerized maps that will show aircrews where they are, where the target is, and where the threats are as the mission progresses.

80. Covault, 61, 63. The Navy increased its use of SPOT imagery during and after the Gulf War. Aircraft carriers are now equipped with a SPOT database.

81. Ibid.

82. Ibid., 64; and Jay Lowndes, "War's Aftermath Tracked," *Space News*, 22–28 April 1991, 8. This is not to imply that the use of one type of imagery data precludes use of another. Each clearly has a unique contribution in warfare. For example, in the aftermath of Hussein's scorched earth policy in Kuwait, *LANDSAT*s broad views guided cleanup crews to oil that remained in the Persian Gulf. At the same time, the high-resolution imagery helped estimate the volume of oil in different locations so that necessary equipment could be on hand.

83. James M. Gifford and Vincent Kiernan, "Military Calls Space Superiority Essential," *Space News*, 6–12 May 1991, 6. In retrospect, some claim that as much as 70 percent of space's assets have tactical value.

84. Air Force Manual 1-1, *Basic Aerospace Doctrine of the United States Air Force*, March 1992, 10.

85. Joint Doctrine, Tactics, Techniques, and Procedures (JDTTP) 3-14, *Space Operations*, 15 April 1992, GL-7.

86. For the text of the Outer Space Treaty, see United States Arms Control and Disarmament Agency, *Arms Control and Disarmament Agreements* (Washington, D.C.: Government Printing Office, 1984). See also Sylvia Maureen Williams, "International Law and the Military Uses of Outer Space," *International Relations*, May 1989, 413.

87. "Sharp Imagery Spurs Call for ASAT," *Space News*, 6–12 March 1995, 17. As recently as March 1995, a senior policy advisor to a Senate Armed Services Committee member said that ASAT capability will be absolutely necessary to protect US troops from being spotted by commercial

satellites. This advisor sees the proliferation of commercial imagery satellites as an indication that ASATs are essential and their employment inevitable.

88. Shutter control refers to the government's ability to block sales of imagery from United States commercial satellites. To implement shutter control, current policy requires a cabinet-level decision during a time of national crisis. Some congressmen believe this policy is too restrictive and are heading a move to empower officials at lower levels to exercise shutter control. Warren Ferster, "Prospect of Policy Review Rattles Imagery Executives," *Space News*, 6-12 March 1995, 9.

89. This lesson is frequently offered to orbital analysts. Several such offerings include the F-15-launched ASAT test, satellite breakups, and space shuttle support operations. The United States' single employment test of its F-15-launched ASAT corroborated the scope of this problem. Orbital analysts assigned to the Space Surveillance Center within a few years of that test are fully aware of the protracted challenge of identifying and tracking those hundreds of pieces of debris. Every time an on orbit satellite breaks up, the challenge begins anew. For orbital analysts charged with the responsibility of space shuttle safety, what was once an infrequent notification requirement has nearly become a full-time job. As the set of orbital objects grows, so does the threat to spacecraft. The scope of this problem significantly increases when considering the limitations of the United States's space surveillance network. The majority of the space surveillance sensors are only capable of tracking objects that have a radar cross section of 10 centimeters or larger. The network daily tracks approximately seven thousand such objects; however, the complete set of orbital objects is much larger than this. Using data derived from some of the network's more sensitive sensors, analysts estimate that there are between 60,000 and one million pieces of orbital debris ranging in size from one to 10 centimeters. Debris objects smaller than one centimeter are estimated to number around the billion mark. The combined efforts of spacecraft shielding and collision avoidance notifications provide some protection for the space shuttle. However, as the 20,000 craters discovered on the Long Duration Exposure Facility after 69 months in orbit indicate, space is a hostile environment. Knowingly taking actions that exacerbate the debris problem places both adversarial and friendly spacecraft at higher risk.

90. One could argue that destruction of a launch facility is necessary when the adversary is preparing to launch an ASAT attack. This unique circumstance would, of course, modify the assumption that one is dealing with an in orbit system and subsequently modify the conclusions. The author, in offering the analysis is not attempting to include all possible scenarios or variables, but rather to disclose a reasoned approach to space control decision making.

91. Wilson, 431. There are currently only 20 spaceports: five in the United States, three in the former Soviet Union, three in China, two in Japan, and one each in France, Italy, Australia, French Guiana, India, Israel, and Brazil.

92. These statements should not be construed as a rationale for the creation of a joint forces space component commander (JFSCC), but rather as an argument that the US's space systems have evolved to the point at which only space systems personnel should be granted authority and responsibility for leadership of space forces. The move to create the position of a JFSCC transfers from Desert Storm the image of the joint forces air component commander who was responsible for coordinating all air operations. The contemporary JFSCC concept is titillating to those who seek position and power, but there is no need for it. The global nature of space systems allows for their centralized command and control from within the United States borders at the hands of USCINCSPACE. The majority of military space assets continue to provide the data they were designed for, namely, strategic intelligence and warning. The space systems that provide tactical data within a theater of operations can be commanded by a field grade officer reporting operationally to the theater CINC and administratively to the respective space command. If the United States's military forces were to reorganize themselves in a manner similar to that chosen by CIS, that is, a military that includes space force and reconnaissance-strike organizational concepts, plus deployment of space weapons, perhaps it would be time to create such a position. Until the United States's space forces have evolved to the point where they play an active role in force application, a JFSCC is unnecessary.

Regarding the leadership responsibilities for space forces, only a person with space systems experience is fully qualified to lead space forces. This follows the same reasoning that has been used since the inception of the United States's independent Air Force to justify that its leadership be restricted to its small set of rated personnel. Air and space are uniquely different media, just as are land, sea, and air. The contributions space systems give to warfare are similarly unique just as a much shorter period of time was sufficient to provide justification for the argument that only airmen lead air forces. Forty years of evolutionary history is sufficient to justify that only space systems personnel lead space forces.

The United States's four decades of military space exploitation has created a large pool of space experts from which to groom and summon the future's space leaders for the future. Many of them possess the cognitive faculties as well as other critical leadership traits; what they may lack is training in the art of warfare. This is a systemic problem that can be overcome.

Injecting rated personnel into space leadership positions only serves to offer auxiliary leadership opportunities to potential air leaders. Concurrently, this hinders the development of future space strategists and leaders who do not gain the experience offered through those leadership positions. This can also impact their selection for subsequent advanced educational and leadership opportunities. Due to their lack of experience with space systems, the rated personnel are placed in the unenviable situation of being responsible for decisions about an unfamiliar medium. In most cases, they do not understand the nature of the US military's role in space, the military space community or its unique subcommunities, or the educational and technical requirements of its people. The level at which this problem exists is obviated by Space Command's 400 percent over manning level for rated officers and its continued ability to hire rated personnel with no space experience to take critical leadership positions. The difficult situation that the Air Force's leadership faces is that it needs the warrior mentality it has given to its top-rated officers yet needs those warriors to have space systems expertise. The solution to this dilemma is a concerted effort on the part of Air Force leaders to, (1) immediately and significantly reduce the number of rated officers in its space commands, (2) open more space systems positions at its warrior training schools, and (3) reserve its space leadership positions for those with space systems expertise. Failing this will perpetuate many of the problems that have stymied the maturation of military space doctrine, policy, and strategy during the United States's first 40 years of military space activity.

Space achievements of the Soviet Union include the majority of firsts: they were the first to place a satellite in orbit; place men and women in orbit; leave orbiting crafts to walk in space; send human artifacts to another planet (Venus); flyby, impact, circle, and orbit a craft around the moon; conduct welding and smelting experiments in space; place an automated lunar rover on the moon; place a space station in orbit; and land spacecraft on Mars. Soviets hold the human endurance records and have the most expertise in scientific investigation. Currently the CIS maintain the only space station. In addition to these firsts, one of their most recent achievements was the formation of an independent space force. The United States also has a few space firsts including the first test of an ASAT weapon (1959), the first and only men to explore the Moon, and the first spacecraft to land like a plane. The United States may not be as ready as the CIS to form an independent space force, but there may come a day when it is the correct decision. In preparation for that day, the Air Force should begin now to wean itself from its reliance upon rated personnel for space leadership. Deferring this decision helps neither the military space mission nor its potential space forces leaders. It will serve the United States well to groom its space experts for space leadership rather than allowing those without space experience to lead US forces into the next century.

To learn more about CIS concepts, see Mary C. FitzGerald, "The Soviet Image of Future War: Through the Prism of the Gulf War," Hudson Institute HI-4145, May 1991; Mary C. FitzGerald, "The Soviet Military and the New 'Technological Operation' in the Gulf," *Naval War College Review,* Autumn 1991, 16–43; Mary C. FitzGerald, "Russian Views on Electronic Signals and Information Warfare," *American Intelligence Journal,* Spring-Summer 1994, 81–87; FitzGerald, "The Russian Military's Strategy for 'Sixth Generation' Warfare," *Orbis,* Summer 1994, 457–76; and Benjamin S. Lambeth, "Desert Storm and Its Meaning: The View From Moscow," RAND R-4164-AF (Santa Monica, Calif.: RAND, 1992). To learn more about the United States's first

ASAT weapon, the *Bold Orion,* see Wilson, 163; and Paul B. Stares, "Deja vu: The ASAT Debate in Historical Context," *Arms Control Today,* December 1983, 2–3.

93. The primary sources for the SPOT technical data are Green, 46–49, and Wilson, 392–94. Even during the cold war, SPOT Image Corporation advertised its intelligence-gathering capabilities and military usefulness through its advertisement entitled "SPOT: The New Way to Win!" advertisement. See *Defense Electronics,* November 1988, 68.

94. This distance is approximately five hundred miles.

95. SPOT's high resolution visible sensors use the push broom scanning technique that utilizes a linear CCD as the active sensor. The camera optics focus the full width of the ground swath onto the CCD array as the spacecraft travels along its orbital path. The CCD is sampled at a specific frequency to provide sequential lines of image data. Beam splitters transfer image data to the spectral CCD detectors to acquire multispectral imagery.

96. The twin imagers can operate independently of each other, in panchromatic or multispectral modes, and at near vertical or variable angles. Each imager can be steered to any of 91 orientations 0.6 degree apart. This results in a capability for a plus or minus 27 degree off nadir view and the ability to view a single area on seven successive passes. SPOT is thus capable of stereo imaging and reattempts when observations are hampered by cloud cover. The oblique viewing capability decreases the actual revisit time from 28 days to 3.7 or 2.4 days depending on where the targeted imaging area is located on the Earth. The ground swath width is 60 kilometers for the panchromatic imagery and 117 kilometers for the multispectral imagery. A SPOT scene will range from a 60-kilometer square for a vertical view angle to a 60-kilometer-by-80-kilometer maximum at a 27-degree viewing angle (the maximum viewing angle).

97. Decommutation is conversion of unidirectional current.

98. Wilson, 393. Level 1 is basic radiometric and geometric corrections. Level 1A is essentially raw data and is useful for stereo plotting and basic radiometric studies. Level 1B is full radiometric and limited geometric corrections and is the basic preprocessing level for photo interpretation and thematic analysis. Stereoscopic pairs data are also available at this level. Level 2 provides rectifications according to a given cartographic projection. Level 2A corresponds to level 2 precision processing but can be implemented without use of map ground control points. Level S scene verification is based on ground control points, ensuring registration with another scene used as a reference to within 0.5 pixels. Level IAP was introduced in 1990 and is optimized for photogrammetric applications using analytical stereo plotters.

99. Ibid. The proposed SDRS at Fucino, Italy, is still under negotiation.

100. de Selding and Lawler, 3.

Bibliography

Burrows, William. "Give Space Reconnaissance Systems a B+." *Space News*, 5–18 August 1991.

Cohen, Eliot. *Gulf War Air Power Survey Summary*. Maxwell AFB, Ala.: Air Command and Staff College, 18 March 1993.

Collins, Capt Brian. "Soviet View of Storm." *Air Force Magazine*, July 1992.

Covault, Craig. "USAF Urges Greater Use of SPOT Based on Gulf War Experience." *Aviation Week & Space Technology*, 13 July 1992.

de Selding, Peter B. "Gulf War Spurs Call for European Spy Satellites." *Space News*, 11–17 March 1991.

de Selding, Peter B. and Andrew Lawler. "SPOT Halts Sales of Gulf Area Imagery." *Space News*, 13–19 August 1990.

Deudney, Daniel. "Unlocking Space." *Foreign Policy*, Winter 1983–84.

Dougherty, Vice Adm William A. "Storm From Space." *Proceedings*, August 1992.

Durham, Guy. "Satellites' Keen Eyes Help Allies See Victory." *Air Force Times*, 3 June 1991.

Dutton, Lyn, et al. *Military Space*. Washington, D.C.: Brassey's, 1990.

Ferster, Warren. "Prospect of Policy Review Rattles Imagery Executives." *Space News*, 6–12 March 1995, 9.

FitzGerald, Mary C. "The Soviet Image of Future War: 'Through the Prism of the Gulf War.'" Hudson Institute HI-4145, May 1991.

———. "The Soviet Military and the New 'Technological Operation' in the Gulf." *Naval War College Review*, Autumn 1991, 16–43.

———. "Russian Views on Electronic Signals and Information Warfare." *American Intelligence Journal*, Spring-Summer 1994, 81–87.

———. "The Russian Military's Strategy for 'Sixth Generation' Warfare." *Orbis*, Summer 1994, 457–76.

Gifford, James M., and Vincent Kiernan. "Military Calls Space Superiority Essential." *Space News*, 6–12 May 1991.

Gilmartin, Patricia A. "France's SPOT Satellite Images helped US Air Force Rehearse Gulf War Missions." *Aviation Week & Space Technology*, 1 July 1991.

Green, William B. *Digital Image Processing: A Systems Approach.* 2d ed. New York: Van Nostrand Reinhold, 1989.

Hough, Harold. "Eyes in the Sky: Satellite Surveillance for the Masses." *Soldier of Fortune*, May 1991.

Hudson, Neff. "Air Force Researching Ground-Based Laser." *Air Force Times*, 3 May 1993.

Kiernan, Vincent. "War Boosts Anti-Satellite Weapons Proponents." *Space News*, 6–12 May 1991.

Lambeth, Benjamin S. "Desert Storm and Its Meaning: The View From Moscow." RAND R-4164-AF. Santa Monica, Calif.: RAND, 1992.

Lowndes, Jay. "War's Aftermath Tracked." *Space News*, 22–28 April 1991.

Matthews, William. "The Digital Warrior." *Air Force Times*, 6 January 1995.

Military Newswire, 28 April 1995.

O'Sullivan, Kate. *The L-Bomb.* London: 1995.

Pelton, Joseph N. *Future View: Communications Technology and Society in the 21st Century.* Boulder, Colo: 1992.

Powell, Gen Colin L. *Report on the Roles, Missions, and Functions of the Armed Forces of the United States.* Washington, D.C.: CJCS, February 1993.

Richelson, Jeffrey T. *America's Secret Eyes in Space.* New York: Harper & Row, 1990.

———. "Implications for Nations Without Space-Based Intelligence Collection Capabilities." In *Commercial Observations Satellites and International Security.* Edited by Michael Krepon, et al. New York: 1990.

Saunders, Renee. "War in Iraq Enhances Value of Commercial Remote Sensing." *Space News*, 21 January–3 February 1991.

SPOT Image Corporation. *The Catalogue of SPOT Products and Services.* 1989.

———. "SPOT: The New Way to Win!" Advertisement. *Defense Electronics*, November 1988, 68.

Stares, Paul B. "Deja vu: The ASAT Debate in Historical Context." *Arms Control Today*, December 1983, 2–3.

Toffler, Alvin and Heidi. *War and Anti-War: Survival at the Dawn of the 21st Century.* Boston, Mass.: Little, Brown, 1993.

United States Air Force. Air Force Manual 1-1, *Basic Aerospace Doctrine of the United States Air Force.* March 1992.

———. *The Space Tactics Bulletin.* Falcon AFB, Colo.: Space Warfare Center.

United States Arms Control and Disarmament Agency. *Arms Control and Disarmament Agreements.* Washington, D.C.: Government Printing Office, 1984.

United States Space Command. Joint Doctrine, Tactics, Techniques, and Procedures (JDTTP) 3-14, *Space Operations,* 15 April 1992.

Wickham, Gen John A., Jr., USA, Retired. "The Intelligence Role in Desert Storm." *Signal,* April 1991.

Williams, Sylvia Maureen. "International Law and the Military Uses of Outer Space." *International Relations,* May 1989.

Wilson, Andrew. *Jane's Space Directory 1994–1995.* Alexandria, Va.: Jane's Information Group, 1994.

Wolf, Capt James R. "Implications of Space-Based Observation." *Military Review,* April 1994.

PART IV

High-Ground Perspectives

Chapter 7

National Security Implications of Inexpensive Space Access

William W. Bruner III

The nation which controls space can control the Earth.

—John F. Kennedy
24 October 1960

There has been a great deal of discussion in the space policy community about the technical challenges of gaining economical and routine access to space. Despite this, there has been little written about the opportunities which exist for the development of new missions for US military space forces. Neither has there been much discussion of the security challenges that any proliferation of access to space may present to the United States and to the established international order. Even the most forward-looking space advocates in the Department of Defense (DOD) assume that access to space will continue to be prohibitively expensive and difficult for the foreseeable future, that a US decision not to take advantage of the military potential of space is deterministic for the rest of the world, and that "navigation, communications, and surveillance activities will likely remain the limits of space-based capabilities" for all countries.[1]

Part of this failure to consider the possibilities of a world radically changed by inexpensive access to space is a reaction to the "expectations gap" set up by the gulf between mankind's collective dreams about its future in space and the realities of its achievements so far. The collective public and political mind has been shaped by powerful and convincing fictional images of

This work was accomplished in partial fulfillment of the master's degree requirements of the School of Advanced Airpower Studies, Air University, Maxwell AFB, Ala., 1996.
Advisor: Maj Bruce M. DeBlois, PhD
Reader: Dr Karl Mueller, PhD

space activities that we are not likely to see for a hundred years. Real world, but slow moving and silent, pictures of earth from space taken from small spacecraft with cramped cabins and short mission duration suffer greatly in comparison to images of robust and operable spacecraft spanning the galaxy at faster than light speeds. A century after the Russian Konstantin Tsiolkovsky conceptually solved most of the problems involved in human space flight, over a third of a century since the Soviet sputnik ushered in the space age, and over a quarter century since the United States left humanity's first footsteps on another celestial body, many thoughtful and technically literate people are conditioned by historical experience to think of access to space as an expensive enterprise that is technically difficult, dangerous, and the exclusive province of huge government and corporate bureaucracies.[2]

This stands in stark contrast with the almost giddy optimism that characterized thinking about humanity's future in space at the beginning of the so-called space age. In a 1959 issue of *Air University Quarterly Review*, for example, a serving Air Force officer submitted an article from Command and Staff College that proposed using lunar craters as ballistic missile silos.[3] Even without the Outer Space Treaty of 1967, it is hard to imagine anyone in today's US Air Force making a similar proposal.[4] This change in outlook, conditioned in part by the "expectations gap" and by changes in the fiscal and political landscape, has shaped thinking on this subject over the past 35 years.

As a result of these diminished expectations, as well as competition with other pressing political and economic issues whose solutions don't seem related to space, the American body politic has concluded that routine civil, commercial, and military access to space is not a national priority; not because it is not technically possible, rather, because the experience of the past 38 years argues against it. This is true even at the end of a century of unprecedented technological change. This lack of practical application for access to space and the relatively small size of today's commercial space industry combine to create uncertainty about where the United States should be headed in space, and because of the bureaucratic and technical complexity of traditional space operations, makes it difficult to set a single long-range direction for the nation's efforts in space. In fact, the uncertainty with which the United States views the new medium is reflected in the fact that there was a national commission chartered to determine what should be done in space every year between the *Challenger* accident in 1986 and 1993. (This streak is still unbroken, the National Aeronautics and Space Administration's [NASA] *Access to Space Study* was again released in 1994 and 1995.)[5]

Political, economic, and technological forces may be converging at this point in history, however, to provide the United States with a way to realistically pursue its national purposes in space. With respect to political forces, there seems to be a growing awareness in the US government that something has to be done to lower the cost of space access. Most of the national reports on space over the past decade either say something like "a coherent national effort to improve launch capabilities is desperately needed," or, "above all, it is imperative that the United States maintain a continuous capability to put both humans and cargo in orbit."[6]

Part of the reason for this new awareness is the high operating costs of the current space launch fleet. As overall space budgets fall, operating costs for old-technology space launchers grow as a percentage of total costs. In fact, space shuttle operations presently consume about one-third of the total NASA budget.[7] This is one of the economic forces that is providing incentives to lower the barriers to space access. The other is the growing commercial space business ($5 billion in 1992 sales and growing at a double digit annual rate) and the possibility that new technology will make space access for profit-making enterprises economical for the first time.[8]

Underpinning these new political and fiscal realities is the maturation of technologies that, together, can solve some of the engineering problems that have traditionally forced spacefaring nations to throw away the largest part of their space vehicles. These new technologies: Lightweight materials from the National Aero-Space Plane (NASP) and National Launch System (NLS) programs, advanced propulsion from the shuttle program and from Russia (in fact, the NASA *Access to Space*

Study bases the propulsion system for its reference reusable launch vehicle on the Russian tripropellant RD-704 engine) as well as new computing techniques from the commercial sector have combined to offer the potential for an order of magnitude reduction in the cost of getting into orbit.[9]

If indeed this important part of President John F. Kennedy's New Frontier becomes more accessible, however, there will not only be new opportunities for the United States; there will also be new challenges and obligations that have not been thoughtfully considered. These issues are considered in the pages that follow.

Forces Reducing the Cost of Space Access

An examination of recent technical literature on space launch, foreign and domestic writings on space policy, and the recent activities of the US government seem to indicate that a confluence of bureaucratic, political, and technological forces may be about to lower the barriers to space access; not just for the United States, but for other nations as well. This expanded space access could have implications for US military doctrine, and more importantly, for US national security.

Since the beginning of the space age with the launch of sputnik in 1957, people who have written and thought about using space for national security purposes have proposed crewed space vessels which did not cost significant fractions of the gross national product (GNP), nor did they require an advanced education in computing and astrophysics to operate. Significantly, official Air Force publications of the late 1950s are full of speculation about the implications of such ideas. They proposed using such manned space vehicles for bombing terrestrial targets (a proposal from a general officer on the Air Staff) or for establishing intercontinental ballistic missile (ICBM) bases on the moon.[10] Even Gen Henry "Hap" Arnold, in his prespace age "Report to the Secretary of War" at the end of World War II predicts manned "space ships" as the weapons with which war would be waged "within the foreseeable future."[11] There is not a lot of this sort of thinking about space in today's military writing. In fact, there is no mention of manned military space flight in joint space doctrine, and the astronaut who returns to Space Command to write doctrine informed by experience in the medium is the exception, rather than the rule.[12]

Because of gradually declining faith in the United States's ability to repeatedly and affordably gain access to space, current thinking has become limited to automated systems with throwaway ballistic-missile-derived launch vehicles that do little more than support traditional terrestrial operations.[13] This declining faith in the potential of space power in warfare is partially traceable to perceived treaty and national policy limitations and partially to the expectations gap described earlier, but it is more fundamentally related to the immaturity of existing technology.[14] It simply has not been physically possible to conduct affordable routine operations in the Mach 18 (suborbital) to Mach 25 (orbital) regime with existing propulsion, materials, or flight control technologies. In addition, the early promise of the space shuttle (dashed with high space shuttle maintenance and launch costs and the loss of *Challenger)*, the realization that air-breathing space planes (such as the late NASP program) are not affordable or technically unachievable in the near term, and large expendable launch costs that stretch far into the future, have combined to make the institutions charged with the responsibility of maintaining US access to space averse to changing the status quo and resistant to proposals that change this calculus because earlier proposals for change have come to naught.[15] Doubting that change is possible, they are loathe to accept new ideas or solutions, even if the technologies required to create General Arnold's ideal "space ship" were to become available. In fact, strong institutional forces have grown up around the established methods of doing business, even if they are demonstrably more expensive in the long run and less operable in the short run. Despite this institutional inertia, however, a conjunction of political, economic, and technological forces in the last few years of the twentieth century may finally bring down the cost and technical sophistication required to get into space, turning this period into General Arnold's "foreseeable future."

Confluence of Political Forces

Now, a quarter century after the first human beings set foot on the moon, there is an understanding at the highest levels of the US government that without repeatable and affordable access to space, it will be difficult, if not impossible, to accomplish national purposes in and from space. This understanding is driven by the poor cost performance of current space access methods (this includes low launch rates, high costs, and lack of reliability) and by the resulting lack of hands-on experience with space which, along with the expectations gap discussed earlier, cripples thinking about what can be done in space.

There are as many proposed solutions to the space access problem as there are players in the space policy debate. Nine national-level studies on the issue in eight years, plus innumerable internal studies in agencies across the government, each with its own solution, are indicative of the lack of a coherent vision for what is possible or desirable to do in space. This incoherence is due in part to the immaturity of space technology, and due in part to the fact that few "experts" have actually been in space (because access is still restricted to the select few by the expense of getting there). It has been due in largest part, however, to the struggle for organizational survival in a world of limited resources.

In the past two years, Congress has attempted to break through the roadblock of diminished expectations and lack of policy direction. There now seems to be congressional understanding that lack of assured access to space prevents the United States from pursuing its national purposes there, but at the same time, Congress has shown itself to be dissatisfied with the solutions proposed by the various agencies of the Executive Branch.[16] Congressional dissatisfaction with Executive Branch space policies has traditionally caused it to do two things: first, to cancel every new expendable launch vehicle (ELV) proposed by NASA and the DOD in recent years (the Advanced Launch System [ALS], NLS, and Spacelifter), and second, to direct a series of studies to address the problem.[17] Immediately after the cancellation of Spacelifter and the effective cancellation of NASP, Congress directed NASA and the

DOD to study space access in the FY93 NASA Appropriations Act and in the FY94 Defense Authorization Act.[18] These studies, released within three months of each other in early 1994, used the same technology base and, in some cases, the same study participants; but came up with diametrically opposed conclusions about the best way to solve the nation's space access problem (perhaps for some of the bureaucratic and organizational reasons outlined above).

The Case for and against Standing Down. With large and continuing requirements for access to space, both the DOD and NASA have little choice but to continue their costly present launch operations as they try to solve this problem. The US government mission model for the next 15 years averages about 30 launches per year, while industry will account for roughly 15 more.[19] These continuing requirements include obligations to our International Space Station (ISS) partners for space station assembly missions, DOD launches of national security payloads, and the replacement of aging communication and sensor satellites to address shortfalls highlighted in Operations Desert Shield and Desert Storm. Although the DOD could use foreign launch services to get its "must carry" payloads into orbit, former US representative Dave McCurdy and his coauthors call such a possibility "truly disturbing" in an article for *Strategic Review* in 1994. Dependence on foreign launch vehicles in time of war or crisis could turn out to be even more costly than the status quo. The private sector, on the other hand, does not mind going offshore for launch services, but with an already negative balance of payments, this poses questions of US economic competitiveness that are also ultimately questions of national security. As recently as 1979, the United States launched 100 percent of worldwide nongovernment satellites. Today, that figure is closer to 40 percent.[20] This situation has deteriorated to the point that Charles Bigot, the chairman of the European launch consortium Arianespace, no longer considers the United States to be a major competitor in the $1 billion commercial launch business because "to develop a really new transportation system you need probably between six and ten years [and] I don't believe that America will do it."[21]

With foreign officials dismissing the United States as unable to compete, with a fiscal vise closing on both NASA's and DOD's launch budgets, and with a continuing national need for sovereign space access, there seems to be a consensus growing in Washington and elsewhere that something has to be done about fixing space launch.[22] The space policy community also recognizes that the United States must simultaneously fly the missions that are necessary to the fulfillment of national policy goals. This is the context within which the following discussion takes place.

The Case for and against the Status Quo. There is always the option of doing nothing to build on the technology developed for the programs that have already been canceled. It would save on the cost of a new space launch vehicle in a time of declining budgets and would decrease the technical risk of developing new spacelift technology when the time finally comes to field a new launch vehicle. However, there are three arguments against this approach.

The first is that the US's foreign competitors are taking more and more of the launch market away. As the Vice President's Space Advisory Board on the Future of the US Space Launch Capability Task Group (the "Aldridge Commission") report put it,

> A decision by the Administration or the Congress not to fund a new, reliable, low-cost operational space launch capability is a de facto policy decision to forgo US competition in the international space launch marketplace, a mandate that the US government will continue to pay higher prices than necessary to meet future government launch requirements, and acceptance of less reliability, less safety, and higher risks for space flight than our technology is capable of providing.[23]

The second argument against the status-quo approach is that the United States has essentially pursued this policy by default after the series of program cancellations discussed earlier. This policy has gotten the nation no closer to solving the problem, but has cost several billion dollars ($2.4 billion for NASP and $600 million for advanced logistics system [ALS] and NLS).[24] If the nation does nothing with the technology from these programs, then this money will have been spent for naught. The third reason, as outlined above, is that the cost of space launch is a large part of both NASA's and DOD's continuing costs. Although the shuttle program is under continuing pressure to cut operating costs, its share of the NASA budget increases as the overall NASA budget decreases. The same can be said for the DOD space budget. As overall budgets decline and launch costs do not, there are not enough resources left over for either organization to carry out its other tasks. This is where much of the political incentive to "do something about space launch" comes from. Both the DOD Space Lift Modernization Plan (SLMP) and the NASA Access to Space Study considered the option of remaining with the status quo. Both concluded that the continuing high cost of their present space launch operations were not supportable. In addition, both concluded that waiting would not, in the end, save money. As the NASA Access to Space Study states, "delaying the decision of which space architecture to select by four or five years but not funding a focused technology phase will achieve nothing, since the lack of a focused technology program during that period will not reduce the risks of developing an advanced technology vehicle. Therefore, the choices available in four to five years would be exactly the same as those we face today."[25] NASA and the DOD both seem to agree that there is nothing to be gained by waiting.

The DOD Space Lift Modernization Plan: the Case for and against Expendables. The DOD study, the SLMP, concluded that pursuing new reusable launch vehicle technology was "controversial" due in part to the risk.[26] DOD recommended, therefore, that it remain committed to the evolutionary development of its present stable of aged Atlas, Delta, and Titan launchers, while investing in incremental technology improvements. The SLMP itself admits that this would deliver little or no per launch or per pound to orbit cost savings.[27]

Despite the DOD's enthusiasm for this new evolved expendable launch vehicle (EELV), however, there is no new money in the president's budget for either new or evolved expendables. Congress has appropriated $40 million in the FY95 Defense Appropriations Bill for an evolved expendable, but that is far from the $2 billion estimated total program cost, so the Air Force plans to take $400 million out of its own budget over the FYDP to fund it.[28] As in the cases of ALS, NLS, and

Spacelifter before it, EELV is a conservative approach based on what is essentially 1950s ballistic missile technology, delivering small savings in per launch costs. It is, in fact, intended to be even more technologically conservative than earlier expendable programs to cap the development cost at $2 billion.[29] Even with such a cap, however, these development costs are still of the same order of magnitude as those for a major weapon system. With this multibillion dollar development cost, the EELV will narrow, not reduce, the range of medium lift costs from $35–$90 million to a projected $50–$80 million.[30] Although standardization of the launch fleet to a single vehicle/contractor combination from the separate and costly Atlas, Delta, and Titan programs will bring some savings, it is impossible to get away from the fact that "staged expendable" means, in effect, building two vehicles every time you fly, mating them meticulously, and sinking both craft in the ocean when the mission is complete. As W. Paul Blase says in the March 1993 edition of *Spaceflight* magazine,

> All current rocket launchers are derived from 1960s era ICBM designs, and man-rating procedures are merely ways of producing man-rated ammunition. Rocket designers are conservative by their nature and the high cost of both the vehicles and their payloads causes them to refine the same basic concepts continuously to finer and finer degrees, taking few risks with radically new ways of doing things. This has resulted in a situation very much like trying to pull a semi-trailer with a racecar. Like a racecar, ICBM-based rockets are designed to get maximum performance from minimum equipment. Technology is pushed to the very brink to wring out that last ounce of thrust. However, it is an engineering truism that when one gets near the theoretical limits of a system, every additional 10 percent increase in performance doubles the systems cost and halves its reliability.[31]

The NASA Access to Space Study—the Case for and against Reusables. The civilians at NASA, using essentially the same data, came to a different conclusion. They believe that neither ELVs nor the shuttle are suitable launch vehicles for the twenty-first century. They believe that the time has come for the nation to move to the next technological level. Accordingly, NASA's Access to Space Study recommended that the United States "adopt the development of an advanced technology, fully reusable single-stage-to-orbit rocket vehicle as an Agency goal."[32] In addition, NASA concluded, "leapfrogging" the United

States into a next-generation launch capability would place the nation in an extremely advantageous position with respect to international competition.[33]

As a result of the separate positions taken by the agencies primarily responsible for the nation's access to space, the Executive Branch has decided not to focus on a single strategy for space access in the twenty-first century. Instead, the new national space policy accepts the NASA position on sprinting ahead to reusable launch vehicle technology while also maintaining a core expendable capability in the interim (managed by the more risk-averse DOD).[34] The language of the new NASA *Implementation Plan for the National Space Transportation Policy* makes this clear. Administration policy, NASA says, "calls for a balanced two-track effort; first, to ensure continued access to space by supporting and improving our existing space launch capabilities, consisting of the Space Shuttle and current ELVs; second, to pursue the goal of reliable and affordable access to space through focused investments in, and orderly decisions on, technology development and demonstration for next-generation reusable transportation systems."[35]

This two-track approach, while it satisfies the competing bureaucracies of NASA and the DOD, and appears to manage risk prudently, does not seem to be fiscally or politically realistic. As outlined above, every expendable launch vehicle that DOD and NASA have proposed in recent years has been terminated by Congress.[36] These cancellations had less to do with the merits of the respective programs than with the limited launch savings over existing launch vehicles and high program costs (relative to those same limited savings) that are characteristic of expendables.[37] With this in mind, a space policy that calls for two new program starts, one of which is an expendable much like those canceled in the recent past, has little likelihood of continued funding from Congress. It seems more prudent, and politically realistic, for the Executive Branch to decide early which track it wishes to pursue, and then to focus its efforts there.

What explains the significant difference between the two recommendations? It is important to answer this question because the political viability of the president's two-track approach depends on the ability of NASA and DOD to convince Congress of the soundness of the reasons underlying their respective recommendations over the lifetimes of the two programs. In an era of limited resources, the recommendation that fails to stand up to the scrutiny of lawmakers will not survive, no matter how strongly its bureaucratic constituency believes in its merits. The rest of this section will attempt to determine the reasoning underlying the two recommendations, and to assess their respective political viability in the Washington of the late 1990s.

The Political Viability of the RLV and ELF. The first question in determining the viability of the respective approaches is whether technology advanced so far between the two reports that reusable launch vehicle development suddenly became more possible and less "controversial." This is not likely. In fact, the NASA report was released first and DOD used the NASA study for purposes of comparison.[38] The NASA report's assessment of the technology's potential to solve the nation's launch problem seems, therefore, to have been driven by some other factor. If the level of technology is acknowledged by both reports as being within striking distance of an operational reusable vehicle, then, to observers in Congress, NASA's choice would appear bold and the DOD's choice suffers by comparison. It would be difficult for DOD to make the "immature and risky" technology argument and maintain the funding level for the old technology EELV when NASA's flying advanced technology demonstrators are competing for the same dollars. (This calculus would change, of course, if either program ran into major technical trouble.)

The second question is whether the two conclusions were driven by differences in the risk tolerances of the two institutions. Perhaps so. The DOD argues, correctly, that the stakes are higher in the national security arena, and that the nation can ill afford another launch hiatus caused by exclusive reliance on high-risk technology (as it suffered after the *Challenger* explosion). NASA argues, also correctly, that risk has been reduced by recent advances in lightweight materials, thermal protection, high speed computing, the attendant flight control and systems integration software, and other technologies. Even though these advances do not reduce the risk of the reusable launch vehicle to zero, NASA, it seems, is willing to take some programmatic risk to protect US competitiveness in the international launch vehicle technology race. Congress is likely to be more sensitive to this concern than to DOD's national security concerns in the wake of the cold war.

Along the same lines, risk tolerance is one thing, but did the two institutions have differing perceptions of the same technical and fiscal risks? On the subject of the same prospective (RLV) technology that NASA considered, the DOD study says, "A fully reusable, single stage to orbit space plane is an exciting concept to all the space sectors and industry alike. It offers benefits of responsiveness, reliability, operability, and very low cost per flight which are universally agreed to be desirable. However, the practicality of achieving those benefits is controversial."[39] NASA, on the other hand, concluded that, "single-stage-to-orbit vehicles appear to be feasible because of reduced sensitivity to engine performance and weight growth resulting from use of near-term advanced technologies (e.g., tripropellant main propulsion, Al-Li [Aluminum-Lithium] and graphite-composite cryogenic tanks, graphite-composite primary structure, etc.). An incremental approach has been laid out to reduce both technical and programmatic risk."[40] Again, with the same information, NASA reaches the more forward-looking conclusion.

NASA may be looking further forward, but did this cause it to manipulate the numbers so that the bold RLV solution was made to look unrealistically inexpensive? The similarity with the DOD figures makes this doubtful. DOD estimated the cost for a reusable launch vehicle program (technology and engineering development) at between $6.6 and $20.9 billion, while NASA estimated the same costs at $17.6 billion.[41] Though the upper end of the DOD range is higher, there does not seem to be a significant enough difference in the estimates alone to cause the wide discrepancy between the two recommendations. If DOD was concerned that it did not have enough money to go it alone (which, given the office of Secretary of Defense (OSD) Bottom-up Review funding levels that were the SLMP's starting point, seems a reasonable assumption), it could have proposed a joint national launch strategy with NASA (as with Spacelifter and NASP), unless of course there were unstated reasons for not doing so.[42] These unstated reasons might include the perception that because cooperation with NASA on ALS, NLS, Spacelifter, and NASP was difficult, and each program ended badly, a DOD-only program might have a better chance of success (although the DOD has managed to get quite a few programs canceled on its own). Unfortunately for the DOD, Congress has a long record of preferring cooperative programs with joint program offices over competing and redundant programs.[43] Unpleasant experiences with previously canceled programs are not a politically palatable justification for the DOD going it alone.

Was there a stronger bureaucratic constituency for expendables than for reusables in the DOD? The answer to this question may lie in the strong institutional tie between the expendable ballistic missile acquisition community at the Air Force's Space and Missile Center in Los Angeles and the Air Force Space Command at Colorado Springs, Colorado. The Space and Missile Center (formerly the Ballistic Missile Office) managed all Air Force ballistic missile acquisition during the cold war. It also managed NLS and is the home of the program office for EELV. Space Command, which was recently assigned responsibility for the peacetime organization, training, and equipage of the ICBM force, has launched the majority of the payloads it now controls on expendables (and the rest on the partially expendable space shuttle), and now is staffed with officers who spent years preparing to carry out the strategic missile mission with expendable rockets. If there is an institutional tie between flying officers and the program offices at Wright-Patterson Air Force Base (AFB) where airplanes are acquired, then there may be a similar tie between the missile officers at Space Command and the Space and Missile Center at Los Angeles, California.

There was a small constituency for RLVs inside DOD who helped in the preparation of the SLMP, but it was confined to the narrow group within Strategic Defense Initiative Organization (SDIO) who had developed the DC-X subscale RLV demonstrators.[44] If there were a single difference between the two studies, this may be the most significant. In contrast with the situation within DOD, there was a strong constituency for RLVs within NASA. In fact, a group of engineers at NASA around 1991 began publishing a number of papers on the feasibility of rocket-powered-single-stage-to-orbit vehicles.[45]

This project is not intended as a study in bureaucratic decision making, it is simply intended to serve as a tool for understanding how bureaucratic forces inside NASA and DOD drove the president to a "two-track" policy, when there were strong political trends favoring one "track" over the other. In fact, a senior administration official has noted strong congressional interest in the RLV.[46] Congress was also willing to back this preference up by voting more money for the RLV subscale demonstrator in the FY95 Defense Appropriations Bill than for initial work on the EELV.[47] The EELV's chances for survival, given the unfortunate precedent of ALS, NLS, and Spacelifter, would not be very good in the best of circumstances, but given the real or perceived competition between an old-technology ELV and a flying RLV advanced technology demonstrator four years hence, Congress is even more likely to cancel the EELV. NASA has scheduled the advanced technology demonstrator RLV to fly no later than July 1999 (the 30th anniversary of the first moon landing, a coincidence to be sure).[48] DOD's EELV, on the other hand, is projected to fly for the first time in 2000.[49] In today's resource-constrained environment, an expendable launch system on the drawing board will find it very difficult to compete for dollars with a flying prototype RLV. The EELV's first flight may very well be a year late and a couple of billion dollars short. As Luis Zea says in the December 1993 issue of *Final Frontier*, "Recycling ideas like the National Launch System and the more recently proposed Spacelifter family of expendable boosters appears to be politically dead."[50] EELV program managers are working hard to prove him wrong, but the weight of history is against them.

Convergence of Economic Forces

Even if RLVs, arguably the precursors of Hap Arnold's space ships, are more politically viable and fiscally realistic than EELVs, they still may not be affordable enough to avoid cancellation themselves. If Congress won't vote $2 billion for an

EELV, why should it vote $20.6 billion, $17.6 billion, $6.6 billion, or even the $5.5 to $6.5 billion figure quoted by former astronaut Pete Conrad for a reusable launch vehicle?[51] Perhaps it would be cheaper to stay with current ELVs or the shuttle. Unfortunately, as discussed earlier, the cost of operating today's launch fleet will not permit that. The DOD's current expendable fleet costs $2.5 billion a year (about 20 percent of the DOD space budget), while NASA launches about eight shuttles a year for $4.3 billion (approximately 31 percent of NASA's budget). This is the source of urgency behind new launch vehicle development. While EELV makes a marginal improvement in per mission and operations and support costs, the RLV promises to bring launch costs down by a factor of five to 50 (to between $1 and $10 million per flight).[52] The cost savings over the life cycle of the single stage to orbit (SSTO) reusable "space ship" would be significant. The DOD estimates the annual operational cost of a fleet of four such vehicles at $0.5 to $1.5 billion (as opposed to the $6 billion plus for today's expendables and the shuttle).[53] In other words, even if the DOD is right about the high up-front investment required, the nation would save at least $4.5 billion per year. NASA conservatively estimates that payback on the initial investment will occur approximately nine years from RLV initial operating capability.[54] If this is accurate, it becomes difficult to make an economic case for remaining with the status quo. The rest of this section tries to determine whether there is a positive economic case for reducing the cost of access to space (in addition to the weaker negative motivation of dissatisfaction with the status quo). The analysis will also attempt to deal with some of the fiscal issues raised by RLV opponents.

The Economic Case for and against RLV. Even people who are skeptical about rocket-powered SSTO understand that the only reason to make the large up-front investment in RLVs is the savings in life-cycle costs. Some opponents of the technology believe that the projected savings in life-cycle costs are too good to be true. There have to be, they believe, some "hidden costs" to SSTO such as; upper stages required to reach geostationary orbit, the inability to carry heavy payloads that will force the DOD to retain the heavy Titan IV expendable for national security payloads, or the expense of building a huge liquid hydrogen storage infrastructure.[55] These criticisms, however, back a conception of new ways of doing business in a world where spacecraft have some of the operability of aircraft. (As will be discussed shortly, this conceptual limitation is even more dangerous in the national security area.)

Analysis of these three charges based on an understanding of how air transport works may be useful in determining whether there are legitimate economic reasons not to proceed with SSTO.[56] The parallel between air transport and reusable space transport operations may not be complete, but it is probably closer than the ballistic missile model in use today.

Charge I. Opponents claim that SSTO RLVs could not carry the significant number of DOD, NASA, and commercial payloads bound for geostationary orbit (22,300 miles equatorial orbit) since the NASA SSTO reference configuration is designed to carry a 25,000 pound payload to the planned international space station orbit at just 220 nautical miles altitude. The critics claim that the SSTO would have to carry an expendable upper stage (adding $16 million to its per launch costs for a total around $26 million, wiping out enough of SSTO's per launch cost advantage, making it uneconomical), or that the government would have to fund a multibillion dollar reusable upper stage to get the per launch costs down to $14–$16 million (with Congress in no mood to fund additional program starts.)[57] Further analysis, however, reveals an answer that is entirely different for three reasons not considered by the critics.

1. On-Orbit Refueling. During the Persian Gulf War, when planners chose targets in Baghdad for aircraft stationed in southern Arabia, a refueling tanker rendezvous was scheduled as a matter of routine. This is what reusable launch vehicles will enable the United States to do in space. Work has already been done on cryogenic fuel transfer in a microgravity vacuum environment, and even the US Air Force has considered increasing the operational availability of space assets by refueling them with ELVs.[58] (Although these ideas never flew because the high cost and long delays of ELV launches made such operations impractical, RLVs could bring them back to life because of their lower cost and greater responsiveness.)

Developing and using these techniques for on-orbit refueling, reusable launch vehicles can themselves become "reusable upper stages" at far less cost than a new program start. The cost for the "tanker" would not be analogous to that of specialized air-breathing tankers for aircraft refueling in the illustration above, and would not require the development of a new vehicle. Instead, it would mean changing out a standard RLV payload for fuel and refueling connections. Developing these new techniques will be difficult, similar to the work involved in making aerial refueling a routine and safe operation. Although ground-based experiments using possible methods of refueling in a microgravity vacuum environment have been conducted, no such experiments have been conducted in space. There are the obvious problems of gaseous venting in vacuum, frozen connections, and unknown propellant flow characteristics in microgravity. Mission needs will drive the development of this capability, not engineering curiosity. If the RLV is as operable as NASA believes it will be (seven-day turnaround with a 0.95 probability of on-time launches), then there will be a strong incentive for civil, commercial, and military operators to exploit the potential offered by that operability.[59] Refueling in space is one way to do this, allowing operators to accomplish missions that are not otherwise possible without developing entirely new vehicles.

Space ship operators would, however, have to ask themselves several essential questions before they proceeded with any refueling modification. Can we do without the ability to get heavy payloads to geostationary orbit (GEO)? Probably not, since the majority of the $5 billion space industry is presently in medium-weight geostationary communications satellites.[60] Can we afford to operate ELVs or partial reusables far into the future? Both the NASA and DOD space access studies say no. Can we afford the billions of dollars that it will take to develop a new orbital transfer vehicle?[61] Probably not, and especially if operators have just spent billions of dollars to buy an RLV. Is there a possibility of extending the range of the RLV to capture medium-weight geosynchronous satellites without the expense of a new program start? There may be, given the encouraging preliminary results of the refueling studies cited above. If so, then a relatively small investment in designing a new payload for an existing RLV seems eminently more sensible than developing an entirely new vehicle for a single purpose. Given these answers, it seems likely that the refueling option will be attractive to RLV operators after their ability to get to low Earth orbit (LEO) routinely has been proven. Again, this modification is not trivial, but engineering studies suggest that it is well within the realm of possibility.

2. Lower Insurance Costs. The ELV is a lot like an artillery shell. Once launched it cannot be recalled. That is why, at every US ELV launch, there is an official at a console monitoring the status of the mission and the ascent trajectory. If the mission deviates a given amount from predetermined parameters, the range safety officer detonates the vehicle's destruct package (if the vehicle hasn't already destroyed itself). RLVs, on the other hand, are intended to land safely after every mission and have built-in mission abort capabilities. The fact that there is no destruct package on the first flying subscale RLV model is a matter of some importance to its program managers.[62] If an engine fails after takeoff, the vehicle executes an emergency landing as the subscale RLV did after an explosion during a test flight in June 1994.[63]

Beyond the obvious material savings, this has enormous insurance implications. At present, payload insurance rates for expendable rockets are a significant part of launch costs for commercial concerns. With insurance rates around 18 percent of the total of satellite cost plus launch cost, any reduction in risk could make for significant savings.[64] Assuming a still relatively new reusable launch vehicle that has demonstrated its intact abort capability at least once, we might guess that satellite insurance companies would give commercial space ship operators an insurance discount, perhaps charging 10 percent of launch value rather than 18 percent.[65] For a $75 million medium-weight geostationary communications satellite on a $60 million expendable mission with the same payload capacity as an RLV to LEO, it turns out to be over $66 million in savings for a single mission which more than covers the cost of up to five RLV "tanker" missions insured for their launch costs at a 10 percent rate.[66] In fact, a $60 million expendable mission launching a $75 million commercial communications satellite to geostationary orbit with $25 million in insurance will cost more than six $10 million reusable missions with one payload carrier and five refueling missions. There would be a total of $13.5 million in insurance costs at 10 percent of satellite plus launch cost for the RLV ($148.5 million total launch, payload, and insurance costs). Of course, to make money, the launch operator would fly as few tanker missions as possible. The amount of fuel brought up by an RLV designed to meet NASA's X-33 requirements on five missions would be far in excess of what was needed to get to GEO. In fact, it would be enough to get to the moon.

In addition, the refueled RLV would be able to take the entire 20,000 pounds to GEO, while the ELV would have to use up some if its payload weight to LEO to get the satellite into a geosynchronous transfer orbit. The numbers outlined above suggest strongly that the enterprises with RLVs would enjoy a significant competitive advantage over those still flying ELVs simply due to insurance savings. This would not directly affect DOD launch costs, but if a significant number of commercial payloads migrate to RLVs, then ELV production rates will slow down and prices will go up. A similar slowdown in Titan IV production has been the principal cause of a 60 percent increase in launch costs.[67]

3. Follow-On Missions. This brings us to the third reason that the "additional cost for upper stages" argument is fallacious. If each of the five-tanker missions in the exaggerated example above brings up 25,000 pounds of fuel, the RLV carrying the payload would not only have enough fuel to deploy the communications satellite, it would also have enough fuel to perform a follow-on mission such as retrieval of the older satellite it is replacing (or even to go to the moon with one more tanker mission).[68]

Using a derivation of the rocket equation, $\Delta\upsilon = g\ I_{sp}\ \ln (M_0/M_E)$, a gross lift-off weight of 1,000,000 pounds; a PMF of 0.90; a resulting vehicle empty weight of 100,000 pounds; space shuttle main engine vacuum I_{sp} of 453 seconds, and an approximate $\Delta\upsilon$ of 12,000 fps required for translunar injection from earth orbit; an RLV could take on six 25,000 pound-fuel loads and reach the moon for a lunar survey mission similar to the Ballistic Missile Defense Office's recent *Clementine* mission.[69] Getting 18,000 more fps (two times lunar escape velocity) for an orbit circularization burn, landing, and takeoff would require 21 more missions (which is less than NASA's projected space station construction mission model using a far less operable spacecraft).[70] This mission also requires a vertical takeoff, vertical landing (VTVL) RLV.

This may seem a massive undertaking for a mission that does not seem to have much national priority, but the operability of the RLV may make such a trip useful for economic reasons to be discussed shortly. That said, when the nation is ready to return to the moon, a $28- to $280-million mission (28 RLV missions at $1 to $10 million each) modifying a vehicle whose cost is recouped in earth-to-LEO operations would be far more cost-effective than paying the development cost for purpose built orbital transfer vehicles, lunar landers, or other specialized vehicles. It is cost competitive with a single Titan IV launch and less expensive than a space shuttle mission. There is no cost comparison with expendables for the retrieval or lunar missions, because no matter how much money is spent on a single ELV mission with present or evolved vehicles, these multiple missions are not possible without developing other specialized expendable vehicles.

This extreme example makes the point: thinking about reusable launch vehicles in the same way as expendables can prevent the analyst from seeing opportunities that will be apparent soon after RLVs become available. As this example also illustrates, it is likely that many more opportunities will arise once the space operability revolution takes place, but these opportunities are so difficult to foresee that they cannot reasonably be used as justification, economic or otherwise, for RLV development. There are, on the other hand, enough possibilities that earthbound analysts at NASA and elsewhere are able to justify the economics of proceeding along this development path if only to reduce today's high operating costs.

Charge II. Opponents also charge that first-generation RLVs will be unable to loft heavy payloads. Where the first charge was that the RLV compared unfavorably with medium-lift ELVs, the second charge is that the RLV cannot compete at all with heavy lifters. On the face of it, this claim is accurate as long as the launch operator limits the mission to a single launch. Today's space community has been conditioned to think of getting satellites into orbit as unitary events, with each launcher custom-tailored to each payload. If a payload weighs 40,000 pounds and its mission is in geostationary orbit, conventional wisdom suggests the need for a heavy-lift vehicle plus a transfer stage to take the whole package there at the same time. Again, this sort of thinking will be inadequate for the age of the reusable launch vehicle. In the RLV world, as in the rest of the transportation world, if the cargo is too heavy to take in one trip, the solution is to put it in two boxes and make two trips. As David C. Webb, president of the International Hypersonic Research Institute and former member of President Ronald S. Reagan's National Commission on Space, suggests in his Aerospace Industries Association of America (AIAA) paper, "Spaceflight in the Aero-Space Plane Era,"

Potentially, the way around this problem is to break the platform up into smaller chunks and launch them on smaller launchers. It would be even less expensive to do this with aero-space planes. [Something he defines as: "aero" because such vehicles utilize the atmosphere, "space" because they go into space, and "plane" because they are operated like airplanes. The SSTO vehicle, therefore, is considered an aero-space plane even though it may not look like an airplane.] It might seem that the large military reconnaissance satellites could not be launched on aero-space planes. However, one possibility could involve splitting the satellite into two modules that are launched separately and assembled in orbit.[71]

If a Titan IV launch costs from $250 to $320 million per launch, then one could theoretically take the payload up as separate components, launching it in 25 to 32 missions at $10 million per trip and still break even. In fact, work-on-line replaceable units for satellites (similar to those in the aircraft world) is presently under way at the US Air Force's Phillips Laboratory. Even though the laboratory is working on modular satellite construction for standardization and cost-savings purposes, some of this work could be directly transferable to the on-orbit assembly idea. Again, the extreme example makes the point. It is poor analysis to make the blanket assumption that a medium-lift RLV will be unable to carry heavy payloads. The operability revolution inherent in RLV technology will enable new solutions to old problems, and create economic and military advantages for the United States in space that are difficult to foresee. This will be addressed in further detail in the discussion of the national security implications of the RLV.

Charge III. Finally, opponents charge that because SSTO requires high I_{sp} fuels, which today means cryogens such as liquid hydrogen, the high cost of the terrestrial hydrogen infrastructure necessary to support robust operations will be prohibitive. This is more an argument against launch sites at every airport than it is against the cost effectiveness of RLVs in replacing the current fleet of expendables and semi-expendables. Many of today's launch vehicles use cryogens, the space shuttle among them. In fact, the shuttle uses the same cryogens that NASA plans to use for its planned RLV demonstrator, the X-33. There will not be large fuel infrastructure costs associated with the transition from the shuttle to RLVs. In fact, as part of the X-33 Cooperative Agreement Notice (CAN), NASA sets out as a program goal that,

> the flight vehicle shall be capable of unplanned landing at alternate landing sites with minimal support equipment/ facilities, e.g.,
> - No existing cryogenic facilities, launch stands/equipment, etc.
> - Self-ferry of flight vehicle between landing and launch sites. . . . Equipment required to repair, process, and return vehicle to launch site shall be transportable.[72]

If indeed the infrastructure requirements for ferry missions are minimal and NASA finds it useful to launch some missions from White Sands Missile Range, New Mexico, for extra energy (because of its elevation), some missions from Florida for eastward equatorial orbits, some missions from Vandenberg AFB,

California, for polar orbits, and some from higher latitudes for higher inclination orbits, then the government is likely to build the skeleton of an infrastructure that private interests can use to begin commercializing the vehicles. Among past examples of infrastructure investment for national purposes that turned out to have enormous commercial implications was the worldwide network of coaling stations for steamships in the late nineteenth century. This network, built by the industry and governments of the great naval powers, became an essential element of national security and a significant factor in the worldwide trade that built the United States's national wealth.

Another example was the infrastructure required to support the automobile. In the early twentieth century, when Henry Ford decided to mass produce the automobile, the infrastructure argument would have gone something like this, "Henry, how do you expect to make any money? There are no roads to run those things on and everyone lives right next door to the store where they work. Even your factory workers are within streetcar distance of your plant. No one will spend the millions and millions of dollars to build the roads or the petroleum-based fuel distribution infrastructure for these things to run on." The critic would have been absolutely right, if Model Ts provided the same amount of productivity per mile as horse carriages.

Similarly, the infrastructure cost critiques would be right if RLVs are only as productive and operable as ELVs. However, if there is money to be made or saved by operating RLVs, then the cost of infrastructure will be amortized through savings and profit, and as the DOD estimate of annual cost savings over expendables shows, those savings are in the billions of dollars per year. If one adds the profit taken from foreign expendable launch operators, one could buy a lot of liquid hydrogen and the infrastructure required to handle it.[73]

The principal economic force acting to drive interest in and funding for the RLV is the desire to reap the benefits of the cost savings inherent in its operability. Launch costs are devouring the NASA and DOD budgets, and both institutions know they have to do something to cut costs in the face of continuing budgetary pressures. So far, this is the principal economic force acting as a stimulus to RLV development, but there are indications that it may not be the only one.

Private Sector Argument for RLVs. Private sector interest in a reusable space launch vehicle and in a possible reusable hypersonic point-to-point (as opposed to earth-to-orbit) cargo carrier is another economic trend working to stimulate RLV development. The US government has attempted to take advantage of this interest by pursuing a unique acquisition approach in the development of the RLV, offering "Cooperative Agreement Notices" rather than traditional requirements statements to begin the acquisition process. NASA, to maximize the private sector's intellectual, entrepreneurial, and financial contribution to the RLV program, has issued a CAN for an experimental flying vehicle, the "X-33," that allows the private sector, for the first time, to propose and include independent research and development as part of their corporate contribution.[74] This new approach is designed to keep NASA engineers from driving RLV design toward a predetermined solution that meets only NASA's needs, and not industry's. In fact, some NASA centers have had difficulty adjusting to the new reality, publishing reports that seemed to favor one RLV solution over another, and earning a written reprimand from NASA headquarters for their trouble.[75] The objective of the CAN, NASA says, is to

> stimulate the joint industry/Government funded concept definition/design of a technology demonstrator vehicle, X-33, followed by the design/demonstration of competitively selected concept(s). The X-33 must adequately demonstrate the key design and operational aspects of a reusable space launch system. As a minimum, the scaleability and traceability of the X-33 airframe, cryogenic tanks, and thermal protection system (TPS) to the corresponding proposed SSTO rocket must be identified.[76]

As of this writing, three prime contractors, Lockheed Martin, Rockwell International, and McDonnell Douglas have entered competitive SSTO concepts. One of their designs is scheduled to be selected by July 1996 for construction and flight as early as possible but not later than July 1999. NASA will make every effort to accelerate this schedule and will assist the selected contractor(s) in any feasible manner to fly the advanced technology demonstrator before July 1999.[77] At least one other private company sees the economic potential of reduced cost access to space and is pursuing RLV technology outside of the CAN process. Kistler Aerospace is using the profit its founders made from their Spacelab venture (a private/NASA cooperative project that has flown on the shuttle) to finance their own reusable launch vehicle. They plan to raise $400 million from private investors and to put up $100 million of their own money to fund the estimated half-billion-dollar program cost. Though industry and government officials give Kistler little chance of success, given estimates of RLV development costs in the billions, the fact that investors are willing to risk $100 million of their own money to pursue the possibility of reusable space ships is another strong indicator that economic forces are in place that are providing a push to the technology.[78]

There are other potential commercial uses for an RLV that have spurred some interest from the private sector. Science and science fiction writers have described intercontinental ballistic passenger and cargo spaceships for years. In Philip Bono and Kenneth Gatland's seminal 1969 book, *Frontiers of Space*, the authors propose a 200-foot-tall intercontinental passenger/cargo carrier for suborbital missions which could haul 1,200 passengers 7,500 miles in slightly over one-half hour. A second idea, *Hyperion*, was a conical VTVL SSTO (much like McDonnell-Douglas's current ideas) that could carry 8,100 pounds to orbit.[79] In the December 1993 issue of *Analog* magazine, science writer G. Harry Stine calls suborbital hops the "hidden market" for SSTO services. As Stine points out, "any SSTO spaceship that can take a payload to orbit can also deliver passengers and cargo to any place in the world in less than an hour."[80]

This could all be dismissed as idle speculation but for the fact that Federal Express (FedEx), one of the leading on-time freight express companies in the world, is giving its support to the design review processes of all three teams competing to develop the X-33.[81] The FedEx interest on its own will not build the space ship, but it does seem to indicate that there are uses beyond access to orbit for reusable hypersonic technology. This could provide an even stronger economic stimulus for the near-term development of single-stage rocket technology in light of the fact that the $200 billion size of the commercial air transport market dwarfs the total worldwide space-market figure of $5 billion.

There are also other missions for RLVs outside of conventional earth-to-orbit NASA/DOD mission models that could drive the market for them. Orbit-to-earth return missions may also turn out to be nearly as lucrative (e.g., space debris cleanup, on-orbit satellite repair and salvage, and what might be called single-stage-to-earth [SSTE] operations). The economics of these missions, however, are difficult to foresee and were already proposed as missions for the space shuttle in the early 1980s (and then turned out poorly). It may, in fact, be so difficult to foresee the cost implications of SSTE missions that they are not useful as economic justification for SSTO. The ability to routinely rendezvous with and retrieve material from space may, however, be an interesting capability that space ships give their operators which has enormous national security implications.

Other possible missions are even more speculative (such as space tourism, deep space exploration, military presence missions); using them as economic justification for RLV development quickly degenerates into an argument over causality. In addition, these missions are not relevant to the debate in the near term. RLV space ships are justifiable on the economic grounds of cost savings to be gained by eliminating ELV and shuttle operating costs, by reducing the need for orbital transfer vehicles and Upper Stage Development programs, and (if FedEx's interest in X-33 is an indication) on the grounds that there are air transport missions they can perform at hypersonic speeds.

Technological Forces

Finally, recent technical advances provide the underpinning for some of the economic and political trends discussed above. Although space ships have been foreseen at least since the advent of the German A-4 rocket (known to the Allies as the V-2) at Peenemünde on the Baltic coast during the Second World War, they have not been technically possible because the weight of the materials and the specific impulse of the rocket engines available did not permit single-stage vehicles to achieve orbit.

As early as 1946, US rocket designers believed that it was possible to build SSTO vehicles with lightweight materials (usually allowing pressurized propellant tanks to double as vehicle structure to save weight as the early Atlas ICBM did) and high specific impulse oxygen/hydrogen engines.[82] Unfortunately, neither lightweight materials nor LOX/LH_2 engines were available in the late 1940s. A LOX/LH_2 engine had to wait until Centaur in the 1960s and the shuttle became the first vehicle to use LH_2 at liftoff in 1981.[83] Early drawings of these prospective single stage vehicles bore an uncanny resemblance to the V-2. Although some successor to the V-2 was arguably what Hap Arnold had in mind when he wrote about "space ships," the V-2, in fact, turned out to be the technological predecessor of the costly expendable rocket approach. The same German rocket engineers who designed the V-2 also developed the Redstone missile for the United States. Alan B. Shepard rode this missile on a 15-minute suborbital hop in 1961 to become both the first American in space and the first American to ride a suborbital hypersonic transport.

The German engineers from Peenemünde then went on to form the nucleus of the design teams that built the Jupiter missile, which led to the Saturn I and, in turn, the Saturn V moon rocket. Offshoots of the Huntsville team include the Titan ICBM, which has become the Titan IV, today's largest and most expensive US ELV.[84] As Stine says in *Confrontation in Space,* "nearly all of the USA space launch vehicle stable stands on the foundation of Peenemünde."[85]

Interestingly, the design heritage of the modern RLV goes back, not to Peenemünde, but to work done by Douglas Aircraft for a nuclear-powered bomber for the US Air Force in the early 1950s. In the late 1950s, a young Douglas engineer named Maxwell Hunter took the engine design for the canceled Air Force nuclear airplane and began to investigate a single-stage-to-orbit nuclear rocket called the Reusable Interplanetary Transport Approach (RITA). After the RITA program ran its course, aerospace engineer Bono came to work for Max Hunter at Douglas and began his long work on the series of VTVL SSTO concepts which he describes in *Frontiers of Space.* Through the 1960s and 1970s, SSTO ideas languished because of materials and propulsion limitations. Serendipitously, US government intervention in the form of the lightweight materials that came out of the NASP and NLS programs revived these discussions in the late 1980s and early 1990s.

At this point, political forces converged with SSTO technology. At the beginning of the President George S. Bush administration, a group of conservative space advocates including Max Hunter and retired Army Maj Gen Daniel Graham, met with the vice president and the National Space Council to advocate a reusable VTVL SSTO rocket vehicle. Given the administration's commitment to former President Ronald Reagan's scramjet-powered SSTO, the National Aerospace Plane, however, it would have been politically difficult to start another NASA/Air Force Joint Program Office to investigate rocket SSTO, so the administration decided that the well-funded SDIO should foot the initial bill. Significantly, General Graham's High Frontier Foundation had been part of the initial impetus for SDI and he remained one of its staunchest supporters. It is not surprising, therefore, that SDIO obligingly funded four aerospace industry study teams to research and design SSTOs capable of launching 10,000 pounds to polar low earth orbit. In 1991, however, Ambassador Henry Cooper, director of SDIO, under funding pressure from Congress and interagency pressure growing out of the perception that SSTO had become a very popular rival to other launch system improvement programs, elected not to assume management of the program beyond suborbital testing of a one-third scale model, the DC-X. The program title was changed to Single Stage Rocket Technology (SSRT), with any additional SDIO funding beyond DC-X contingent upon a derivative of DC-X meeting SDIO's suborbital launch requirements. As a result, and with the 1993 dismemberment of SDIO, SSRT became an institutional orphan.

Not content with cutting Air Force follow-on funding for the technology, agencies with competing agendas actively worked to dismiss the possibility of rocket SSTO. In 1991, Martin Marietta (makers of the Titan IV ELV) cast doubt on the economics of rocket-powered SSTO and the Air Force space acquisition community dismissed the technology in a 1992 NLS decision brief to the secretary of the Air Force.[86] A primary back-up chart from the briefing reflects this position in a quote from the Aldridge Report, "NASP, SSRT, and High Speed Civil Transport (HSCT) are not in competition with or a substitute for NLS since these technologies are not sufficiently mature to risk 'leap-frog' development."[87]

Despite this Air Force and contractor nay-saying Dan Goldin, the NASA administrator, became interested in the idea of a reusable single stage-to-orbit launch vehicle after seeing the DC-X fly.[88] He saw the possibility of an advanced technology program building on the knowledge gained from DC-X that would restore US leadership in space and perhaps solve the nation's access to space problem. This was the genesis of NASA's sponsorship of the subscale advanced technology demonstrators that are now flying, and arguably, the beginning of NASA's interest in the X-33 idea.[89]

This idea did not spring up overnight. It has a long technological and engineering history and significant backing inside and outside of the space technology community (there are even three Internet home pages dedicated to RLVs and to political activism on the technology's behalf).[90] With private sector interest inside and outside of the NASA CAN process, with public advocacy groups developing briefings for members of the public to show to their members of Congress, and with a real national need to solve the access to space problem, there now seems to be a significant impetus for the RLV to change how the United States operates in space.

This moment in history is unique in American development of the space frontier. The combination of the political, fiscal, and technological forces that are driving the RLV idea seem to add up to the possibility, and perhaps even the probability, of significant near-term change in our ability to access space. What will that mean for US national security? That is the topic of discussion in the remainder of this work.

Military Implications of Inexpensive Space Access

As already outlined, the lack of routine civil and commercial access to space militates against the development of robust methods for using the medium for commercial purposes. For similar reasons, it may also work against the development of robust methods for using the medium for national security purposes. In fact, the current difficulty in accessing space is a fundamental reason for the limited perceptions of what it is possible to do there.

The state of present joint US military space doctrine as the United States lowers the barriers to space access is a case in point. Joint doctrine assumes that one of the "operational characteristics" of space cited in "Joint Doctrine, Tactics, Techniques, and Procedures" (JDTTP) 3-14, *Space Operations*, is "difficult access."[91] Any doctrine that assumes that access to the medium it addresses is going to be difficult and infrequent is also likely to assume that operations which require robust and continuous access (such as protracted combat or logistic resupply) will not take place there. If the conditions underlying the doctrinal assumptions change, however, then the doctrine derived from those assumptions is not likely to be prepared for the changed conditions. This happened on the Western Front when Great Power assumptions about the density of fire on the World War I battlefield proved incorrect, it happened to the French during the Battle of France in 1940 when assumptions about the speed of armored maneuver coordinated with airpower changed, and it happened to the Iraqis during the Persian Gulf War in 1991 when assumptions about the effectiveness of airpower changed. This section attempts to determine whether this sort of doctrinal discontinuity is likely in the next few years if the RLV programs called for by the president's new National Space Transportation Policy are developed and access to space is made much less "difficult."

Despite the limiting assumption of "difficult access," there are nevertheless ideas in present joint space doctrine and objectives in the president's National Security Strategy (NSS) that will be useful in the RLV era. The 1994 NSS, for example, says that two of the United States's main policy objectives in space are, "continued freedom of access to and use of space" and "maintaining the US position as the major economic, political, military and technological power in space."[92] The draft joint armed forces space doctrine, although written two years earlier in support of Bush-era space policy, supports the objectives of the 1994 NSS in this regard with the recognition that there are certain strategic locales in space that have to be controlled in order to maintain access, what Joint Pub 3-14 calls "decisive orbits."[93] It also posits that space forces should consider capabilities to "control" these orbits by force, but then, in a "Tactics, Techniques, and Procedures" manual, it provides no tactics, techniques or procedures for doing so.[94]

To be fair, access to space has heretofore been difficult and, in part because of that difficulty, few people on the Joint Staff have had to think about how realistically to control "decisive orbits." Nevertheless, as General Arnold said of Air Forces in 1945, "National safety would be endangered by an Air Force whose doctrines and techniques are tied solely to the equipment and processes of the moment. Present equipment is but a step in progress, and any Air Force which does not keep its doctrines ahead of its equipment, and its vision far into the future, can only delude the nation into a false sense of security.[95]

The same might be said today for "any space force" or for any service that claims as its mission the defense of the United States through the control and exploitation of their respective realms. If, in fact, access to space is about to become much less difficult, then it behooves military thinkers and doctrine writers to determine what the deficiencies in their doctrines are before the fundamental assumptions underpinning them are invalidated (or at least to think far enough ahead not to be blindsided when it does happen).

That said, the next section builds on the technological possibilities previously discussed to determine what doctrinal deficiencies a possible "space operability revolution" will reveal in US joint space doctrine, and what new doctrines might be required in a proliferated space access world. Before proceeding, however, it is necessary to challenge some shibboleths about the military uses of space.

Political Sensitivity of the "Militarization" of Space

The Outer Space Treaty of 1967 prohibits several specific activities in space. It prevents signatories from stationing weapons of mass destruction anywhere in space and forbids the construction of military bases on the moon. Article II says that "Outer space is not subject to national appropriation by claim of sovereignty" and Article V says that the Moon and other celestial bodies shall be used exclusively for peaceful purposes. There are no prohibitions, however, against reconnaissance, surveillance, military communication, navigation, or other uses that support terrestrial military operations. These uses, whose value to the United States and its Coalition partners was demonstrated in the Persian Gulf War, create tension between the "no national appropriation" rule and reality. The war demonstrated that there are orbits and force structure in space that the United States must be able to control and protect in time of war to fight successfully. This is the origin of the "decisive orbits" idea in the 1992 draft joint space doctrine as well as the statement that force may have to be used in order to secure them. On the face of it, this statement is a violation of the spirit, if not the letter of the Outer Space Treaty, but the president's National Security Strategy echoes this sentiment when it speaks of "freedom of access" (similar language with respect to freedom of the sea has been the basis for a good part of the development of the US Navy).

The very existence of the "space control" mission, in joint as well as Air Force doctrine, is an acknowledgment that the United States has equities in space that it cannot afford to lose in time of conflict, the Outer Space Treaty notwithstanding. As a result of the new higher stakes in space, it has been suggested that military space operations could see the same progression from observation and signaling to pursuit and bombardment that aviation made during the course of World War I.[96] Since early airplanes were relatively inexpensive, the armed forces could afford to experiment with various types and to determine their capabilities under combat conditions. A few aircraft losses while trying to work out the details did not threaten the air program as the loss of *Challenger* threatened the space program. Another analogy may also be useful, that of the development of submarine warfare before World War II. Submarine warfare, after the political and moral opprobrium aimed at the Germans for sinking troop ships and merchant men in World War I, could not be politically justified based on the Corbettian idea of the commerce raider.[97] Nevertheless, the Navy was able to buy submarines and field them in a world where new technology and doctrine had to be developed in a hostile political climate to set the stage for American success in World War II. In fact, the submarine's threat to the battleship Navy led to its misapplication in war games and to the promotion of conservative skippers who had to be replaced by a more aggressive breed in 1941. One submarine captain put it this way, "The minds of the men in control were not attuned to the changes being wrought by advancing technology. Mahan's nearly mystical pronouncements had taken the place of reality for men who truly did not understand but were comfortable in not understanding."[98]

This example shows that it is possible for the US armed forces to field new technologies that give them the edge in future wars without clear positive national policy goals (and even in the face of some political and senior military resistance). As we have seen, the NSS already reflects American national interests more than it does the spirit of the Outer Space Treaty. If and when RLVs begin to fly, policy makers can reasonably be expected to use them to further the national interests of the United States, as they did with the submarine in the 1940s, and as any nation will if and when it builds its own space ships.

Traditional Military Missions in Space

Some of the possibilities for reaping the economic rewards of increased operability in space have already been discussed. Using some of these economically useful capabilities, this section will explore some possibilities that space operability offers for national security.

Current joint and air doctrine divides military operations in space along the same lines as current US armed forces doctrine. These four broad functions are force enhancement, force application, space control, and space support.[99] Today's doctrine lists activities such as communications, navigation, intelligence and surveillance, environmental monitoring, mapping, charting, and warning processing and dissemination as part of force enhancement. Within force application are ballistic missile defense, aerospace defense, and power projection. In space control, protection, negation, and surveillance of space are listed. Space support consists of launch, satellite control, and logistics.

As with much of terrestrial US armed forces doctrine, this speaks very much to the nuts and bolts of how military power is used in warfare, but does not say a great deal about what it is used for. It also is deficient in describing uses for military power outside of the context of a shooting war. There is usually a diagram at the beginning of US doctrine manuals that outlines the tie between the National Security Strategy of the United States, the national military strategy (NMS), and the doctrine in question, but the logic flow between the boxes or circles in the diagram is not clearly spelled out.[100] For example, when the same four pillars of the National Military Strategy of the United States (deterrence, forward presence, crisis response, and reconstitution) can support both former President George Bush's NSS and President William "Bill" Clinton's new National Security Strategy of Engagement and Enlargement without significant change, it is reasonable to suspect that there is little real deterministic relationship between the NMS and the grand strategy it is supposed to support. The military has simply divided warfare into four parts and tied it to the NSS at only the most superficial level. What is the logical tie, for example, between President Clinton's new national objective of "promoting democracy" and the combat-oriented strategy of "deterrence, forward presence, crisis response, and reconstitution?"[101] As a result, when the president wants to use military forces to achieve precise political effects that don't involve combat, the armed forces are often reluctant, pressing instead for either overwhelming force or non-involvement. Unfortunately for the Department of Defense, achieving precise political effects (not involving combat) is what the armed forces are called upon to do much of the time. In the first 45 years of the US Air Force's existence, for example, it was called upon for "air movements of national influence" hundreds of times, as opposed to only a few combat operations.[102] American military forces are often used in situations where "force" and "control" (as in force enhancement, force application, and space control) are not acceptable. Humanitarian operations and operations other than war are good examples of this language's failure to describe the full range of possible military operations in support of national policy objectives.

Joint doctrine's inadequate treatment of these subtleties in terrestrial operations is a handicap, but not a fatal one, because policy makers can conceive of and implement uses of the terrestrial military for noncombat policy purposes without the help of military doctrine. The blockade of Cuba during the 1962 missile crisis is a good example. Even though traditional US Navy blockade procedures were not followed (sometimes over the vociferous objections of flag officers), the blockade was conducted as the president wanted it, not in accordance with traditional naval practice. Similarly, in 1993, President Clinton directed a reluctant US Air Force to begin night food-pallet drops to Bosnian civilians to directly achieve specific national policy objectives. If this sort of operation, which often characterizes the exercise of US power in both the cold war and post-cold-war periods, continues to be prevalent, then space doctrine as well as terrestrial doctrine should reflect this reality.

However, doctrine's inadequate treatment of this type of operation in space may be a more serious handicap in the coming RLV era. This is because decision makers will find it much more difficult to conceive of the possibilities for using newly operable space power to implement their policies. Missions such as enforcement of today's ongoing terrestrial sanction regimes or air exclusion zones, blockade of other groups' access to space, repositioning space forces over a target state or group's territory as a demonstration or to provide presence over a given region or in a specific "decisive orbit," or providing rapid humanitarian relief using the suborbital lift technique discussed previously could be extraordinarily useful politically, but they are likely to be outside of the cognitive schema of most military leaders, let alone civilian policy makers.[103]

New Space Missions in the RLV Era

The RLV space ship's characteristics would make it not only possible, but affordable and politically feasible to use military space forces to "move national influence" in the same way that air and sea power do today. In other words, operable space forces could participate in military missions that directly support the achievement of national policy goals not necessarily in direct support of a combatant commander on earth in ways that today's few and fragile space forces cannot.[104] Some of the contributions of the space operability revolution that would enable such participation would be timely logistic resupply, rapid maneuverability, and on-scene human judgment. All three are to be discussed here, with no particular significance to the order in which they are presented. Relationships between the three will become evident in the discussion. As each is discussed, trade-offs with current terrestrial methods, some possible strategic circumstances under which these capabilities might be useful, and some tactics, techniques, and procedures for using them are also addressed.

Logistics. There have been a number of US Air Force and NASA studies of refueling and refurbishment of on-orbit force structure.[105] Many of these studies were predicated, however, on expensive and unresponsive expendable launch vehicles to bring refueling and servicing payloads up to target satellites from earth. As a result, these studies never progressed past the paper stage. With reusable space ships, however, the calculus changes. As previously discussed, RLVs make it economical to replace and retrieve the current generation of satellites. It also becomes possible to refurbish satellites that are designed for on-orbit servicing, thus avoiding the cost of new satellite design and construction. Reconnaissance and warning satellites could have their sensor packages upgraded with the latest technology using line replaceable units (like those the Air Force's Phillips Laboratory is developing today), rather than becoming obsolete. In today's context, with RLVs and modular satellite design, the debate over the Defense Support Program (DSP) follow-on would have a simpler answer. Rather than asking Congress for a new program start (such as the canceled Follow-on Early Warning Satellite or the controversial DSP II proposal), the United States could replenish station-keeping fuel, replace sensors, and upgrade the communications and data-processing equipment aboard existing spacecraft.[106] No longer, for instance, would this nation's reconnaissance and surveillance architecture require tens of billions of dollars invested in lump sums to wholly replace on-orbit capability. Rather, individual spacecraft could be updated or replaced without entire constellation replacement. The mean mission duration (lifetime) of these national assets would be significantly extended at a great savings.

Such logistic resupply (especially of oxidizers for propellant) could actually be easiest using a base on the moon. The 9,000 feet per second (fps) change in velocity (_υ) required to escape the moon's gravity is a lot cheaper than the 31,000 fps required to get to LEO from earth, even assuming a 12,000 fps _υ to get back to the moon. For GEO and high earth orbit (HEO), the advantages are even greater. In fact, the energy required to bring materials from the moon to HEO is less than a twentieth of that needed to lift an equal mass from earth to such an orbit.[107] Since oxygen is about 40 percent by weight in lunar soil, it would be fairly simple to extract. In fact, some have called the moon a "tank farm" in space.[108] Although hydrogen is in low concentration at the Apollo landing sites, its relatively higher concentration in fine-grained lunar soils may allow for its extraction as well.[109] Just as building RLVs would save billions of dollars every year in continuing launch costs, building an automated lunar extraction facility and geostationary satellite resupply base would save a significant amount in propellant costs over time. Since it takes 1/20th as much fuel to get to HEO from the moon than it does to get to HEO from earth, we would burn 6,429 pounds of hydrogen and 38,571 pounds of oxygen ($18,000 in fuel at current prices of $0.05/pound for oxygen and $2.50/pound for hydrogen) to get to HEO from the moon (with the notional 100,000-pound dry weight, 0.90 PMF vehicle).[110] This saves 122,142 pounds of hydrogen and 732,858 pounds of oxygen compared to launch from earth (with 900,000 pounds of fuel at a 6:1 oxygen/hydrogen ratio, which would cost $360,000). That is a total fuel cost savings of $342,000 per mission (which becomes significant if per mission cost is as low as $1 million), with the added benefit that such a logistic base would be even more useful to the numerous commercial and civil satellite operators than to the military.[111] The downside is, of course, the infrastructure investment in building such a facility. In addition, there is the cost of semipermanent stationing of RLVs on the moon that would not be available for earth-to-orbit launch services. The savings and profits from such an enterprise would have to be tremendous to justify such an investment.

If, however, there are hundreds of US flights per year leaving earth to refuel and refurbish high-altitude satellites, then the United States, as the only space power capable of such a project in the near term, could improve its balance of payments by selling propellant resupply and on- orbit repair services to the rest of the world at premium prices. The continuing high cost of lifting fuel out of earth's deep gravity field (sometimes described as a "gravity well") could convince RLV operators to make the investment in a lunar base to lower their operating costs, just as NASA is investing in the RLV itself to lower large and continuing operating costs. Such a base, essentially civilian in nature, would also provide enormous treaty-compliant strategic advantages.[112]

Rapid Maneuverability. Although spacecraft governed by the laws of orbital mechanics move at five miles per second with respect to the surface of the earth, they are not very maneuverable from orbit to orbit. ELV-era space operability does not allow the United States to position its space forces where it wants them when it wants them there. At present, with a limited and unreplenishable amount of maneuvering fuel in orbiting satellites, it is not a trivial matter to reposition them to influence or even monitor events on earth. Although the details of defense satellite fuel-states are not releasable, the laws of physics suggest that the unexpected movement of today's unrefuelable DSP missile warning satellites to cover the Arabian Gulf during the 1991 war undoubtedly reduced their on-orbit lifetime and reduced the US's flexibility in responding to future emergencies. If RLVs gave us the ability to refuel sensors such as DSP and other satellites (as discussed in the preceding section), they could be repositioned to cover any area of interest without posing the danger of future station keeping fuel shortfalls. Later, smaller, less capable, but less expensive and more numerous sensors could be deployed in orbit in response to a crisis. With the RLV and a supply of such sensors ready to be launched on short notice, this could be done in a matter of hours rather than the months that are currently required for a launch campaign.

Today's maneuverability shortfall also limits thinking about nondestructive inspection of unknown satellites. Instead, we inspect satellites that we want to know more about by taking pictures of them from the ground, which is hundreds of kilometers away and blanketed by the distorting interference of the earth's atmosphere. After the space operability revolution, reusable space ships or satellites resupplied by them could close the minimum distance permitted by international treaty in peacetime and inspect unidentified satellites and their payloads (by optical, radar, and other means) up close without the distortion of the atmosphere. In a period of escalation short of a shooting war, RLV space ships would intercept unidentified traffic and inspect it for hostile capabilities or intent. If no such capabilities are found, the satellite could be released to go on its way. If hostility is suspected or confirmed, or in accordance with policy-driven rules of engagement, the RLV would have a wide range of options. It could capture the offending satellite, jam it, or disable it (preferably using nondestructive means that would enable the use of the disabled satellite for leverage in negotiations, which would have the added advantage of not worsening the space debris problem). Contrast this with today's space doctrine. The neutralization of hostile space forces by nonlethal technical methods is currently the only method of space control short of destruction. The United States is limited to these techniques (such as eclipsing adversary solar panels or jamming uplinks), however, because rendezvous with, and capture of, hostile satellites is considered a rare, expensive, and risky operation. This will not be the case after the operability revolution, when rendezvous and capture are practiced on a routine basis in the course of repairing and retrieving friendly satellites. There are also fewer simple countermeasures to physical capture. Jamming originating from the earth can be overridden and satellites can maneuver on battery power to escape an artificial "eclipse." It will, on the other hand, be much more difficult for an adversary to avoid capture by a grappling arm guided by human intelligence in real time. In addition, a captured asset can be used to coerce or deter some space-faring adversary from a hostile course of action. Leaving the satellite on-orbit, as done with today's disabling schemes, however, gives the adversary time to devise a technical countermeasure to the disabling technique. Capture puts an end to such hopes.

The maneuverability of RLV space ships would also make them useful for missions that are more accurately described as denial than destruction. They could mine decisive orbits (as could ELVs), but they could also conduct mine-clearing operations, soft landing the cleared mines for storage back on earth, something an ELV could not. These mine fields could be laid in a crisis and cleared afterward, giving new flexibility to national policy makers. RLVs would also be able to respond to crisis situations with all of these capabilities more quickly than the ELV due to launch preparation times that are forecast to be months shorter.[113]

The increased mobility provided by the RLV would enable the United States to move its forces to decisive orbits in space or over any trouble spot on earth more quickly (typically 31,000 feet per second with reference to the earth's surface) than any form of terrestrial military power.[114] Threatened uses of force or nonlethal inspection of enemy forces (space or terrestrial) could work to achieve policy objectives without firing a shot.

As the president's *National Security Strategy of Engagement and Enlargement* puts it, "all nations are immediately accessible from space."[115] It follows that when space itself becomes immediately accessible to the United States, then the United States will have immediate access to all other nations. This access can mean the ability to observe, or it can mean the ability to influence. The movement of space forces to threaten on-orbit force structure have been discussed, but RLV space ships would also allow the United States to deliver destructive or nonlethal power to any point on earth less than an hour after launch.

Although many of the missions made possible by the RLV's maneuverability discussed to this point are not captured in present space doctrine, the idea of force application from space is. Although the perception exists that force application from space is prevented by international treaty or US policy, it is not. Joint Pub 3-14 puts it this way, "international law . . . allows the development, testing, and deployment of force application capabilities that involve nonnuclear, nonantiballistic (ABM) weapon systems (i.e., space-to-ground kinetic energy weapons)."[116] Because it has been difficult to access space, however, it has been difficult to develop any such concept beyond the idea stage. Concepts such as Sandia National Laboratory's Winged Reentry Vehicle Experiment, a ballistically delivered, nonnuclear, long range, precision-guided kinetic energy penetrator flew three times on the front end of ICBMs before it ran out of funds.[117] Many other studies never got past the paper stage. Studies with acronyms such as data analysis control (DAC), program management plan (PMP), independent cost estimate (ICE), BRIM, and GPRC spent hundreds of thousands of dollars and produced stacks of reports without really demonstrating any technology.[118] With reusable space ships and routine access to space, however, research payloads can be flown on operational missions without waiting for rare ICBM test launch opportunities. Separation tests would be scheduled similar to current scheduling for US Air Force Seek Eagle weapons carriage and separation tests for air breathers.

The RLV could also deliver nonlethal payloads such as ground-based sensors, radio and television transmissions, and humanitarian relief supplies (via suborbital lift into secure areas or via shielded reentry containers in denied areas) to places that may not be accessible even to airpower (due to threat, distance, or overflight restrictions). If fuel costs for an orbital mission are $360,000 and overall launch costs can fall to $1 million, then suborbital missions requiring less _υ and therefore less fuel should cost even less. These missions could be cost competitive with military aircraft. A 1991 Air Force regulation says that in FY92, the DOD would have had to charge NASA $403,132 for a 28-hour, 450-knot average speed, 12,500 nautical miles, non-stop C-5 mission.[119] In the RLV era, if NASA has priority cargo to transport to its few remaining overseas tracking stations, it might be smarter to pay the same or similar costs and cut the trip time by 27 1/2 hours.

Such a capability would allow the United States to protect its interests, on earth or in orbit, at times and places of its choosing, without having to consider the risk of loss to enemy action. States or other groups with nascent ballistic missile or space programs will soon have primitive ASAT capability in the form of sounding rockets carrying kinetic energy submunitions (as simple as sixpenny iron nails) launched in the path of an oncoming satellite in a predictable orbit.[120] These ASATs, a threat to any satellite in a predictable LEO, are of limited utility against an RLV space ship launched on a suborbital or fractional orbital trajectory. There is very little possibility that nonspace-faring nations or groups could detect launches from US sovereign territory. At present, only the United States has a publicly disclosed missile warning satellite, although the Russians have reconnaissance satellites and are likely to have missile warning satellites left over from the cold war as well. If these nations detect launchers, they do not have the data-processing infrastructure to predict and disseminate suborbital trajectories and impact points to space weapon defense forces. While making a case for an independent European satellite reconnaissance capability in the wake of the Gulf War, former French foreign minister Pierre Joxe acknowledged the "supremacy of the US space surveillance machine with its range of missile early warning, ocean surveillance, photographic and radar reconnaissance, electronics eavesdropping and weather satellites . . . with its massive supporting processing and communications chain."[121] France's and Britain's $1 billion investment in military spacecraft could not match the $200 billion US military space machine during the war, and it is not likely that many other nations on earth could do so in the foreseeable future.[122]

That said, it does not take a lot of money to buy sixpenny nails. Low technology ASATs would, however, be difficult to use against an RLV changing its orbit from revolution to revolution. Even the United States would have a great deal of difficulty engaging hypersonic maneuvering reentry vehicles (which would be very similar to the strategic defense problem).

On-Scene Human Judgment. The "difficult access" paradigm has also worked to keep space doctrine notably free from references to the idea of military personnel in space. Even White House policy makers recognize the Department of Defense's aversion to the idea of manned space flight. Richard Dalbello, assistant for aeronautics and space in President Clinton's Office of Science and Technology Policy says, "policy recognizes that DOD has little current interest in human spaceflight."[123] This could be related to the fact that there is a "manned military space expectations gap" that goes along with the overall launch expectations gap. This part of the expectations gap is also a result of dashed hopes and unsatisfactory reality. The dashed hopes can be traced to events such as President Richard Nixon's cancellation of the Air Force Manned Orbiting Laboratory, the shutdown of the Air Force's space shuttle launch facility at Vandenberg AFB, in 1986 after the *Challenger* accident, and, in both cases, the subsequent disbanding of military astronaut groups who had been screened and selected through an arduous board process.[124] The disillusionment (or, at least, the skepticism) concerning the role of military man in space is evident from the deafening silence on the subject in Air Force doctrine, in joint doctrine, and in even the most forward-leaning research papers and projects such as the US Air Force's recently completed *Spacecast 2020* study. This has led to an almost universal assumption in the US space community that most DOD space missions can be performed by robots; some contend that any requirement for human judgment in space can be fulfilled today by unmanned systems and tomorrow by telepresence or virtual reality.

There may nevertheless be a case for military personnel in space. The experience of land, sea, and air warfare seems to indicate that the judgment and initiative of the human being on the scene is critical to success in battle against a reacting enemy. It is not obvious that this pattern will be repeated in the new space medium, but history suggests that the presence of military personnel could help with the continuous tactical improvement and adaptation that has traditionally made for victory in war. As John Collins of the Congressional Research Service says in *Military Space Forces: The Next 50 Years,* "sizable manned contingents probably should deploy in space, because commanders and staff far removed from crises seldom can assess the situation and take appropriate actions as well as on-the-spot counterparts."[125] Commanders and staff on the ground may also have their links with RLV ships disrupted or jammed, while it is much more costly for the enemy to break the man-machine link in a piloted vehicle. There are also complexities in military operations that may not lend themselves well to remote control. As with the submarine, a complex vehicle with multiple missions in a challenging and dynamic physical environment with a reacting enemy, it is very difficult to imagine a remote crew of operators coordinating rendezvous, grappling, defensive countermeasures, damage reporting and control, and all of the subtasks implicit in those operations simultaneously, whether under the sea or in space.

To adapt to such rapidly changing situations, military man on earth has had to have repeatable and regular familiarity with the medium in which combat operations take place. This repeatable and regular familiarity with the medium is what the RLV operability revolution will provide that is now missing from current space doctrine. Without personal experience with the medium, it is arguable whether sound doctrine can be devised for operating there. It is difficult to imagine that the Navy could have gained enough experience in subsurface warfare before World War II to enable it to sink over five million tons of enemy shipping in the Pacific if all subsurface operations before the war had been conducted by remotely controlled undersea robots.[126]

It can be argued that the same results would have been obtained with submarines controlled from shore via twenty-first century telepresence or virtual reality. The complexity of submarine combat suggests otherwise. Damage control and loading torpedoes in combat situations would have to be done by onboard robots. Torpedo misfires would also have to be cleared by such robots. Software would have to be written to fuse sonar inputs and onboard ambient noise so that the teleoperator could monitor both for damage cues and situation awareness. It seems that the added level of complexity required for a teleoperated combat submarine would be significant and might outweigh the advantages of removing man from the scene. In any case, the pre-World War II US Navy overcame the inherent hostility of the undersea environment and a clear lack of political enthusiasm for undersea warfare and put men to sea on submarines.[127] A similar case may be made for the manned combat RLV in space.

Policy makers, too, may also be reluctant to trust unmanned or teleoperated warships even though the teleoperated RLV would be like any weapon on earth, a machine executing a decision made by man just as a firearm does when a soldier pulls the trigger. There should, therefore, be the same amount of trust in the teleoperated RLV as in the soldier's rifle. The difference is, however, that when the soldier's rifle misfires, he is on the scene to unjam it or fix the bayonet. In the event of onboard failure, link jamming, or battle damage, the unmanned or teleoperated RLV would have no trained soldier on the scene to make sure that high-stakes political missions are carried out successfully.

In addition to the arguments outlined above, there is also a simple physical argument against remote or virtual reality (VR) piloting of space vehicles in wartime or crisis situations: the speed-of-light delay inherent in the long slant ranges that would be involved. It would take an earth-based operator at a console or in a VR environment, 0.25 seconds to send a command to a refueled RLV intercepting a maneuvering adversary satellite in geostationary orbit and perceive that the vehicle was responding (22,300-mile orbit, 186,000 miles-per-second speed of signal, two-way trip). This assumes that the vehicle is directly overhead the operator. If the space ship is inspecting a satellite in geostationary orbit on the other side of the planet, the signal is likely to be relayed via two or more geostationary satellites. The round trip in this case is over 1.00 light seconds and begins to be problematic even for cooperative targets. Speed-of-light delay is acceptable when sending instructions to unmanned deep-space probes, but, just as in air-to-air refueling at 0.70 Mach, rendezvous would be much more difficult and dangerous with a one-second flight control delay as would maneuvers in close proximity to another spacecraft at Mach 25. This would be especially true if the target spacecraft were itself maneuvering.

Automation, VR, and telepresence would reduce vehicle cost and complexity since there would be no need for life support and a reduced need for vehicle reliability. There would, therefore, be military missions that machines or telepresence can perform perfectly well (e.g., routine reconnaissance, space station resupply, satellite deployment). The Russians have been resupplying their manned and politically valuable *Mir* space station for years via automated docking with the unmanned *Progress* resupply rocket.[128] But in general, high-stakes missions in which failure would be politically disastrous, especially in an international crisis, argue for man's presence, even if this increases the risk to RLV crews.

Although the weight and complexity of the generic RLV might be reduced through teleoperation, the necessity for combat vehicles to operate in degraded modes, the onboard maintenance often required in dynamic situations, and the coordination required for multiple missions would seem to argue for the restriction of teleoperation and automation to relatively benign environments. Man should not be excluded from space simply because he requires added vehicle complexity in the form of life support. What he brings to the game in terms of degraded operations, jam resistance, and damage control may be worth the extra weight. This, however, is not the approach of today's US space policy and doctrine. People sitting at consoles on earth sending inputs to robots in space are the US armed forces' space officers, who are the experts qualified to write space doctrine. It may be useful to remember how unsophisticated early air doctrine, created by people without much flying experience, seems today.[129] Space doctrine developed in institutions that assume away routine manned operations in space may not stand the test of time much better.

The preceding discussion of potential missions and arguments for and against manned RLVs highlights interesting parallels with undersea warfare. Given long duration inspection and/or presence missions in an extremely hostile environment, multiperson crews performing specialized tasks, and the ability to maneuver in three dimensions, the best model for the fighting RLV may be the submarine. Missions requiring presence over adversary territory or near adversary space facilities through the course of a terrestrial political crisis, long inspection patrols to survey other nations' satellites, confinement in small pressurized spaces for long periods, and specialized crew functions appear to fit the submarine paradigm more than any other.[130] This is not to say that there are not significant analogies to air operations as well, but there are many things about military operations in space, especially those that have to do with control of the medium, that seem to be closely analogous to submarine operations.

It is when space power acts to affect political outcomes on earth that the tie to airpower roles and functions is strongest. If airlift (as suggested by the "suborbital hop" idea), strategic attack, interdiction, and perhaps even close air support are possible from space, then these missions, more than space control or presence, are where military power from space might have real leverage on political outcomes on earth.[131] That said, space operations will require an infusion of naval as well as air "culture" and doctrine. This will be discussed further in the next few sections.

Building on the Joint Doctrine of "Decisive Orbits"

After the discussion of what the RLV revolution will allow the United States to do in space, it may be useful to explore the physical nature of the earth-moon system and why certain places in it have military advantages over others. The doctrine of decisive orbits touches on this point, but the RLV space ship could make control of these orbits even more decisive, especially if it makes them more usable.

Physical Characteristics of Decisive Orbits. Before proceeding with how decisive orbits in space should be used, however, it is necessary to define their physical characteristics. It is also necessary to understand how the physical characteristics of space fit into air, land, and sea doctrine.[132]

Some space doctrine writers focus on the physical differences between operating in space and operating in the atmosphere to emphasize the point that air and space are distinct military media.[133] The organization and doctrine of forces designed for operating in one medium are not appropriate, these writers believe, for the organization and doctrine of forces in the other. These writers focus on the physical differences of astrodynamics versus aerodynamics rather than on whether the effect of an action in or from space is the same as actions taken in or from the air. This could be called doctrine with a focus on engineering, rather than doctrine focused on what one is trying to do to the enemy. Air and space vehicles do require different sorts of engineering, but the effect of a destructive strategic attack from space (given good intelligence and similar accuracy) is likely to be the same as a destructive strategic attack from the air (allowing for the greater energy inherent in orbital energy states). The reason for the similarity of effect is the similar nature of the advantages that air and space power hold over terrestrial forces and political entities. US Air Force doctrine says that speed, range, and flexibility are among the characteristics of airpower. It seems that a case can be made for these as characteristics of space power as well.

Both air and space power have the advantage of elevation (with its corollaries, superior viewing, and energy advantage) over terrestrial forces. This difference between air and space forces on the one hand and terrestrial forces on the other unites air and space power in a very fundamental way. It means that no matter what its physics, flight is still flight, and that the "control and exploitation of air and space" should be performed for very similar political purposes. If the advantages and uses of the two media are the same or similar, it does not seem to make a lot of doctrinal sense to try to separate them.

That said, there are physical characteristics of operations in the space medium that make the methods for gaining control of the medium very different from the "air superiority" mission. First, there are certain energy-states in earth orbit that are of particular utility in conducting space operations. These energy-states are associated with certain orbits that have been proven to be militarily useful. Among these, and cited by Collins of the Congressional Research Service as "key terrain" in *Military Space Forces: the Next 50 Years*, are geostationary and other equatorial earth orbits.[134] Second, these orbits can be controlled by occupation or other forms of denial in ways that have no analogues in air operations. It is necessary to send up several multiship formations of air superiority fighters in more than one combat air patrol (CAP) "orbit" to prevent enemy aircraft from entering friendly airspace. It is only necessary to occupy an equatorial geostationary orbit with a single long-duration "fighting RLV" at a given longitude to prevent anyone else from putting a spacecraft there (just as with terrestrial power, blocking avenues of approach by occupying key terrain is possible in space where it is not possible in the air). Circumstances are somewhat different for orbits that are not fixed with respect to the earth's surface, which describes virtually all other Earth orbits. For these orbits, multiple spacecraft are necessary to provide global coverage. Third, and related to the previous point, the laws of orbital mechanics allow spacecraft to persist in these decisive orbits with very little expenditure of energy. As a result, spacecraft on blockade or blocking missions could stay on station without refueling significantly longer than the two to three hours characteristic of fighter CAPs because one can maintain an orbit above the drag of the atmosphere with the expenditure of little or no energy. In simple terms, the air-to-air fighter's engine is running the whole time it is on patrol, the RLV's is not.

Geostationary orbits are obviously critical to terrestrial forces because they provide stationary "relay towers" in the sky for communication and other purposes, and may therefore qualify as "decisive." There are other militarily useful orbits that may also qualify for this distinction. Among these are the polar orbits flown by many reconnaissance satellites. As Collins notes, "reconnaissance and surveillance missions inclined 90 degrees sooner or later loop directly over every place on Earth."[135] That is why he counts these orbits as "key terrain" as well, which leads one to believe that they may also be "decisive" even though it would take many more spacecraft to occupy them.

The RLV will play in this military geography of earth orbital space in four ways. First and foremost, it gives the United States routine access to these orbits for peaceful purposes, for political signaling and other nonlethal propaganda purposes, as well as for military purposes. One of these purposes will be to take unimpeded advantage of one of the corollaries of space power's elevation, superior view. A space-faring power's awareness of what is going on on earth is far superior to that of nonspace-faring nations. A nation with routine access to space will multiply that advantage with the ability to access any orbit at will. Second, as noted above, the RLV will be able to occupy these orbits to prevent others from using them. Third, it will allow the United States to engage adversary space forces at times and places of its choosing from a position of energy advantage. Fourth, it will allow the United States to engage adversary ground, air, and sea forces and political entities at times and places of its choosing from a position of energy advantage. As mentioned above, one of the corollaries to the elevation of air and space power is the energy advantage of superior altitude (what fighter pilots call "God's G"). This discussion naturally leads to a concept which may be most useful in understanding the importance of this energy advantage to space doctrine in the RLV era.

The "Gravity Well." The earth, with its relatively strong gravitational field, "bends" space in its vicinity to create an attraction to nearby objects. That attraction decreases as the inverse square of the distance from the earth. What this means is that objects farther away from earth ("higher up" in the gravity well) have more gravitational potential energy than those below. This has obvious military implications. Collins points this out when he says,

> Military forces at the bottom of Earth's so-called gravity well are poorly positioned to accomplish offensive/defensive/deterrent missions, because great energy is needed to overcome gravity during launch. Forces at the top, on a space counterpart of the "high ground," could initiate action and detect, identify, track, intercept, or otherwise respond more rapidly to attacks. Put simply, it takes less energy to drop objects down a well than to cast them out. Forces at the top also enjoy more maneuvering room and greater reaction time. Gravitational pull helps, rather than hinders, space-to-Earth flights.[136]

The military implications of the physical facts have long been recognized, but again, the high cost of doing anything about them has made force application from space problematic. As mentioned earlier, this is less a problem of policy than a lack of a realistic and affordable way to take advantage of the leverage that space provides. Space-to-earth kinetic energy weapons that would achieve the same effects as air-delivered weapons do not merit multibillion dollar investments (current Air Force concepts of permanently orbiting space strike weapons are unmanned and can be launched on expendables).[137] Space strike weapons developed incidentally to highly profitable RLV operations (that will go on with or without those weapons) may, on the other hand, merit the relatively small investment required. An example is Gen William "Billy" Mitchell's development of antiship bombing techniques in the early days of aviation. The US Army did not set out to take advantage of the energy advantage of the airplane over the surface ship when it bought its first airplane for the Signal Corps. Despite this, once aircrews gained practical experience with the "reusable air vehicle," experimenting with it and finding out what it could do became part of the airman's culture. A similar course for the development of the RLV is logical and desirable.

Nature of Space Doctrine in the RLV Era. This discussion leads to at least three possible conclusions about what the RLV will mean to the broad outlines of space doctrine. First, it may mean that space doctrine should become more naval, with emphasis on the protection of US economic interests in space and protection of free access to space lines of communication. This would tend to emphasize the control of the medium. Second, it may mean that space doctrine should become more aerial, focusing on the earth as the seat of political purpose and space as a place from which to affect those purposes. In the language of the US Air Force, that would be "exploitation" of the medium. The third possibility is that there is some intermediate position between the first and second ideas, some merging of air and naval culture and doctrine that would be most useful for space. A comparison of the relative merits of all three options may shed some light on how doctrine writers should approach space doctrine in the RLV era.

1. ***Space Doctrine More Naval.*** As outlined earlier, there are strong arguments to support this position. The physical characteristics of orbital space, the nature of possible operations there (blockade, mining), the ability to conduct long duration patrols, and the enormous national and commercial investment on station in orbit all lend themselves to naval thinking. Satellites on orbit are much like commercially valuable islands or oil platforms in strategic locations at sea. In addition, once in space, the RLV is far closer to a ship than to an airplane in terms of the amount of effort required to stay "afloat." Aircraft must be continually "flown," ships float more or less of their own accord. Even at five miles per second, the similar characteristics of the space ship will give the crew time to devote its attention to other things, including interaction with other vessels. The RLV, unlike the airplane, can rendezvous with other spacecraft and exchange crew members or cargo other than fuel, and doesn't have to destroy or even disable adversary spacecraft to control the medium. Control of the sea or of space does not necessarily mean using lethal firepower to destroy an adversary (as it usually does for the airman). It can also mean interposing oneself between adversary forces and the objective, occupying the objective, or nondestructive inspection backed up with the threat of force (as in the Gulf War maritime intercept operation). Mastering such operations would take a tremendous amount of time, doctrine development, and training. If they were the priority missions of a "space force" as a result of maritime tradition or service culture, there might not be much time left over for other important tasks that may also be done from space.

2. ***Space Doctrine More Aerial.*** Although counterintuitive, it seems fair to say that space forces become more aerial as they look toward the earth. The fundamental elevation advantage of both air and space forces over terrestrial forces is the underpinning of this assertion. Because most policy objectives for the foreseeable future will be aimed at adversary terrestrial decision makers, strategic operations (nonlethal and lethal) from space aimed at the center of the enemy's decision-making apparatus (food drops and propaganda broadcasts to target national populations, high probability of strikes against leadership and national-level command and control as well as other targets) are most like air operations. At the operational level, space power will be able to conduct air interdiction and counterair missions, and with enough affordable force structure in space (provided by the advent of the RLV), terrestrial forces should be able to call in all of the close support they need to accomplish tactical objectives.

This leads us to the important advantage of space power over other forms of military power. This advantage is the previously cited corollary of air and space power's elevation: higher energy states. The energy states inherent in orbital and suborbital spacecraft can provide an enormous amount of firepower for a relatively small investment in the size of a given vehicle or weapon. As Collins notes, "Offensive kinetic energy weapons (KEW) plummeting from space to Earth at Mach 12 or more with terrific penetration power have a marked advantage over defensive Earth-to-space counterparts that accelerate slowly while they fight to overcome gravity."[138] Space forces will look very much like air forces to those who are at the receiving end of their effects on earth. They will also look very much like air forces at their terrestrial bases. They must, after all, traverse the atmosphere in order to get into space. In this respect, they are much like air forces, vulnerable and useless while on the ground. The compensating factor is their range. American military RLV bases are likely to be far from the US coastline and secured against terrorist attack. This is beyond the strategic reach of most nations on earth. They will, however, (within the limits of RLV response time and dispersability) be vulnerable to intercontinental, submarine-launched, or space-launched hypersonic strikes. If such an attack were launched, though, with or without nuclear weapons, the United States would have larger concerns than RLV survivability.

The demonstrated ability to strike any target on earth with precision and discrimination could, in fact, be a potent deterrent to or factor in conflict. This deterrent, unlike nuclear weapons, could be used against nonnuclear powers without the collateral damage and the negative moral and political fallout of nuclear weapons use.

A notional case may be useful in developing this argument. Assuming RLVs in orbit that are able to employ 30-pound kinetic energy weapons using the same techniques as ICBM bus separation, precision guidance of the type employed on the DOD's information network system (INS)/global positioning system (GPS) guided joint direct attack munition (JDAM), and a global communications system (i.e., the proposed Iridium or Teledesic cellular systems), a US ambassador anywhere in the world would have a "flying gunship" that could support him or her with precise and discriminate force when necessary.[139] Unmanned space-to-earth strike platforms similar to ICBM reentry vehicle buses could be employed quickly in times of crisis, as in the mine example discussed earlier, and cleared when not needed. Putting these platforms in orbit should be no more difficult than the civil satellite deployment for which the RLV is being designed. This would also allow the United States to upgrade the platforms on the ground in the periods between crises, and would reduce their vulnerability to ASATs, unlike permanent stations in orbit.

With such a capability before the Gulf War, the American ambassador's meeting with Saddam Hussein might have gone a little differently. With platforms launched in the preceding weeks passing overhead every few minutes (assuming little or no cross-range for their weapons, 32 space ships in 90-minute orbits would be in employment range every 45 minutes), the ambassador could have made a case for Iraqi vulnerability to US power by looking at her watch, making a phone call, and asking Hussein to step to the window to watch a demonstration. (Admittedly, this example may not ring true because of the low probability of State Department use of strategic strikes on foreign territory.) Perhaps an example of sea control from space may seem more politically plausible. Again, assuming little or no cross-range for the orbit-to-earth weapon, it would take 128 orbital weapons employed by RLVs in a crisis to revisit a maritime exclusion zone every 11 minutes. United States or allied naval vessels enforcing international sanctions could order threatening or suspicious vessels to heave-to with the knowledge that they were supported with precise firepower from space. Hypersonic projectiles could create impressive warning shots across the bows of recalcitrant ships. If such a situation escalates, sinking the ship from space is not only physically possible, but could also be much more politically palatable than the first scenario.[140]

3. Space Doctrine as a Combination of Naval and Air Doctrine. The preceding discussion seems to show that operations for control of the space medium are more nautical, while the leverage it provides in accomplishing the most important national policy objectives is more like airpower.

Between the two emphases, it seems clear that in high stakes conflict, US objectives will likely be tied to some outcome on earth rather than in space. That said, the strategic view of the airman, whose culture and doctrine is more consonant with such ideas, seems to be best suited to carry them forward into space. If, on the other hand, humanity's political centers of gravity move outward into space, then control of the medium and the lines of communication between these new political entities will become most crucial. For the foreseeable future, however, the United States is most concerned with what happens in the international system here on earth. This seems to argue fairly strongly for airmen to lead the US armed forces into space. These airmen must, however, adapt to the naval nature of the new medium. This may mean discarding many of the things that make airmen unique. The destructive offensive counterair model as the best way to gain control of the medium may have to be deemphasized, as may the role of the solitary pilot. If launch and landing are automated (which is the NASA CAN requirement) and orbital mechanics allow the vehicle to keep on station without much intervention, there is little need for a pilot who is continuously at the controls.[141] Again, the terrestrial analogue is the ship captain who is rarely in direct physical control of the helm. He or she has more important things to do. The ability to command a crew rather than hand-eye coordination may become the yardstick by which space combat officers are measured. These new ship captains must, however, remember that their mission is to directly affect adversary decision making on earth in accordance with national political objectives, not simply to fly around in orbit. In this, they will be more akin to airmen than to sailors.

This section has attempted to show the changes in US space doctrine that will be the outgrowth of reduced barriers to space access. It has outlined the assumptions in current doctrine that will be shaken and drawn parallels between what the RLV will mean for civil operators and what it will mean for military operators. It has also tried to use the physical characteristics of space and the capabilities of the RLV to outline a rudimentary space doctrine. The reasoning here is handicapped, however, by the same problem besetting the overwhelming majority of all space doctrine. It is written by someone who has not left Earth. Nevertheless, this outline, based on the assumption that space access will soon be routine and inexpensive, may more closely reflect the realities of the RLV era than doctrines which do not.

Summary and Conclusions

After determining that the United States is making steady political, economic, and technical progress toward fielding an affordable reusable launch vehicle, this study has attempted to induce the economic and military implications of such a development. From this, a few key themes and conclusions can be drawn.

1. The United States is developing an RLV that will lower the cost of access to space early in the twenty-first century.
2. RLV operations will have significant economic impact on the cost of today's commercial space activities and foster the development of new ones.
3. The RLV will have a significant impact on joint US military space doctrine.
4. The RLV will make space operations much more analogous to present-day naval and air operations.
5. Of the two analogues, the similarity to air operations will have the greatest impact on terrestrial political structures in the immediate future.

A short discussion of each conclusion may help to provide direction for thinking about these issues as the United States and the world enter the RLV era.

The RLV is Coming

The first conclusion that this study suggests is that the RLV is coming soon. The president's new Space Transportation Policy indicates that the US government is serious about building a fully reusable launch vehicle that will reduce the cost of access to space.

The idea has growing support in Congress and in the space policy community, if not in DOD. There is a confluence of political, economic, and technological factors creating an environment conducive to the development of a reusable rocket ship.

Economic Impact of the RLV

The first order economic consequence of the advent of the RLV will be reduced cost access to space and reduced demand for expendable launch vehicles. The ultimate result of reduced ELV production would be increased prices for ELV launches, reducing demand and production even further. Eventually, prices would rise to an uneconomic level. This could presage the end of the throwaway rocket industry, both in the United States and abroad.

There would be at least two other economic consequences of low-cost access to space. The first would be improvements in the US's economic competitiveness and balance of payments. The second would be an even further reduction in the cost of access to space after the amortization of the cost of the RLV. In such a case, DOD would find resisting RLV technology more difficult, especially with the concomitant reduction on operating costs. This would allow the US armed forces to achieve the US's national objectives of assured access to space and maintenance of its military advantage there using technologies whose cost was recouped in the private sector.

Military Impact of the RLV

The high "sortie rate" of the RLV will rapidly fill orbital space with billions of dollars worth of politically and economically important manned platforms, civil and commercial remote sensors, cellular communications satellites, and other objects. Conflicts over orbital position (which have already arisen over the desire of poor equatorial nations to "own" the geostationary orbits over their territory) will become more frequent as the number of satellites increases.

Space-faring nations flying RLVs will have the ability to monitor, threaten, sabotage, disable, or destroy the space investments of other states using techniques very similar to those used in commercial operations. If the United States sees the possibility of such operations, then other powers may as well. If so, the assumptions underlying US space doctrine (difficult access to space, no role for man in space) would become dangerously out of date.

Military Space Operations More Aerial than Naval

Space operations even in the near-term RLV era will have many characteristics of naval operations. Most of these characteristics will have to do with control of the space medium. Where military space operations intersect with terrestrial forces and political structures, space power will have more of the characteristics of airpower. These operations, especially at the strategic level, will be more decisive than the missions with naval analogues.

Conclusions

The energy advantage of RLV-equipped space-forces will be their most significant military characteristic in the context of the present international system. As orbital energy-states become more accessible to larger numbers of people and groups for commercial reasons, they will also become more accessible for military reasons. That said, a world in which any state or political group can buy an RLV whose cost has been amortized by years of routine operations may be a world where there are new and larger threats to US security than terrestrial dictators and intercontinental missiles.

Notes

1. Michael R. Mantz, *The New Sword: A Theory of Space Combat Power*, Air University Research Report AU-ARI-94-6 (Maxwell Air Force Base [AFB],

Ala.: Air University Press, May 1995), 11; and Richard Szafranski, "When Waves Collide: Future Conflict," *Joint Forces Quarterly*, Spring 1995, 80.

2. Carl Sagan, *Cosmos* (New York: Random House, 1980), 110-11.

3. S. E. Singer, "The Military Potential of the Moon," *Air University Quarterly Review*, Summer 1959, 35.

4. This treaty prohibits stationing weapons of mass destruction in space, or any military base on the moon.

5. Dana J. Johnson, *Issues in United States Space Policy*, RAND PM-141-AF/A/OSD (Santa Monica, Calif.: RAND Corp., 1993), 3-4.

6. Vice President's Space Policy Advisory Board, *A Post Cold War Assessment of US Space Policy: A Task Group Report* (Washington, D.C.: Government Printing Office [GPO], 1992), 26. National Commission on Space, *Pioneering the Space Frontier* (New York: Bantam Books, 1986), 109.

7. Vice President's Space Policy Advisory Board, 7.

8. Ibid., 29.

9. *NASA Access to Space Study, Summary Report* (Washington, D.C.: Office of Space Systems Development NASA Headquarters, January 1994), 48.

10. Homer A. Boushey, "Blueprints for Space," *Air University Quarterly Review*, Spring 1959, 18; Singer, 35.

11. Henry H. Arnold, *The War Reports of General of the Army George C. Marshall, Chief of Staff, General of the Army H. H. Arnold, Commanding General, Army Air Forces, and Fleet Admiral Ernest J. King* (New York: J. B. Lipincott Co., 1947), 463.

12. Joint Doctrine, Tactics, Techniques, and Procedures (JDTTP) 3-14, *Space Operations*, final draft, 15 April 1992.

13. Szafranski, 80.

14. The Outer Space Treaty prohibits weapons of mass destruction in space and military bases on the moon, but does not prohibit the use of space for military purposes. This perception exists even though there are no US policy limitations on force application from orbit. In fact, pages 11–21 of DOD Joint Pub 3-14 says, international law "allows the development, testing, and deployment of force application capabilities that involve non-nuclear, non-ABM weapons systems (i.e., space-to-ground kinetic energy weapons)."

15. *Report of the Task Force on the National Aero-Space Plane (WASP) Program* (Washington, D.C.: Defense Science Board, November 1992), 23.

16. "DOD Space Launch Modernization Plan Executive Summary," draft, April 1994, iv.

17. Ibid.

18. *NASA Access* to *Space Study Summary Report*, 1; "DOD Space Launch Modernization Plan," 1.

19. DOD Space Launch Modernization Plan, 30.

20. Ibid., 20.

21. "Ariane Chief Says Eastern Nations, Not US, Will Be Future Competition," *Aerospace Daily*, 4 October 1993.

22. "DOD Space Launch Modernization Plan," 10-11.

23. "Air Force Position on the Future of Space Lift: An Approval Briefing," 2.

24. Memorandum for the Special Assistant to the Secretary of Defense (Patty Howe) Reference the Clinton Transition Team Tasking of 17 December 1992, Attached NASP Paper (Unclassified); and "Air Force Position on the Future of Space Lift: An Approval Briefing," 14.

25. *NASA Access to Space Study Summary Report,* 72.

26. "Air Force Position on the Future of Space Lift: An Approval Briefing," 29.

27. "DOD Space Launch Modernization Plan," 13, 18.

28. "DOD Appropriators Boost NASA Reusable Rocket Effort," *Space Business News,* 11 October 1994, 3; "Air Force Wants Less Oversight and More Insight for EELV Program," *Space Business News,* 2 March 1995, 4; and Lieutenant Colonel Smith, EELV Program Manager, telephone interview with author, May 1995.

29. Smith interview.

30. "DOD Space Launch Modernization Plan," 13, 18.

31. "The First Reusable SSTO Spacecraft," *Spaceflight,* March 1993, 91.

32. *NASA Access to Space Study Summary Report,* 72.

33. Ibid., 70.

34. NSTC-4.

35. "Revised NASA Implementation Plan for the National Space Transportation Policy," November 1994, 3.

36. "DOD Space Launch Modernization Plan Executive Summary," 1994, iv.

37. Marcia S. Smith, *Military Space Programs in a Changing Environment: Issues for the 103d Congress,* CRS Issue Brief 92879 (Washington, D.C.: GPO, 1992), 5.

38. "DOD Space Launch Modernization Plan Executive Summary," 10.

39. Ibid., 29.

40. *NASA Access to Space Study Summary Report,* 57.

41. "DOD Space Launch Modernization Plan Executive Summary," 19; *NASA Access to Space Study Summary Report,* 62.

42. "DOD Space Launch Modernization Plan Executive Summary," 11.

43. Marcia S. Smith, 4–5.

44. This is evident from the list of participants cited on the last page of the SLMP. Only two, Maj Jess Sponable and Colonel Worden, had worked in offices where there was RLV work in progress.

45. Alan W. Wilhite, *Advanced Technologies for Rocket Single-Stage-to-Orbit Vehicles,* Aerospace Industries Association of America (AIAA) Paper 91-0540 (New York: AIAA, January 1991).

46. "Conversations," *Aerospace America,* May 1995, 12.

47. "DOD Appropriators Boost NASA Reusable Rocket Effort," *Space Business News,* 11 October 1994, 3.

48. "Revised NASA Implementation Plan for the National Space Transportation Policy," 7.

49. "Air Force Wants Less Oversight and More Insight for EELV Program," *Space Business News*, 2 March 1995, 5.

50. "Sailing Space on the Clipper Ship," *Final Frontier*, December 1993, 52.

51. Charles "Pete" Conrad, *Beyond the NASA Shuttle*, remarks to the Washington Roundtable on Science & Public Policy, George C. Marshall Institute, Washington, D.C., 8 November 1994, 8.

52. William A. Gaubatz, "Designing for Routine Space Access," McDonnell Douglas Corp., briefing, 10.

53. "DOD Space Launch Modernization Plan Executive Summary," 19.

54. *NASA Access to Space Study Summary Report*, 57.

55. Martin Marietta Corp., "Propulsion and Cost Considerations for a Single Stage to Orbit Launch System," proprietary briefing, 6 December 1993, 5; and *NASA Access to Space Study Summary Report*, 60.

56. There are other charges against SSTO technology, but these are representative of the thinking of the opponents.

57. Briefing, Martin Marietta Corp.

58. David I. Chato and William J. Taylor, *Small Experiments for the Maturation of Orbital Cryogenic Transfer Technologies*, NASA-TM-105849, E-7295, NAS 1.15:105849, September 1992; William J. Bailey and Douglas H. Beekman, *Ullage Exchange—An Attractive Option for On-Orbit Cryogen Resupply*, AIAA Paper 90-0511 (New York: AIAA, January 1990); and Air Force Human Resources Laboratory, *On-Orbit Refueling* (Wright-Patterson AFB, Ohio: Air Force Human Resource Laboratory, February 1993).

59. "Reusable Launch Vehicle (RLV) Advanced Technology Demonstrator X-33"; "NASA Cooperative Agreement Notice," 12 January 1995, A-5.

60. "DOD Space Launch Modernization Plan," 4.

61. Briefing, Martin Marietta Corp.

62. Strategic Defense Initiative Organization, "Single Stage Rocket Technology, Program Status and Future Systems Feasibility Assessment," briefing, August 1992, 15.

63. "Rocket on a Round-Trip," *Discover*, May 1995, 54.

64. Russell Hannigan and David Webb, *Spaceflight in the Aero-Space Plane Era*, AIAA Paper 91-5089 (New York: AIAA, December 1991), 6.

65. Ibid.

66. For the ELV, 18 percent of $135 million is $24.3 million for a total cost of $159.3 million. For the RLV, 10 percent of a total launch cost including payload of $85 million is $8.5 million. This gives a total cost of $93 million—$66.3 million less than the ELV mission. Part of the payload to LEO for the ELV is the upper stage required to get the satellite to GEO. For a 20,000-pound-class payload in LEO, about 8,000 pounds gets to GEO.

67. "DOD Space Launch Modernization Plan," 1994, 5.

68. Dietrich E. Koelle and Michael Obersteiner, *Orbital Transfer Systems for Lunar Missions*, IAF Paper 91-181, October 1991.

69. Where *g* is the acceleration due to earth's gravity, I_{sp} is rocket engine vacuum specific impulse in seconds, M_O is starting mass, M_E is ending mass, and PMF is propellant mass fraction or the fraction of the gross vehicle weight that is fuel as opposed to structure—the critical number for SSTO and why breakthroughs in lightweight materials have been necessary to make it a possibility; Strategic Defense Initiative Organization, briefing, 28; Robert M. Zubrin and Mitchell B. Clapp, *An Examination of the Feasibility of Winged SSTO Vehicles Utilizing Aerial Propellant Transfer*, AIAA Paper 94-2923 (New York: AIAA, June 1994), 1.

70. Elske A. P. Smith and Kenneth C. Jacobs, *Introductory Astronomy and Astrophysics* (Philadelphia, Pa.: W. B. Saunders Co., 1973), 529; and Hannigan and Webb, 3.

71. Hannigan and Webb, 5–6.

72. "Reusable Launch Vehicle (RLV) Advanced Technology Demonstrator X-33"; "NASA Cooperative Agreement Notice," A–5.

73. NASA's Access to Space Study suggests that RLVs "would lower launch costs so dramatically that US industry could underprice all competitors," *NASA Access to Space Study Summary Report*, 68.

74. "Revised NASA Implementation Plan for the National Space Transportation Policy," 13.

75. "Marshall Engineers Barred From X-33 Effort," *Space News*, 8–14 May 1995, 1.

76. "Reusable Launch Vehicle (RLV) Advanced Technology Demonstrator X-33"; "NASA Cooperative Agreement Notice," iii.

77. "Revised NASA Implementation Plan for the National Space Transportation Policy," 7.

78. "Can a Reusable Launch Vehicle Be Developed without NASA's Money?" *Space Business News*, 30 March 1995, 2.

79. Philip Bono and Kenneth Gatland, *Frontiers of Space* (New York: Macmillan, 1976), 182–83.

80. "Single Stage Technology, Space Policy for Opening Up the Space Frontier; Single-Stage Rocket VTOL," *Spaceflight*, July 1994, 219.

81. "Federal Express Offers Consulting Services to RLV Teams," *Space Business News*, 13 April 1995, 6.

82. Andrew Wilson, *The Eagle Has Wings: The Story of American Space Exploration, 1945–1975* (London: British Interplanetary Society, 1982), 8.

83. Ibid.

84. G. Harry Stine, *Confrontation in Space* (Englewood Cliffs, N.J.: Prentice-Hall, 1981), 29.

85. Ibid.

86. Martin Marietta Corp.

87. "Air Force Position on the Future of Space Lift: An Approval Briefing," 3 December 1992, primary backup chart 1.

88. "Rocket on a Round Trip," *Discover*, May 1995, 56.

89. "Revised NASA Implementation Plan for the National Space Transportation Policy," 11.

90. The X-33 Home Page for Space Activists, http://www.contrib.andrew.cmu.edu/usr/fj04/x33.html; The NASA/Marshall Space Flight Center RLV Home Page, http://rlv.msfc.nasa.gov/rlv_htmls/rlvl.html; and Delta Clipper/SSRT Program, http://gargravarr.cc.utexas.edu/delta-clipper/title.html.

91. Joint Pub 3-14, 11–8.

92. *A National Security Strategy of Engagement and Enlargement* (Washington, D.C.: The White House, July 1994), 10.

93. Joint Pub 3-14, I-7, II-11.

94. Ibid.

95. Arnold, 455.

96. Robert F. Futrell, *Ideas, Concepts, Doctrine: Basic Thinking in the United States Air Force,* vol. 2, *1961–1984* (Maxwell AFB, Ala.: Air University Press, 1989), 691; John M. Collins, *Military Space Forces: The Next 50 Years* (New York: Pergamon-Brassey's, 1989), 53.

97. Kenneth J. Hagan, *This People's Navy: The Making of America Sea Power* (New York: Free Press, 1991), 273–75.

98. Quoted in Kenneth Hagan, 275.

99. Joint Pub 3-14, III-1.

100. Ibid., III-4.

101. *A National Security Strategy of Engagement and Enlargement,* 18.

102. Secretary of the Air Force Staff Group, *"45 Years of Global Reach and Power: 1947–1992"* (Washington, D.C.: GPO, 1992).

103. See Tom Blow, *Defending Against a Space Blockade,* CADRE Report AU-ARI-CP-89-3 (Maxwell AFB, Ala.: n.p., December 1989).

104. Unclassified on-orbit satellite force structure numbers, limited by high launch and payload costs, range from two for DMSP weather satellites to five for fleet composite communications system (USN) (FLTSATCOM) naval communications satellites. National asset program numbers would be limited by the same factors. These forces are not very maneuverable and are vulnerable to primitive ASATs as will be described later, *ergo,* "few and fragile."

105. Scott Smith, *Preliminary Analysis of the Benefits Derived to US Air Force Spacecraft from On-Orbit Refueling,* Air Force Logistics Command Study, February 1993; William I. Shanney, *On-Orbit Servicing for US Air Force Space Missions-A Phased Development Approach,* proceedings of the International Symposium on Artificial Intelligence, Robots and Automation in Space, Kobe, Japan, November 1990.

106. House, *Strategic Satellite Systems in a Post-Cold War Environment: Hearings before the Subcommittee on Legislation and National Security of the Committee on Government Operations,* 103d Cong., 2d sess., 1994, 48–49.

107. The National Commission on Space, *Pioneering the Space Frontier* (New York: Bantam Books, 1986), 7.

108. Collins, *Military Space Forces: The Next 50 Years* (Washington, D.C.: Pergamon-Brassey's, 1989), 43.

109. *Pioneering the Space Frontier,* 85.

110. Telecon with Maj Jess Sponable, DC-X Program Manager, 14 December 1992.

111. This assumes final cost of extracted lunar oxygen and hydrogen is similar to present cost (infrastructure excluded).

112. This complies with Outer Space Treaty of 1967 which the United States has signed, and does not comply with the Moon Treaty of 1979, which the United States has not signed. This is not to say there are no political issues to be considered. If the president's policy is to remain the major economic power in space, this is one way to do so.

113. "RLV Advanced Technology Demonstrator X-33"; "NASA Cooperative Agreement Notice," A-5.

114. Zubrin and Clapp, 1.

115. *A National Security Strategy of Engagement and Enlargement,* 10.

116. Joint Pub 3-14, II-21.

117. Air Force Space Command briefing.

118. Ibid.

119. Air Force Regulation (AFR) 173-13, Attachment A15-1, "Aircraft Reimbursement Rates Per Flying Hour, US Government (Excluding DOD)" 30 November 1991.

120. Stine, 82.

121. Michael J. Muolo, AU-18, *Space Handbook. A Warfighter's Guide to Space* (Maxwell AFB, Ala.: Air University Press, 1993), Vol. 1, 85; Sir Peter Anson and Dennis Cummings, "The First Space War," in *The First Information War,* Alan D. Campen, ed. (Fairfax, Va.: Air Forces Communications and Electronics Association [AFCEA] International Press, 1992), 128, 130.

122. Anson and Cummings, 121.

123. "Conversations," *Aerospace America,* May 1995, 12.

124. Air Force Space Command briefing, 5.

125. Collins, 123.

126. R. J. Overy, *The Air War 1939–1945* (Chelsea, Mich.: Scarborough House, 1991), 96.

127. This was partly because of its association with the politically unpleasant memory of World War I's "unrestricted submarine warfare."

128. "Beam Me Down, Scotty," *Time Daily News Summary,* 3 February 1995. Interestingly, the Russians were reluctant to let a possibly defective US space shuttle flown by a human pilot within 30 feet of *Mir* on the first shuttle-*Mir* rendezvous mission.

129. There is very little mention of weather, for example in Giulio Douhet's *Command of the Air.* Some attribute this to the fact that Douhet was not himself a flier.

130. The RLV could perform ship-to-satellite reconnaissance, ship-to-earth reconnaissance, orbital intercept and rendezvous, satellite grappling and capture, strategic attacks, and a myriad of other tasks that seem to require some human intervention.

131. As laid out in three "Axioms of Space Combat Power" by Lt Col Michael R. Mantz in *The New Sword: A Theory of Space Combat Power*

(Maxwell AFB, Ala.: Air University Press, 1995), "Axiom 1. Space strike systems can be employed decisively by striking earth forces, both independently and jointly. . . . Axiom 2. Space-strike systems can be employed decisively in war when the enemy's essential means for waging war (industry, transportation, and communication) are vulnerable to attack from space. . . . Axiom 3. Space-strike systems can be employed decisively by striking at the decision-making structure (leadership and command and control) of the enemy." Mantz, 75–76.

132. Clausewitz was more fortunate in *von Krieg* because most of his readers understood how horses and the firearms of the day worked. The contrast with modern doctrine is stark. Today, lengthy technical explanations are necessary before even getting to subjects as important as the subject of policy.

133. For example, see Michael Wolfert, "Concept Paper on Space Organizational Options" (paper for the DOD Roles and Missions Commission, 12 December 1994), 9.

134. Collins, 23.

135. Ibid., 24.

136. Ibid., 23.

137. Mantz, 20.

138. Collins, 56.

139. Warheads of this size were tested recently in hypersonic weapons effects tests at Sandia National Labs. Orbital versions of these weapons would have to weigh more than 30 pounds when launched because each would need a small rocket motor to impart about 7,450 fps of _*v* to deorbit.

140. Mantz, 48.

141. "Reusable Launch Vehicle (RLV) Advanced Technology Demonstrator X-33"; "NASA Cooperative Agreement Notice," A-5.

Bibliography

"45 Years of Global Reach and Power: 1947–1992." Washington, D.C.: Secretary of the Air Force Staff Group, 1992.

A National Security Strategy of Engagement and Enlargement. Washington, D.C.: The White House, July 1994.

AFR 173-13, Attachment A 15-1. "Aircraft Reimbursement Rates Per Flying Hour." 30 November 1991.

Air Force Human Resources Laboratory. *On-Orbit Refueling.* Wright-Patterson AFB, Ohio: AFHRL, February 1993.

"Air Force Position on the Future of Space Lift: An Approval Briefing." 3 December 1992.

Air Force Space Command. "The Evolution of National Security Space Policy." Briefing, 9 October 1992.

———. "Force Application Study." Briefing, 9 May 1991.

"Air Force Wants Less Oversight and More Insight for EELV Program." *Space Business News,* 2 March 1995.

Anson, Sir Peter, and Dennis Cummings. "The First Space War." In *The First Information War.* Edited by Alan D. Campen. Fairfax, Va.: AFCEA International Press, 1992.

"Ariane Chief Says Eastern Nations, Not US, Will Be Future Competition." *Aerospace Daily,* 4 October 1993.

Arnold, Henry H. *The War Reports of General of the Army George C. Marshall, Chief of Staff, General of the Army H. H. Arnold, Comanding General, Army Air Forces, and Fleet Admiral Ernest J. King.* New York: J. B. Lipincott Company, 1947.

Bailey, William J., and Douglas H. Beekman. *Ullage Exchange-An Attractive Option for On-Orbit Cryogen Resupply.* AIAA Paper 90-0511. New York: AIAA, January 1990.

"Beam Me Down, Scotty." *Time Daily News Summary,* 3 February 1995.

Blow, Tom. *Defending Against a Space Blockade.* CADRE Report AU-ARI-CP-89-3. Maxwell AFB, Ala.: n.p., December 1989.

Bono, Philip, and Kenneth Gatland. *Frontiers of Space.* New York: Macmillan, 1976.

Boushey, Homer A. "Blueprints for Space." *Air University Quarterly Review,* Spring 1959.

"Can a Reusable Launch Vehicle Be Developed without NASA's Money?" *Space Business News,* 30 March 1995.

Chato, David J., and William J. Taylor. *Small Experiments for the Maturation of Orbital Cryogenic Transfer Technologies.* NASA-TM-105849, E-7295, NAS 1.15:105849. September 1992.

Collins, John M. *Military Space Forces: The Next 50 Years.* New York: Pergamon-Brassey's, 1989.

Conrad, Charles Pete. "Beyond the NASA Shuttle, Remarks to the Washington Roundtable on Science & Public Policy, George C. Marshall Institute," Washington, D.C., 8 November 1994.

"Conversations." *Aerospace America,* May 1995.

Delta Clipper/SSRT Program Home Page, http://gargravarr. cc.utexas.edu/delta-clipperititle.html.

"DOD Appropriators Boost NASA Reusable Rocket Effort." *Space Business News,* 11 October 1994.

"DOD Space Launch Modernization Plan." Draft briefing. April 1994.

———. Draft executive summary. April 1994.

"Federal Express Offers Consulting Services to RLV Teams." *Space Business News,* 13 April 1995.

"First Reusable SSTO Spacecraft." *Spaceflight,* March 1993.

Futrell, Robert F. *Ideas, Concepts, Doctrine: Basic Thinking in the United States Air Force.* Vol. 2, *1961–1984.* Maxwell AFB, Ala.: Air University Press, 1989.

Gaubatz, William A. "Designing for Routine Space Access." McDonnell Douglas Corporation briefing.

Hagan, Kenneth J. *This People's Navy.* New York: Free Press, 1991.

Hannigan, Russell, and David Webb. *Spaceflight in the Aero-Space Plane Era.* AIAA Paper 91-5089. New York: AIAA, December 1991.

Howe, Patty. Special Assistant to the Secretary of Defense. Memorandum. Subject: The Clinton Transition Team Tasking of 17 December 1992.

Joint Doctrine, Tactics, Techniques, and Procedures (JDTTP) 3-14, "Space Operations." Final draft. 15 April 1992.

Koelle, Dietrich E., and Michael Obersteiner. *Orbital Transfer Systems for Lunar Missions.* IAF Paper 91-181, October 1991.

Mantz, Michael R. *The New Sword: A Theory of Space Combat Power.* AU Research Report AU-ARI-94-6. Maxwell AFB, Ala.: Air University Press, May 1995.

"Marshall Engineers Barred From X-33 Effort." *Space News,* 8–14 May 1995.

Martin Marietta Corporation. "Propulsion and Cost Considerations for a Single Stage to Orbit Launch System." Proprietary briefing. 6 December 1993.

Muolo, Michael J. AU-18, *Space Handbook: A Warfighter's Guide to Space.* Maxwell AFB, Ala.: Air University Press, 1993.

NASA Access to Space Study Summary Report. Washington, D.C.: Office of Space Systems Development, NASA Headquarters, January 1994.

NASA Cooperative Agreement Notice. "Reusable Launch Vehicle (RLV) Advanced Technology Demonstrator X-33." 12 January 1995.

NASA/Marshall Space Flight Center RLV Home Page, http://rlv.msfc.nasa.gov/rlv_htmls/rlvl.html.

National Commission on Space. *Pioneering the Space Frontier.* New York: Bantam Books, 1986.

Overy, R. J. *The Air War 1939–1945.* Chelsea, Mich.: Scarborough House, 1991.

Report of the Task Force on the National Aero-Space Plane (NASP) Program. Washington, D.C.: Defense Science Board, November 1992.

"Revised NASA Implementation Plan for the National Space Transportation Policy." November 1994.

"Rocket on a Round-Trip." *Discover,* May 1995.

"Sailing Space on the Clipper Ship." *Final Frontier,* December 1993.

Shanney, William I. *On-Orbit Servicing for US Air Force Space Missions—A Phased Development Approach.* Proceedings of the International Symposium on Artificial Intelligence, Robots and Automation in Space, Kobe, Japan, November 1990.

Singer, S. E. "The Military Potential of the Moon." *Air University Quarterly Review*, Summer 1959.

"Single Stage Rocket Technology; Program Review of Future Systems & Applications." Strategic Defense Initiative Organization briefing. January 1993.

"Single Stage Rocket Technology; Program Status and Future Systems Feasibility Assessment." Strategic Defense Initiative Organization briefing, August 1992.

"Single Stage Technology; Space Policy for Opening Up the Space Frontier; Single-Stage Rocket VTOL." *Spaceflight*, July 1994.

Smith, Elske A. P., and Kenneth C. Jacobs. *Introductory Astronomy and Astrophysics*. Philadelphia, Pa.: W. B. Saunders Company, 1973.

Smith, Marcia S. *Military Space Programs in a Changing Environment: Issues for the 103d Congress*. CRS Issue Brief 92879. Washington, D.C.: GPO, 1 December 1992.

Smith, Scott. *Preliminary Analysis of the Benefits Derived to US Air Force Spacecraft from On-Orbit Refueling*. Air Force Logistics Command Study, February 1993.

Stine, G. Harry. *Confrontation in Space*. Englewood Cliffs, N.J.: Prentice-Hall, 1981.

Strategic Satellite Systems in a Post-cold War Environment: Hearings before the Subcommittee on Legislation and National Security of the Committee on Government Operations. US House of Representatives, 103d Cong., 2d sess., 1994.

Szafranski, Richard. "When Waves Collide: Future Conflict." *Joint Forces Quarterly*, Spring 1995.

Wilhite, Alan W., et al. *Advanced Technologies for Rocket Single-Stage-to-Orbit Vehicles*. AIAA Paper 91-0540. New York: AIAA, January 1991.

Wilson, Andrew. *The Eagle Has Wings*. London: British Interplanetary Society, 1982.

Wolfert, Michael. "Concept Paper on Space Organizational Options." Paper for the DOD Roles and Missions Commission, 12 December 1994.

The X-33 Home Page for Space Activists, http://www.contrib.andrew.cmu.edu/usr/fj04/x33.html.

Zubrin, Robert M., and Mitchell B. Clapp. *An Examination of the Feasibility of Winged SSTO Vehicles Utilizing Aerial Propellant Transfer*. AIAA Paper 94-2923. New York: AIAA, June 1994.

Chapter 8

Concepts of Operations for a Reusable Launch Space Vehicle

Michael A. Rampino

The objective of NASA's technology demonstration effort is to support government and private sector decisions by the end of this decade on development of an operational next-generation reusable launch system.

The objective of DoD's effort to improve and evolve current ELVs is to reduce costs while improving reliability, operability, responsiveness, and safety.

The United States Government is committed to encouraging a viable commercial U.S. space transportation industry.

US National Space Transportation Policy
5 August 1994

Introduction

On 18 May 1996, the United States took another small step toward maturity as a space-faring nation. Under the scorching sun of the New Mexico desert, an attentive media corps readied their cameras. Ground and flight crews monitored consoles and waited for the latest global positioning updates to be received and processed. At 0812:02, a small, pyramid-shaped rocket, the McDonnell Douglas Aerospace (MDA) DC-XA, rose from its launch mount on a column of smoke and fire. Unlike today's operational spaceships, this one landed on its feet after a 61-second flight with all its components intact. This ninth flight of the Delta Clipper experimental rocket was no giant leap for mankind given the limited capabilities of the

This work was accomplished in partial fulfillment of the master's degree requirements of the School of Advanced Airpower Studies, Air University, Maxwell AFB, Ala., 1996.
Advisor: Maj Bruce M. DeBlois, PhD
Reader: Dr Karl Mueller, PhD

vehicle, but it proved once again that reusable rockets are a reality—today.[1]

The US military must be prepared to take advantage of reusable launch vehicles (RLV) should the National Aeronautical Space Administration (NASA)-industry effort to develop an RLV technology demonstrator prove successful.[2] The focus of this study is an explanation of how the US military could use RLVs, by describing and analyzing two alternative concepts of operations (CONOPS).

The most recent National Space Transportation Policy assigned the lead role in evolving today's expendable launch vehicles (ELV) to the Department of Defense (DOD). It assigned NASA the lead role in working with industry on RLVs.[3] The United States Air Force (USAF), as the lead space lift acquisition agent within the DOD, is an active participant in RLV development but with limited responsibility and authority since it is a NASA-led program.[4] USAF leadership has maintained interest in the program but has focused on ensuring continued access to space without incurring the technical risk of relying on RLV development. The USAF's initiation of the evolved expendable launch vehicle (EELV) program reflects this approach.[5]

As of this writing, the USAF, on behalf of DOD, is formulating and defining DOD requirements for an RLV in an effort to plan for a possible transition from ELVs to RLVs. Specifically, the NASA-USAF integrated product team (IPT) for Space Launch Activities is currently examining "operational RLV DOD requirements."[6] In addition, the USAF's Phillips Laboratory started a Military Spaceplane Applications Working Group in August 1995 which may indirectly help identify DOD's RLV needs.[7] This research is intended to contribute to the ongoing process by describing how the US military should use RLVs.

To help remedy the lack of specific DOD requirements for an operational RLV, this study identifies CONOPS for military use of such a vehicle. Obviously, identifying CONOPS requires addressing other issues along the way. For instance, the attributes of an operational RLV must first be identified to facilitate development of the alternative CONOPS. If there are new missions enabled by the vehicle's reusable nature, missions which are not feasible using ELVs or the Space Transportation System (STS) (also known as the shuttle), they must be identified as well. Given the timeline of the RLV program, the year 2012 is a reasonable estimated date for the fielding of an operational system. This date will serve as the basis for analysis in this study.

Four assumptions are worth mentioning at the outset. First, the estimate that RLV technology could become operationally feasible by 2012 is reasonable. Second, a fiscally constrained environment will continue. Third, the US government will continue to support growth and development of the US commercial space lift industry and encourage dual-use, or perhaps triple-use, of related facilities and systems.[8] Fourth, the US government's national security strategy will continue to emphasize international leadership and engagement to further American political, economic, and security objectives.

Given the assumptions of fiscal constraint and a government policy of cooperation with and encouragement of the US commercial space lift industry, any military RLV acquisition strategy will do well to take maximum advantage of possible dual-use or triple-use opportunities and economies of scale. For instance, the US military could pursue development of a military RLV which would share design similarities (i.e., hardware components) with commercial RLVs to the greatest practical extent, minimizing military-unique design requirements and thereby lowering costs. Such an approach would also take advantage of the economies of scale possible if the commercial space lift industry were to operate an RLV similar to the one manufactured for the military. Of course, this assumes there is a need for a military-unique RLV—not just military use of a commercially produced and operated RLV.

Military RLV Requirements

One answer to the research question proposed earlier might be that the DOD does not need RLVs. There may be no requirement for them. One way to confirm or deny this assertion is to examine the relevant requirements documentation.

Space Lift Requirements. An Air Force Space Command briefing on mission area plan (MAP) alignments and definitions lists four functions for a "reusable spacecraft for military ops": strike, transport, space recovery, and reconnaissance.[9] However, the most recent space lift MAP takes a more conservative approach. Using the strategies-to-tasks methodology, the MAP documents five tasks of space lift derived through the mission area assessment process: launching spacecraft, employing the ranges to support these launches, performing transspace operations, recovering space assets, and planning and forecasting government and commercial launches.[10] Prioritized space lift deficiencies are determined through mission needs analysis. These nine deficiencies are mainly cost-related concerns but also include two capability related deficiencies: "cannot perform transspace operations," and "no DoD capability to perform recovery and return."[11] The mission solution analysis concludes that the EELV is the number one priority in the midterm (within 10 years) although RLVs, orbital transfer vehicles (OTV), and a space-based range system are desirable in the long term (within 25 years).[12] The five key space lift solutions are developing the EELV, completing range upgrades, cooperating with NASA in their RLV program, developing advanced expendable and reusable upper stage systems, and fielding space-based range systems.[13] Although potential RLV applications in other mission areas such as reconnaissance and strike are discussed, these are seen as long-term (10–25 years) capabilities.

The fact that the USAF's MAP for space lift (DOD's by default) does not aggressively pursue the potential of RLVs is not surprising. Being based on the strategies-to-task framework, the MAP process will not identify a deficiency or state a requirement when there is no existing higher-level objective or task calling for that capability. Further, the National Space Transportation Policy clearly identifies NASA as the lead agency for RLV technology demonstration, not the DOD (USAF). Finally, the USAF's low-risk approach is understandable given the very real need to ensure continued access to space in support of national security requirements. The last time our country put all of its space lift eggs in one basket, the

STS, major disruptions in access to space for national security payloads resulted when the basket broke. The 1986 *Challenger* accident combined with our national policy to emphasize use of the STS over expendable launch vehicles created a situation USAF space lift leaders never want to see repeated.[14] Given these factors, it is laudable that the space lift MAP identifies transspace operations and recovery and return as capability deficiencies and foresees the use of RLVs in reconnaissance and strike missions. These two deficiencies will not be satisfied by EELV development, but they could be used to derive requirements for military use of an RLV.

Commander in Chief, USSPACE Command Desires. It is interesting to note that a different approach to generating requirements, a revolutionary planning approach, has identified RLVs as promising for broader military applications and sparked the interest of America's most senior military space commander.[15] In a 1995 message discussing implementation of the conclusions and recommendations of the Air Force Scientific Advisory Board's *New World Vistas* study, Gen Joseph W. Ashy, commander in chief of United States Space Command (CINCSPACE) and commander of Air Force Space Command, identified reusable launch vehicles as one of the most important technologies cited in the findings of this revolutionary planning effort.

General Ashy identified the capabilities to "take-off on demand, overfly any location in the world in approximately one hour and return and land within two hours at the take-off base" as desirable. He further suggested reconnaissance, surveillance, and precision employment of weapons as potential RLV applications.[16]

Requirements Identified. For the purpose of exploring military RLV concept of operations, this study identifies spacecraft launch and recovery, transspace operations, strike (in and from space), and reconnaissance as potential RLV tasks. The first two tasks flow from the space lift MAP.

The second two tasks are not identified as tasks for space lift in the MAP, probably because of the inherent near-term emphasis of the MAP, but may prove feasible with RLVs. Further, as shown above, they have been identified as potential RLV applications by the CINCSPACE.

Project Overview

Before developing and analyzing CONOPS for military use of RLVs, current RLV concepts and attributes are summarized and hypothetical attributes of a notional RLV for use in military applications are suggested in the next section. Identifying these notional RLV attributes is a necessary step in the process of answering the research question; they are not intended to be the final word on military RLV design.

Following the discussion of RLV concepts and attributes, another section presents two CONOPS. The two operations concepts are intended roughly to represent military space plane advocates' visions in the first case and to be a logical extension of the current RLV program's goals in the second case. An analysis of these concepts of operations is provided. The criteria used in the analysis include capability, cost, operations efficiency and effectiveness, and politics. The last section in this chapter summarizes significant conclusions and recommends a course of action for the US military to pursue with respect to RLVs.

RLV Concepts and Attributes

To facilitate CONOPS development and analysis, this chapter summarizes current RLV concepts and attribute, and suggests hypothetical attributes of a notional RLV for use in military applications. These notional RLV attributes are not intended to serve as the final word on RLV design, as an endorsement of any particular company's concept, or as a recommendation regarding whether an RLV should take off or land vertically or horizontally. Describing the attributes of an RLV is simply required to provide a basis for the subsequent analysis.

Before stating these attributes, this section first presents an overview of the three RLV concepts proposed by Lockheed Advanced Development Company (LADC), MDA, and Rockwell Space Systems Division (RSSD), as well as the Black Horse transatmospheric vehicle (TAV) concept made popular by Air University's *SPACECAST 2020* project. Next, RLV attributes are discussed in terms of the requirements introduced earlier. Finally, this chapter presents the attributes of a notional RLV to be used for further analysis.

Representative RLV Concepts

Definitions. The lexicon associated with RLVs can be confusing. Often, the term *RLV* is used interchangeably with terms like *SSTO*, for *single-stage-to-orbit*; *TAV*, for *transatmospheric vehicle*; and *MSP*, for *military spaceplane*. Unfortunately, there doesn't appear to be a consensus that these terms are interchangeable. RLV is not interchangeable with SSTO. A one-piece expendable rocket might also achieve orbit with a single stage, and a completely reusable multistage vehicle could be constructed. TAV tends to be used in connection with winged, aircraft-like vehicles that operate substantially in the atmosphere while maintaining some capability to reach orbit. MSP appears to be more general, including RLVs and TAVs used for military applications. For the sake of clarity, RLV is used here to refer to a completely reusable vehicle which is capable of achieving earth orbit while carrying some useful payload and then returning.

RLV Concepts. Three companies are currently participating in Phase I of the NASA-industry RLV program, the concept definition and technology development phase. One of these three will be selected to continue developing its RLV concept in Phase II of the program, the demonstration phase. NASA has scheduled source selection to be complete by July 1996.[17] The winner of this source selection will develop an advanced technology demonstration vehicle, the X-33, which will be used to conduct flight tests in 1999. The focus of this second phase will be to demonstrate aircraft-like operations and provide enough evidence to support a decision on whether or not to proceed with the next phase in the year 2000. Phase III of the RLV program would include commercial development and RLV operations.[18] The decision to enter Phase III will be a complex one. It will depend on Phase II results as well as many other contextual factors bearing on decision makers at the turn of the century. In keeping with the recommendations of NASA's *Access to Space Study*, all phases of the RLV program are to be "driven by efficient operations rather than attainment of maximum performance levels."[19]

All the RLV concepts are currently focused on satisfying the requirement to deliver and retrieve cargo from the International Space Station, Alpha (ISSA). This, perhaps artificially, drives a certain payload requirement (table 35).[20] All three concepts use cryogenic propellants, liquid oxygen and liquid hydrogen (LOX/LH_2), to achieve high specific impulse.[21] Other common attributes are based on objectives of the RLV program, such as the mission life and maintenance requirements.[22] The required thrust-to-weight ratio (F/W), specific impulse (I_{sp}), and mass fraction are based on current estimates and analysis.[23] Current cost estimates are based on paper studies. The estimates vary widely and are affected by the size of the RLV, the number built, whether or not they are certified to fly over land, the basing scheme, other aspects of the concept of operations, and the acquisition strategy, to name just a few of the factors involved.[24] For example, a smaller, lighter, and less capable (with respect to payload) RLV would presumably prove cheaper to build and face less technical risk in development.[25]

Where the three RLV concepts diverge is in their propulsion systems and takeoff and landing concepts. Lockheed Advanced Development Company's RLV would be a lifting body using linear aerospike rocket engines as opposed to more traditional rocket engines with bell-shaped nozzles.[26] The vehicle would take off vertically and land horizontally (VTHL). McDonnell Douglas Aerospace's RLV would be a conical reentry body using traditional bell-shaped nozzle rocket engines. The vehicle would takeoff vertically and land vertically (VTVL). Rockwell Space System Divisions RLV would be a winged body using traditional bell-shaped nozzle engines.[27] Like the Lockheed concept, Rockwell's is a VTHL vehicle (see fig. 11 for an artist's concept of all three vehicles).

Black Horse. The Black Horse TAV concept was identified by Air University's *SPACECAST 2020* as the most promising space lift idea evaluated by the team.[28] The Black Horse is

Figure 11. Current RLV Concepts

included here for comparison because it continues to be of interest to military spaceplane advocates and provides an interesting contrast to the concepts being explored under the NASA-led RLV program. However, this is comparing apples and oranges to a great extent.

The Black Horse does not come close to achieving the RLV payload capability (see table 35).[29] Also, some analysts have doubts about its technical feasibility.[30] Even if Black Horse were technically feasible, the market for small payload launchers is highly competitive and includes the most operationally responsive of all expendable vehicles.[31] This would likely limit Black Horse's utility to only military missions, and perhaps just a subset of those.

Discussion of Requirements

Officially stated requirements for the RLV concepts currently being proposed do not include conducting military operations such as reconnaissance and strike (in and from space). As discussed earlier, there is growing support for de-

Table 35

Attributes of Proposed RLV Concepts and One Popular TAV Concept

Vehicle	Payload	Propulsion	Mass Fraction	Takeoff & Landing Concept	Cross-range Capability	TurnaroundTime	Mission Life	Nonrecurring Costs	Recurring Costs
Lockheed Advanced Development Company RLV	25K lbs. to ISSA orbit (50x244 NM x51.6 deg.) 40K lbs. to 100 NM circular orbit	LOX/LH_2 Linear Aerospike engines F/W = 75 I_{sp}=440 (vacuum I_{sp})	0.10-0.11	VTHL Lifting Body Runway req. for landing (8K feet)	500-600 NM	Nominal: 7 days Contingency: 2 1/2 days	100 (Depot maintenance after 20 missions)	$5 - 20B	Annual costs $0.5 - 1.5B (4 vehicles) $1K/lb.
McDonnell Douglas Aerospace RLV	25K lbs. to ISSA orbit 40K lbs. to 100 NM circular orbit	LOX/LH_2 (tri-prop opt.) Bell nozzle rocket eng. F/W = 75 I_{sp} = 440 (vacuum I_{sp})	0.10-0.11	VTVL Conic reentry body Propulsive landing on 150x150 foot grate	11-12K NM	1-2 days	100 (Depot maintenance after 20 missions)	$5 - 20B	Annual costs $0.5 - 1.5B (4 vehicles) $1K/lb.
Rockwell Space Systems Division RLV	25K lbs. to ISSA orbit 40K lbs. to 100 NM circular orbit	LOX/LH_2 (tri-prop opt.) Bell nozzle rocket eng. F/W = 75 I_{sp} = 440 (vacuum I_{sp})	0.10-0.11	VTHL Winged body Runway req. for landing (10K feet) Erect for takeoff	Nominal: 800 NM Contingency: 1100 NM (with TPS degradation)	Nominal: 7 days Contingency: 3 1/2 days	100 (Depot maintenance after 20 missions)	$5 - 20B	Annual costs $0.5 - 1.5B (4 vehicles) $1K/lb.
Black Horse TAV	1K lbs. to 100 NM circular orbit (4x5x6 feet)	Jet fuel & H_2O_2 Rocket eng. I_{sp}=323–335	0.05 in theory, but aerial refuel allows 0.08 mass fraction	HTHL Runway req. for takeoff & landing (3,150 feet)	1,800 NM+	12 hours - 1 day	Unknown	$700M "X" program cost= $150M	$260K/sortie Annual costs $100M (8 vehicles) $1K/lb.

veloping a system that is capable of accomplishing these missions. It will be a great challenge to identify a system that can meet these military requirements, does not require a great increase in the military space budget, and also satisfies civil (non-DOD government) and commercial needs.

Payload. The payload capability required of an RLV is a very important attribute. Determining the desired payload weight and size capability based on anticipated requirements for delivering and retrieving satellites and other cargo to and from orbit, flying reconnaissance payloads to space and back, and delivering weapons on the other side of the earth is not enough. Determining the desired payload weight and size must also be tempered by the technical risks, monetary costs, and operational costs which might be incurred as a result of establishing the payload requirement. The payload requirement drives the vehicle's physical size, engine performance requirements, development cost, and other attributes. There is general, although not complete, consensus that a smaller RLV than currently conceived by NASA may be more feasible. An argument for the smaller vehicle can be made based on three factors.[32]

First, the National Research Council's 1995 assessment of the RLV Technology Development and Test Program indicated that scaleability of structures from the X-33 test vehicle to a full-scale RLV is an area of uncertainty.[33] The report also concluded that "an increase of 30 percent or more" in current rocket engine performance will be required for the full-scale RLV.[34] The X-33 engine will not satisfy full-scale RLV performance requirements, so development of a new engine will be required. The report estimates it will take a decade to develop.[35] The report does not comment on the feasibility of developing a full-scale RLV but identifies the necessary engine development as a "difficult challenge."[36] These conclusions suggest that developing an RLV closer in size to the X-33 would minimize potential scaleability problems and reduce the requirement for increased engine performance. The result would be less technical risk.

Second, incurring less technical risk may also directly contribute to incurring less financial risk. If RLV development can avoid the need to develop engines with thrust-to-weight ratios of more than 75, then nonrecurring costs may be reduced. Cost is an important consideration for both government and commercial funding. Reducing the cost of access to space, not performance, is the primary driver for the RLV program.

Third, the greatest demand for launch services is not in the area of delivering 40,000-pound payloads to low earth orbit (LEO). Recent forecasts show the greatest demand to be in the medium and small payload class, not more than 20,000 pounds to LEO, and less than 10,000 pounds to geosynchronous transfer orbit (GTO).[37] These forecasts may indicate that sizing an RLV to compete in this market is more likely to result in a successful commercial development. Developing a less expensive vehicle that can satisfy commercial requirements as well as the majority of government requirements has the greatest potential for economic development. Of course, a larger RLV could deliver smaller payloads, perhaps more than one at a time, but it is not at all clear that using the larger RLV would be more efficient. The *Titan* IIIC, a large space lifter originally designed to support launches of the Dyna-Soar spaceplane, never quite caught on as a commercial vehicle. The *Ariane* 5 was originally designed to launch the Europeans' Hermes spaceplane which has since been canceled.[38] It remains to be seen if the heavy-lift *Ariane* 5 can become a commercial success without government assistance.[39]

An argument against developing a smaller vehicle can be made based on the fact that it would not satisfy all the government's requirements. For instance, it might not be able to deliver the necessary cargo loads to the space station or launch the largest national security payloads. Some suggest that even commercial payload size is on the increase.[40] This deficiency could be addressed in several ways. First, a large RLV could be developed after the smaller version, allowing more time for technology maturation and the development of an experience base with the smaller RLVs. In the interim, the large government payloads could be delivered using existing systems or the heavy-lift version of DOD's EELV projected to be available in 2005.[41] Second, the large payloads could, in theory, be made smaller, by taking advantage of miniaturization or by assembling modular components in orbit. Making payloads smaller may not be a panacea, especially for space-station loads, but there is some evidence that the DOD is moving in this direction.[42] Third, a technique referred to as a pop-up maneuver may be used to deliver large payloads with a smaller RLV. This would entail flying the RLV on a suborbital trajectory to deploy larger payloads into LEO than it would be possible to deploy if the RLV itself had to achieve orbit.[43] The pop-up maneuver requires the physical dimensions of the payload bay in the smaller RLV to be sized to accommodate the largest payloads the vehicle is planned to fly. It also forces the RLV to land downrange and be flown back to the primary operating base.

Cargo area dimensions for an RLV are under study, and recommendations vary considerably. NASA's *Access to Space Study* considered payload bay lengths of 30 and 45 feet—large enough for space station cargo but still too small for some national security needs.[44] The USAF's Phillips Lab has proposed a 25-foot-long payload bay to satisfy military requirements.[45] One RLV competitor, Rockwell, believes a 45-foot payload bay is needed even to accommodate "future generations of commercial satellites and their upper stages."[46]

Propulsion and Mass Fraction. Propulsion and mass fraction are important attributes of an RLV but are not stated as desired attributes here. The appropriate figures would result from design of an RLV to meet other requirements.

Takeoff and Landing Concept. An RLV's methods of takeoff and landing are significant to the extent that they affect its operations. Obviously, the need for a runway limits basing and delivery access options. The VTHL vehicles will also require some means for erection prior to launch. On the other hand, even a VTVL vehicle will require some unique basing support, such as a 150-foot square grate.[47] Both approaches require cryogenic fuel facilities which are not typically available at most airfields. Perhaps more important than whether an RLV lands vertically or horizontally is the overall ease and simplicity of operations achieved through its design.

Cross-range Capability. The term *cross-range capability*, as used here, refers to the ability of an RLV to maneuver within the atmosphere upon its return from space. This does not include the ability of an RLV to change its orbital path while in space.[48] The ability of an RLV to maneuver within the atmosphere can be a significant advantage during contingencies requiring an abort while ascending or a change in landing location while returning from a mission. This capability could also prove useful in military applications. For ascent contingency purposes, 600 nautical miles (NM) is adequate.[49] If the vehicle must land at the same base from which it took off after one revolution around the earth, then a cross range on the order of 1,100–1,200 NM is required.[50] The cross-range capability requirement for certain military missions could potentially be higher.

Turnaround Time. For commercial and civilian applications, this attribute is primarily an efficiency question. It will contribute to determining how many RLVs are needed and the nature of launch base facilities. For military missions, this attribute is not only related to efficiency but effectiveness as well. Reconnaissance and strike missions in particular could be facilitated by shorter turnaround times.

Related to turnaround time is the issue of responsiveness, how long it takes to prepare an RLV for launch. Again, military missions are likely to demand quicker response times.

Mission Life. This attribute is closely related to costs. Given the current uncertain state of RLV technology, it is hard to predict what a reasonable mission life would be, so the figure of one hundred has been established. Some think a five hundred-mission life is a reasonable expectation.[51] The frequency of required depot maintenance is also difficult to anticipate.[52]

Other Attributes. There are several other attributes not yet addressed which can significantly affect RLV operations, such as the ability to operate in adverse weather conditions and crew size. Today's space lift vehicles are severely constrained by weather, from lightning potential, to winds at altitude, to winds on the surface.[53] Delays due to weather can add to the cost of operations and dramatically decrease responsiveness. A truly operational RLV, especially one which will conduct military missions, should be able to operate in all but the most extreme weather conditions. A truly operational RLV should also require smaller operations crews than are required by current systems. Today, thousands of people are employed in STS launch base operations at the Kennedy Space Center. Unmanned, expendable launch vehicle operations at Cape Canaveral Air Force Station require hundreds of people to launch a vehicle. These figures should be well under one hundred for an operational RLV.[54] Finally, all payloads should use standard containers and interfaces to facilitate operations efficiency and responsiveness.[55]

Desired Attributes for a Notional RLV

A review of current concepts under study and development in support of the RLV program provides reasonable bounds for requirements or desired attributes for a notional RLV which could be used to support military missions. At the same time, one of the assumptions underlying this study is continued fiscal constraint. This assumption is the basis for a desire to maximize dual or triple use (i.e., military, civil governmental, and commercial use) of an operational RLV to the greatest extent practical. If more user requirements can be satisfied, especially those of commercial operators, it is more likely that funding will be available and that economies of scale can be achieved. Of course, trying to satisfy too many requirements with one vehicle could lead to failure. Defense procurement history is filled with programs that attempted to satisfy so many users that they failed to stay within budget, stay on schedule, or deliver the desired operational capability. With this caution in mind, the attributes of a notional RLV to be used as the basis for analysis are described below.

The notional RLV should be able to deliver 20,000 pounds to a circular LEO with an altitude of one hundred NM (table 36). This payload weight capability should also allow the vehicle to deliver commercial communications satellite-sized payloads to GTO, carry reconnaissance payloads on orbital or suborbital missions, and deliver significantly more weapons payload than today's F-16 and F-117 fighter aircraft or as much as an SS-18 heavy intercontinental ballistic missile (ICBM).[56] Its propulsion system's attributes are not described or stated as requirements, but based on current RLV concepts the assumption is that cryogenic rocket engines will be used.

Table 36

Summary of Attributes of a Notional RLV

Attribute	Value
Payload Size and Weight	20K lbs. to 100 NM circular orbit (due east) 30-foot-long cargo area
Propulsion	As necessary (LOX/LH2 propellant rocket engines based on current concepts)
Mode of Takeoff and Landing	As necessary (assume 10K foot airfield required at any RLV base)
Required Runway Length	10K feet maximum (if necessary at all)
Cross-range Capability	1,100 NM minimum
Turnaround Time	1-day nominal, 12-hour contingency (6-hour response)
Mission Life	100 minimum Depot maintenance after 20+ missions
Development Cost	$4–13B
Cost Per Mission	Annual Costs: $0.50B for 4 RLV squadron <$1Klb.

The method of takeoff or landing is also not specified. To provide a basis for analysis, it will be assumed that any RLV operating base will need no longer than a 10,000-foot runway. If a VTVL vehicle is pursued, this requirement might still exist in practice if it is necessary or desirable to supply an operating base rapidly using large transport aircraft. In any case, this assumption should not constrain choices of operating bases too severely. An RLV used for military applications must have shorter turnaround and response times than what might be necessary or desired for commercial and civil applications, but a nominal one-day turnaround, 12 hours for contingencies, and a six-hour response time do not seem unreasonable based on current concepts. Standard payload containers and interfaces would be used for all missions. Finally, mission life and costs are essentially accepted from the current concepts with one exception. Given the choice of an RLV with less payload capability, the cost figures are estimated to be in the lower end of the range established for a full-scale RLV.

RLV Concepts and Attributes Summary

The concepts being proposed for a full-scale RLV under the NASA-industry RLV program are driven by requirements which may not be completely compatible with requirements for a military RLV. The large, full-scale RLV may not target the space lift market in the most economically viable way. Given the potential to reduce technical risk, save money, and more effectively target the vast majority of user requirements, these attributes for a notional RLV can serve as a basis for CONOPS development and further analysis.

Concepts of Operations

This section presents an outline of two concepts of operations. The first concept, CONOPS A, is intended to be representative of military space plane advocates' visions. It uses the notional RLV described in table 36. CONOPS A makes the fullest military use of the roughly one-half scale RLV to accomplish not only traditional space lift missions but also the additional missions of returning payloads from orbit, transspace operations, reconnaissance, and strike (in and from space). The second concept, CONOPS B, is intended to represent a logical extension of the current RLV programs' goals. It is based on the full-scale vehicle concepts currently being proposed under the RLV program (table 35). CONOPS B also makes expanded use of RLVs. The capabilities of each RLV used for analysis are summarized in table 37.[57]

New systems, weapons, and technologies are usually fielded without the ultimate utility or best application (CONOPS) having been elaborated—the RLV may show its greatest application to have been unanticipated. An RLV may have to be built and operated for some time before its greatest utility is appreciated or the best methods of employment are discovered.[58] In spite of this reality, describing a CONOPS for RLVs at this early stage is vital. Without defining how an RLV force is to be fielded, organized, and operated, its development is bound to

Table 37

CONOPS A and B RLV Capabilities

RLV	Fleet Size	Turnaround time (hours)	Payload	Sorties/day	20K lb. weapons/day
CONOPS A	6	Nominal: 24 Contingency: 12 Response: 6	20K lbs. to LEO	12	96
CONOPS B	4	Nominal: 48 Contingency: 24 Response: 12	40K lbs. to LEO	4	64

be unguided by practical considerations and its utility is guaranteed to be limited.

Each concept of operations is intended to conform to the same fiscal environment since they both have the same budget. Due to this constraint, and as a result of cost estimates presented earlier, the two concepts of operations have different numbers of RLVs available. Since CONOPS A uses the half-scale RLV developed with less technical and financial risk, six are available for employment. Since CONOPS B uses the larger RLV developed with more technical and financial risk, four are available for employment. These figures are based on the development cost estimates presented earlier (tables 35 and 36).[59]

Each concept of operations is described in terms of its mission, systems, operational environment, command and control, support, and employment. The missions of space lift (to and from orbit), transspace operations, reconnaissance, and strike (in and from orbit) contribute to the broader military missions of space superiority, precision employment of weapons, global mobility, and achieving information dominance.[60] The systems description includes not only the RLVs but also their associated ground systems and payloads. The operational environment addresses threats and survivability issues while command and control deals with command relationships as well as authority and responsibility for the mission and the people. Support addresses the numerous activities required to conduct successful operations. Finally, the employment discussion illustrates concepts of how the systems may be used throughout the spectrum of conflict, from peace to war and back to peace.

CONOPS A

Missions. The missions of the RLV force are to conduct space lift, transspace, reconnaissance, and strike operations. Space lift operations include deployment, sustainment, and redeployment of on-orbit forces—earth-to-orbit, orbit-to-earth, and intraspace transportation. Transspace operations involve delivering material through space, from one point on the earth's surface to another. Reconnaissance missions are not limited to the earth's surface, but include inspection of adversary space systems as well.[61] Similarly, the strike mission may be accomplished against surface, air, or space targets. Strikes within space will likely be accomplished with directed energy, high power radio frequency (HPRF), or information weapons rather than explosive or kinetic impact weapons to minimize the chance of debris causing fratricide.[62]

In peacetime, routine launch and recovery of spacecraft and reconnaissance will be the primary occupations of RLV forces. Exercises, training missions, and system tests will also be accomplished. During contingencies, requirements for responsive launch, transspace operations, and more frequent and responsive reconnaissance are likely.[63] Contingencies may also include the need for heightened readiness to accomplish strike missions. During wartime, the full range of missions must be anticipated. Actions to achieve control of the space environment, such as reconnaissance and strike against adversary space systems, as well as surge launch and transspace operations will be conducted.[64] RLVs may be called upon to accomplish prompt strikes against surface targets early in a conflict in an attempt to disrupt an adversary offensive.[65] Once hostilities have passed the opening stages, RLV operations would continue, complementing the capabilities of forces from other environments. For example, strikes from space may enable attacks on targets which would otherwise be beyond the reach of air, land, and sea forces. Strikes from space may also enable attacks against targets deemed too heavily defended for nonspace forces. Once hostilities have ceased, RLV forces may be called upon to conduct reconnaissance missions and provide a deterrent force so air, land, and sea forces may redeploy. RLV strike readiness could be maintained to ensure a prompt response if an adversary decided to take advantage of force redeployment and resume hostilities.

Systems. Six RLVs with the attributes described earlier are available (tables 36 and 37). Payload capabilities include a wide range of systems all using a standard container and interface.[66] Spacecraft, reconnaissance payloads, and weapons dispensers use the same standard container to ensure simplicity and ease of RLV operations. For surface attack, weapons options include maneuverable reentry vehicles which may contain a variety of munitions and guidance systems depending on the nature of the targets to be struck.[67] For strikes within space, weapons options include directed energy, HPRF, and information munitions.

In-flight vehicle operations and control may be affected remotely; however, the vehicle is capable of executing all missions based on programs loaded prior to takeoff. The ability to operate autonomously helps minimize the force's vulnerability to electronic warfare and enhances in-flight security. Communication for purposes such as in-flight operations and control and payload data transfer is available throughout the mission primarily through space-based tracking and data relay spacecraft, though line of sight communication with ground stations is possible.

RLV self-defense capabilities include its ability to use maneuver and speed to avoid threats, and onboard electronic and optical countermeasure systems which can operate autonomously and through remote control. The vehicle's thermal protection system gives it some inherent passive defense against lasers. As with vehicle operations and control, in-flight payload operations and control may be affected remotely. The payload functions can also be executed based on programs loaded prior to takeoff.

The two primary operating bases are located in Florida and California.[68] Four alternate bases may be used as necessary. Two of the alternate bases are located on the coasts—one each on the East and West Coasts. The other two alternate bases are located in the US interior. The alternate bases may be used in the event of contingencies such as those related to system malfunction, extremely severe weather, or threats to primary base physical security. RLV units and personnel also have the capability to establish a contingency base at virtually any airfield in the world with a runway length capable of accommodating large jet-powered aircraft. Other space systems necessary for RLV operations besides the tracking and data relay satellites already mentioned include communications satellites, warning satellites, and space surveillance systems.

Operational Environment. The operational environment of the RLV currently contains few direct threats. However, the proliferation of technology, particularly rocket, spacecraft, and directed-energy technology, combined with the increasing importance of space operations to war-fighting success indicates that more threats are likely to develop. It would be tempting to follow Giulio Douhet's example from the 1920s and predict there will be no way to defend against an RLV attack, but this is not likely to be the case.[69] The world's leading space-faring nations, the United States and the former Soviet Union, have already demonstrated the capability to attack spacecraft using ground-based and air-launched kinetic impact weapons as well as coorbital kinetic impact systems. Lasers and other directed energy devices may also present threats in the RLVs operational environment.[70]

When in flight, the RLV's onboard defensive systems and inherent maneuverability and speed make it difficult for adversary weapon systems to prevent mission accomplishment. The fact that an adversary has to detect the RLV's launch, predict its orbit, pass that information on to its defense force, and then execute an anti-RLV mission would require a high degree of technological sophistication and operational capability. Striking an RLV will be more complicated than a typical antisatellite (ASAT) mission where the spacecraft's orbit is well established, predictable, and less likely to be altered.

However, even if an RLV in flight poses a difficult target for an adversary, its associated command and control centers, communications links, and bases are very vulnerable to enemy attack. This vulnerability drives the need for warning and other intelligence support, an autonomous operations capability, active and passive operating base defenses, and redundant systems. Secure, antijam, low-probability-of-intercept, communications connectivity provides some measure of protection for in-flight vehicle and payload operations and control when autonomy is not acceptable.[71] Assuming vehicle autonomy and security measures for necessary communication links are achieved, the system's greatest vulnerability will be at the operating base. The existence of alternate bases and the capability to establish contingency bases mitigates this vulnerability when combined with active and passive base defense measures.

Command and Control. RLV forces are divided between military and commercial organizations. During peacetime, four of the six RLVs available are operated by a commercial organization engaged primarily in providing space lift services. This company also provides commercial remote sensing services. The remaining RLVs are operated by the US military under the combatant command (COCOM) of the commander in chief, United States Space Command.[72] The military RLVs conduct very little space lift during peacetime to avoid any real or perceived competition with the US commercial space lift industry.[73] They primarily conduct reconnaissance while training for and exercising their strike and transspace missions.

During times of heightened tension or war, the National Command Authorities may direct mobilization of some or all of the commercial RLV fleet based on existing government-industry agreements.[74] These RLVs may then be modified as necessary to conduct military missions. This mobilization of commercial RLVs is necessary to avoid requiring commercial organizations and their employees to accept the increased risk, hardship, and discipline required of military RLV missions. In a war, RLVs used in direct military action or in support of military operations, along with their associated systems, facilities, and personnel, will likely be targeted by the enemy. When CINCSPACE is acting as the supporting commander in chief (CINC) to a geographic CINC, RLV forces may be put under the tactical control (TACON) of the joint force commander (JFC) to ensure the most effective use of these systems in direct support of the theater campaign plan.[75] For air and surface strike missions, the joint force air component commander will normally direct the use of RLV forces.[76] CINCSPACE directs the use of RLV forces supporting the campaign for space superiority and conducting transspace missions. RLV forces may be used to help wage a campaign for space superiority by conducting strikes and reconnaissance within space, space lift, and strikes against surface-based elements of an adversary's space force. The JFC resolves any disputes over apportionment and allocation of RLV forces.

Support. Intelligence support for RLV forces covers a broad range of requirements. Operating base threats must be assessed and threat information provided continuously. Such information will drive defense status and relocation from prime to alternate bases or deployment to a contingency base. RLV surface strike missions will require extensive intelligence support, similar to that required to accomplish precision strikes with today's air forces or missiles. Strikes in space will require extensive space surveillance support. Some space surveillance information may actually be collected by the RLV itself, but it will require support from systems or a network with broader and continuous coverage of the near-earth environment. Mission planning will require not only the information just described but very capable computer hardware and software to process planning information inputs and to generate mission programs for in-flight payload and vehicle operations.

Security of operating bases is paramount. The greatest threats may come from terrorists or an adversary's special forces. In this regard, security requirements will be similar to today's requirements to protect high-value assets at DOD bases in the continental United States except that the threats will have evolved by the year 2012. Logistics support is simplified to the greatest extent practical. Organizational-level maintenance actions at the operating bases are accomplished by military enlisted maintenance technicians organic to RLV units. The primary RLV base on the East Coast is home to RLV unit headquarters.[77]

Employment. During contingencies and war, RLV operations consist of three phases: readiness planning, mission planning, and execution. Readiness planning requires being responsive to world events and direction from higher headquarters to maintain a specified readiness posture. At the highest state of readiness, RLVs may be maintained on alert to respond within six hours for surge space lift, transspace, reconnaissance, or strike missions.

The RLV force's ability to execute specific missions within six hours may be constrained by factors beyond the control of the RLV force. For instance, orbital dynamics may dictate an appropriate launch time for a particular spacecraft deployment, space strike, or space reconnaissance mission that falls beyond the six-hour response time—the RLV may be available, but physics will require waiting longer to execute the mission. Maintaining alert at the highest state of readiness impacts RLV availability to conduct routine missions. Mission planning is conducted once a hypothetical or actual mission tasking is received. Mission planning is conducted by the RLV unit, nominally within one hour for any mission, taking full advantage of the support outlined above. Mission planning includes payload selection and generation of mission programs to be loaded prior to takeoff, assuming the specified mission has not been previously planned and stored for later use.

The execution phase of RLV operations includes final launch preparations, launch, flight operations, and recovery. Recovery is normally at the base from which the sortie generated. System malfunctions, extremely severe weather, or threats to base security may drive recovery at another base. Transspace operations may require establishment of a contingency base and operations from that location. RLV recovery is followed by immediate preparation for subsequent missions. Deployment to an alternate or contingency base may be directed by higher headquarters or the local RLV unit commander.

CONOPS B

Missions. The missions of the RLV force are to conduct space lift, transspace, reconnaissance, and strike operations. The CONOPS B RLV force of four full-scale vehicles is commercially operated. Given the full-scale RLV's longer turnaround time relative to the notional CONOPS A RLV, its utility for reconnaissance and strike missions during contingencies and war is diminished but not eliminated. Further, its completely commercial operation complicates use of the RLV fleet in direct military actions.[78] Nevertheless, this CONOPS include strike operations for completeness and to provide a basis for subsequent analysis.

During peacetime, routine launch and recovery of spacecraft and remote sensing will be the primary occupation of the RLV fleet. During contingencies, requirements for responsive space lift, transspace operations, and surface reconnaissance are likely. Actions to achieve control of the space environment, such as reconnaissance and strike against adversary space systems, are also likely to be required. During war, surface strike missions may be conducted. Once hostilities have ceased, RLV forces may be called upon to conduct reconnaissance missions and maintain some level of strike readiness.

Systems. Four RLVs with the attributes described earlier are available (tables 35 and 37). Payload capabilities are similar to those described for the CONOPS A RLV in that they all use a standard container and interface, but the weight and size of CONOPS B payloads is larger. In-flight vehicle operations, communications, self-protection systems, and payload operations are the same.

The basing scheme includes the same two primary operating bases. There are no designated alternate bases, but the operators have the capability to establish a contingency operating base at virtually any airfield in the world with a runway length capable of accommodating large jet-powered aircraft.

Operational Environment. The operational environment of the RLV is much as described under CONOPS A, except it is less hostile. The apparently civilian, and thus less threatening, nature of peacetime RLV operations would minimize the provocation of hostile action against the vehicles by potential adversaries.

Refraining from exercising the RLV fleet in strike operations during peacetime could help to de-emphasize any potential military applications. Exercising strike operations would obviously hurt the RLV fleet's peaceful appearance, although it would undoubtedly improve the operators' proficiency to execute the mission. Unfortunately, regardless of whether or not the RLV fleet is used for strike missions, threats from ASAT-like systems as described above for CONOPS A are still likely to exist. Further, as long as the RLV fleet is used in even indirect support of military operations (e.g., surge launch of spacecraft used to support military surface or air operations), it will be a potential target of enemy action.

Command and Control. The RLV force is owned and operated entirely by a commercial organization.[79] The company provides space lift and remote sensing services for government and commercial customers.

US government agreements with the RLV operator include a measure of military oversight and involvement to ensure the RLV force is ready and available to conduct missions in support of national security objectives in peace and in war. The systems are never operated by military personnel, but mobilization agreements allow for close military direction of activities during contingencies and war. The secretary of defense (SECDEF) may approve mobilization of the RLV fleet during contingencies and war for the purposes of conducting space lift and transspace operations in support of national security requirements. The president must approve any use of the RLV fleet for strike missions. When mobilized, CINCSPACE exercises COCOM over RLV assets. CINCSPACE also retains operational control (OPCON) and TACON of all RLVs given the fleet's high value and few numbers.[80] When strike operations are to be conducted, military personnel must be present to provide a measure of positive control.

Support. Intelligence support to RLV forces is much the same as under CONOPS A. Logistics support requirements are less stringent due to decreased readiness required for deployment and mission accomplishment. Maintenance actions are accomplished entirely by civilian personnel. There is no requirement for military personnel to be trained and certified in maintenance or operations tasks. Military personnel simply develop tasks and oversee their execution by the commercial civilian operators.

The only exception is with respect to strike missions. Military personnel working with RLV operators must be trained and proficient in implementing positive control measures for RLV strikes. Military personnel are assigned to a detachment collocated with the RLV operator's headquarters.

Employment. During contingencies and war, RLV operations will be responsive to national security requirements. If directed by the secretary of defense, the RLV fleet will be mobilized to conduct surge space lift and transspace operations at a cost that compensates for lost commercial revenues. These operations would be conducted in the same fashion as peacetime RLV operations, but with close military coordination. SECDEF mobilization of the RLV fleet will require the civilian operators to meet contingency turnaround and response times of 24 and 12 hours, respectively.

CINCSPACE will direct tasks and priorities for the fleet once mobilized. CINCSPACE, in conjunction with the supported CINC if CINCSPACE is playing a supporting role, will determine whether or not RLV strike operations are warranted and request presidential approval as appropriate. If use of the RLV fleet for strike missions is approved, measures will be taken to ensure military control of these operations.

Summary of CONOPS A and B

This section presents an outline of two concepts of operations. The first concept, CONOPS A, attempts to make the fullest military use of the roughly half-scale notional RLV to accomplish not only traditional space lift missions but also the additional missions of returning payloads from orbit, transspace operations, reconnaissance, and strike (in and from space). CONOPS A is intended to represent military space plane advocates' visions.

The second concept, CONOPS B, based on the full-scale vehicles currently being proposed under the RLV program, also attempts to make expanded use of RLVs, but their application is inhibited by design attributes and completely commercial operation. CONOPS B is intended to represent a logical extension of the current RLV program's goals.

Analysis

The criteria used to analyze the concepts of operations described in this study include capability, cost, operations efficiency, operations effectiveness, and politics. Capability analysis includes all the required mission areas: space lift, reconnaissance, strike, and transspace operations. Cost analysis addresses operating base, ELV augmentation, and transspace operations costs, as well as the potential for technology maturation to reduce development costs. Operations efficiency and effectiveness analysis emphasizes the impact of using cryogenic propellants, deployment operations, and overall system reliability. Political analysis examines the suitability of each CONOPS in both the international and domestic environments.

Capability

Each concept of operation was intended to satisfy all RLV mission requirements: spacecraft launch and recovery, reconnaissance, transspace operations, and strike (in and from space). Each CONOPS meets these requirements but, as a result of the differences in the attributes of the vehicles used in each CONOPS and the way in which they are organized, deployed, and employed, their capabilities in each mission area vary to some degree. This variation in the extent to which each CONOPS satisfies mission requirements is examined below.

Space Lift. Both CONOPS provide dramatically improved space lift capability from a responsiveness perspective. The most responsive of today's space lifters requires a minimum of two months from call-up to launch compared with less than a day for either RLV described here. [81] However, when considering space lift payload capability the two RLVs are not equal. The half-scale RLV used in CONOPS A (RLV-A from here forward) may not necessarily meet all users' needs from a payload weight and size perspective. If a smaller RLV is developed, an alternative lift means might be required, such as a heavy ELV, if a particular payload cannot be downsized.

At 8.5 meters (28 feet) long and 2,724 kilograms (kg) (about 6,000 pounds) unequipped, the US components of the ISSA would fit within the dimensions of RLV-A. Not to mention that they will have already been deployed long before the first operational flight of an RLV.[82] However, NASA is concerned about minimizing the number of visits to the space station to avoid disrupting microgravity materials processing work. NASA also has concerns about accommodating the crew module envisioned for transporting US astronauts to and from the station. These concerns appear to be driving a desire for the large payload capability of current RLV program concepts.[83] Another factor behind the large payload requirement is the desire to capture the large national security payloads that currently fly on the Titan IV expendable rocket in the interest of pursuing further reductions in life-cycle costs.[84] It is difficult to predict whether or not these payloads will be lighter and smaller in the future. However, if we plan on building vehicles big enough to carry the largest payloads, it is easy to predict that payload designers will take advantage of the capability.

If large national security payloads cannot, or will not, be downsized, they could be lifted on the heavy version of the DOD's EELV, predicted to be available in 2005. If large space-station payloads cannot, or will not, be downsized, they could be lifted on the heavy version of EELV as well. Large Russian rockets could also be used.[85] In fact, launching into the ISSA orbit from the Baykonour cosmodrome in the former Soviet republic of Kazakstan instead of Cape Canaveral, the planned launch base for American ISSA missions, provides more than a 35 foot-per-second velocity advantage to the relatively high-inclination orbit, 51.6 degrees.[86] This higher-inclination orbit is the same as that currently used by the Russian space station *Mir*, which was launched and is resupplied out of Baykonour. Another alternative might be launching large space station payloads on the *Ariane 5*. The Europeans plan to develop their own manned crew transfer vehicle as part of their participation in ISSA.[87] The *Ariane 5* will be able to lift 18,000 (about 39,600 pounds) to LEO, which is comparable to the payload capacity of the Titan IV.[88]

A final, but not least significant, consideration is the need to return large payloads from orbit. While the Russians or French might happily provide return from orbit services using their Soyuz capsule or crew transfer vehicle, respectively, will they be large enough for the loads coming back from the station? As stated above, they might if we plan on using these vehicles and size the return payloads from ISSA appropriately, but certainly won't if we plan to use a larger vehicle.

Reconnaissance. Some may question the need to use an RLV for reconnaissance given the US ability to perform space-based reconnaissance of the earth's surface using satellites.[89] However, there may be times when the element of surprise is desired and not likely to be obtained using on-orbit assets. It is conceivable that a potential adversary might have enough information about US space-based reconnaissance systems to effectively implement operations security measures and avoid detection.[90] Another motivation for using an RLV for reconnaissance might be the need for responsiveness. For a fast breaking contingency, RLVs may provide a quick response not attainable with on-orbit spacecraft, manned aircraft, or unmanned aerial vehicles (UAV). For instance, a low-orbiting remote sensing spacecraft might not have a given location on the earth's surface within its field of view until several orbits have passed. Manned aircraft and UAVs may not allow overflight of a location deep within the target country's territory.

With respect to reconnaissance within space, one might pose a similar question about the utility of RLVs. There are undoubtedly other systems which can perform space surveillance. Paul B. Stares, in *The Militarization of Space*, claims that the USAF attempted to develop a satellite inspection system (SAINT) in the earliest days of the space age.[91] It was canceled in 1962, but Stares suggests the US ability to survey space was not degraded since advances in ground-based sensors made by the mid-1960s facilitated the gathering of a great deal of data. This may be true, but on-orbit reconnaissance may allow for more detailed as well as active inspection of spacecraft in LEO. Reconnaissance of payloads in higher orbits, such as geosynchronous earth orbit (GEO) or Molniya orbits, may require reducing the reconnaissance payload weight or may have to be conducted from a greater distance. This reconnaissance capability might also support strike missions in space with prestrike target information and poststrike battle damage assessment inputs.

Strike. Accomplishing strikes using RLVs is technically feasible. However, to be militarily useful, the vehicles should be able to deliver significant weapon payloads. With respect to surface strike, it appears RLV-A can deliver as much payload as a typical modern fighter. RLV-B can deliver as much weapons payload as a B-2 Spirit stealth bomber.[92]

Obviously, there are additional considerations besides payload weight when analyzing surface strike capability. Response and turnaround times have a dramatic effect on the usefulness of RLVs for surface strike missions. Both RLVs could deliver initial strikes earlier than B-2s. Due to RLV-A's quicker response time and shorter turnaround time, it compares favorably with the strike capability of a cost-equivalent number of B-2s conducting strikes over a two-day period even though RLV-A's payload capability is roughly half that of the B-2 (table 38 and fig. 12).[93] RLV-B, on the other hand, cannot compare as favorably through this same period despite its relatively large payload capability. The B-2's strike capability exceeds that of both RLVs over a three-day period. Strike in space using RLVs is also technically feasible.

Both concepts include the capability to strike adversary spacecraft. The means used and type of strike are only limited by the creative development of strike mission payloads. For instance, RLV space strikes might be accomplished in a manner which minimizes debris and affects only a specific subsystem on board the target spacecraft. Information strikes causing disruption of adversary communications and command and control, or aimed at deception, might also be conducted. Strikes against spacecraft in high earth orbits, such as GEO or Molniya orbits, may require reducing the strike payload weight or be conducted from a greater distance.

Table 38

Cumulative 2,000-Pound Weapons Delivery within Three Days

Time (hours)	RLV-A (6 RLVs)	RLV-B (4 RLVs)	B-2 Spirit (10 B-2s)
6.75	48	0	0
12.75	48	64	0
20.25	96	64	0
21	96	64	160
33.75	144	**64**	160
38.25	144	128	160
47.25	192	128	160
60	192	128	320
60.75	240	128	320
63.75	240	192	320
72	240	192	320

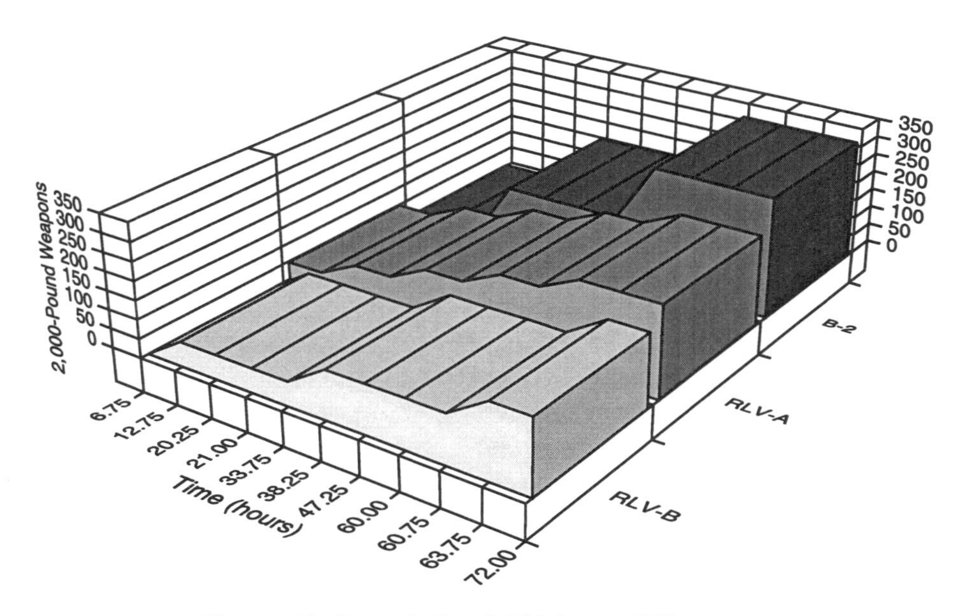

Figure 12. Cumulative 2,000-Pound Weapons Delivery within Three Days

Transspace Operations. The requirement for this capability is not very well defined.[94] One might easily doubt its feasibility except that any RLV capable of recovering and returning payloads from orbit will have an inherent capability to deliver cargo from one location on the earth to another.

Putting aside cost considerations for the moment, a major factor in assessing the feasibility of transspace operations is the ability to establish an RLV operations base at the pickup and delivery points. Experience with the subscale, suborbital McDonnell Douglas Aerospace DC-X can only hint at what an operational RLV operating base might look like. The base established for DC-X (now called the DC-XA in its modified form) operations at White Sands Missile Range, New Mexico, includes propellant facilities, electrical power facilities, vehicle control systems, and connections. The propellant facilities include liquid oxygen, liquid hydrogen, gaseous helium, and gaseous nitrogen storage tanks, transfer lines and control systems. The vehicle control systems include ground control systems and a "real-time data system" to collect, store, and display vehicle data centrally before, during, and after flight. The real-time data system also provides a means for operator intervention, if necessary, and allows for receipt, processing, and loading of autonomous flight operations programs.[95]

While an operational RLV design should include operations efficiency considerations, any RLV operating base will certainly require very large propellant facilities and associated equipment. Given that a full-scale RLV, such as RLV-B, will require about 100 times more propellant than the DC-X, the propellant facilities will not necessarily lend themselves to quick and easy transport. In this sense, RLV-A may have some advantage in that its propellant facilities would be smaller than RLV-Bs. Obviously, RLV-A also has less payload weight capability. Without a clear definition of requirements for transspace operations, it is difficult to evaluate this tradeoff between the two CONOPS.

Cost

Operating Base Costs. CONOPS A is sensitive to operating base costs. CONOPS A includes two primary and four alternate bases as well the capability to establish a contingency base. CONOPS B simply has two primary bases with a capability to establish contingency bases.

Launch base costs for today's fleet of expendable rockets may not be a good indicator of future RLV launch base costs given the objectives of the RLV program. This is fortuitous since today's launch bases are expensive to operate. Operating the US's largest and busiest launch base, Cape Canaveral, and its associated range, costs about $160 million a year. Experience with the DC-XA is also difficult to use as a basis for estimation since the vehicle is very much smaller and less capable than an operational RLV. The DC-XA launch base also uses existing facilities and equipment at the White Sands Missile Range.[96]

Nevertheless, industry sources estimate it will cost roughly $50 million to setup an RLV operating base, at a minimum. Using this figure, CONOPS A's operating base costs may be estimated at $200 million more than CONOPS B's.

ELV Augmentation. CONOPS A may also require ELV augmentation if large space station and national security payloads are not downsized. The Moorman Study reported that simply shrinking the size of the RLV payload bay from 45 to 30 feet might cost an extra $26.6 billion in ELV costs through the year 2030.[97] Employing foreign heavy lift vehicles could reduce this cost.

Transspace Operations. It is unclear that transspace operations will be economical. If one accepts the program goal of delivering payloads to orbit for $1,000 per pound, then the same estimate may be used for the cost per pound of delivering cargo from one point on the earth's surface to another using an RLV. Costs for sending cargo internationally, say from New York to Seoul, using an express package delivery service range from about $50 per pound for a one-pound box to $5.70 per pound for a 100-pound box.[98] Sending loads by military airlift is less expensive, but takes longer. For example, shipping a 20,000-pound load on military airlift from Dover AFB, Delaware, to Ramstein Air Base, Germany, will cost $1.079 per pound and take 3.1 days, if the cargo is given the highest priority.[99]

Such costs make it unlikely that RLV cargo delivery will be economically attractive. Whether or not RLV cargo delivery will be militarily useful remains to be seen.

Technology Maturation. A recurring theme in studies related to the RLV program is the idea that program costs can be reduced through technology maturation.[100] A technology development program targeted against specific high-risk areas executed before system development and acquisition can mitigate the technical and financial risks. Advancing the technology readiness of a system from "concept design" to "prototype/engineering model" prior to entering full-scale development can lower development costs by more than 40 percent.[101] Phase I of the RLV program is intended to include demonstration of the maturity level of candidate technologies.[102] The X-33 flight demonstrations at the end of Phase II represent an attempt to demonstrate technological maturity levels.

However, the major recommendations of the National Research Council's recent review of RLV technology indicate a need for more vigorous development of propulsion technology.[103] There is a government-industry effort currently under way that can help address this issue. The integrated high payoff rocket propulsion technology (IHPRPT) program has goals for booster, orbit transfer, spacecraft, and tactical propulsion systems. Noteworthy booster cryogenic propulsion goals include achieving a "mean time between removal" or "mission life" of 20 for reusable systems by the year 2000, an improvement of 3 percent in I_{sp} by 2010, and an increase of 100 percent in the thrust-to-weight ratio by the year 2010.[104] Unfortunately, funding levels for this program have not increased in spite of the start of the RLV program and recommendations by high-level studies to increase funding in this area.[105] Given the critical nature of propulsion technology development to the success of the RLV program and US space lift competitiveness in general, it is surprising IHPRPT funding has not been raised to the recommended levels.[106]

Operations Efficiency and Effectiveness

Cryogenic Propellants. Cryogenic propellants are not ideal for operations efficiency and effectiveness. A good historical basis for this assertion is the Atlas missile's short life as an ICBM. It was relegated to a space lift only role in 1965 after being an operational ICBM for less than six years because it was not well suited to the responsive operations and reliability required of an ICBM. The extreme caution needed in fueling the missile immediately before launch kept it from ever meeting its required 15-minute reaction time. It also suffered from a host of reliability problems, many related to its propulsion system.[107] The Atlas was quickly followed by the Titan and Minuteman. The Titan ICBM, using hypergolic propellants, could stay propellant loaded since hypergolics did not need constant refrigeration. The Minuteman, using solid propellants, provided outstanding responsiveness and reliability.[108]

The legacy of the Atlas missile's operational life as an ICBM may provide a caution when contemplating the development of an operational RLV with a goal of high reliability and low operations costs. It may be even more relevant when considering the military use of an RLV that drives quicker turnaround and response times. Today's Atlas space lift vehicle outfitted with a cryogenic Centaur upper stage requires cryogenic propellant loading about two hours prior to launch, well within the response time specified for either RLV-A or RLV-B.[109] During test flights in July 1995, the DC-X required a similar time line for propellant loading and was prepared to demonstrate an 11-hour turnaround time.[110] While these time lines seem to bode well for an operational RLV, there is no denying the relative complexity of cryogenic propulsion systems compared with hypergolic or solid alternatives. This complexity will make achieving RLV operations efficiency and effectiveness goals a challenge.

Deployment. The nature of cryogenic propellants also drives complexity in the RLV operating base. This complexity will challenge designers and operators faced with the problem of how to build, deploy, and operate an RLV contingency base. Ideally, such a base will be deployable by air. This is particularly true in CONOPS A, where dispersion for security and increased responsiveness for military missions is required. This contingency base capability will also be a key to transspace operations. Power and propellant systems are likely to comprise the majority of the weight and bulk required to be moved. Lessons may be learned from efforts within the USAF to develop multifunction support equipment for aircraft maintenance.[111]

Being able to reduce the number of operating base support equipment pieces, as well as their size, could ease mobility requirements. It could also lead to a decrease in the number of personnel required to deploy and reduce the cost of outfitting a contingency RLV operating base. Winston Churchill once said, "except in the air [the Royal Air Force] is the least mobile of all the armed services."[112] If the deployability of the RLV force is neglected, it might suffer a similar criticism.

Reliability. Air Force Space Command's *Draft Operational Requirements Document (ORD) for the EELV*, dated 31 March 1995, defines reliability as "the ability of the space lift system to successfully accomplish its intended mission."[113] The *ORD* defines the terms *reliability of the schedule* or *dependability* separately as "the ability of the system to consistently launch . . . when planned."[114] The Moorman Study identified three factors which affect space lift system reliability: complexity, flight rate, and design stability.[115] After considering these factors, one can see evidence of their impact in today's space lift systems. The Delta II, the least complex system, has the highest reliability, 100 percent over the last five years, compared to 84.2 percent and 85.7 percent for the Atlas and Titan, respectively.[116] The Delta also has the highest flight rate and the most stable design of today's expendable systems. The message for RLV development is clear: keep system complexity down, flight rate up, and design stable. The second item, flight rate, may be achieved by capturing the largest share of the launch market practical and/or capitalizing on military applications. The third item, design stability, is aided by requiring standard payload containers and interfaces. Happily, current RLV program competitors already include a standard payload container and interfaces as a key design element.[117]

Unhappily, the National Research Council's warning about the need for vigorous propulsion system development may indicate danger ahead for RLV reliability. One of the reliability problems today's space lifters face is their lack of performance margin.[118] A robust design approach in the RLV program could avoid this pitfall and lead to increased reliability. Rather than focusing on eliminating variation in performance, a requirement when operating a system with no performance margin, a robust design approach would minimize the effects of variation in performance. If the RLV is designed to be a high-performance system without any performance margin, then the operators will be in the same position as today's space lift operators: reliability goals simply will not be achieved. This would seem to indicate the desirability of building a system with plenty of propulsion power for its intended operations. As current RLV concepts plan on milking existing engine (SSME or RD-0120) derivatives for their last ounce of capability, this will result in fielding a full-scale RLV always operating at its performance limits.[119] In this respect, RLV-A may offer some advantage, as the smaller vehicle is not likely to push propulsion performance requirements to the same extent as a full-scale vehicle.

Any potential lack of reliability is also directly related to cost in that the cost of failure is typically high in the space lift business. The ability of an RLV to abort and land back at its base during ascent or descent may minimize the cost of failure in flight. However, any unreliability can cause delays which increase costs, although they do so less dramatically than a catastrophic in-flight failure. If an in-flight RLV failure does occur, its cost will be considerably higher than that of losing one of today's expendable space lifters.

Finally, as highlighted by the Atlas missile's ICBM experience, reliability is a key attribute of military weapon systems. As much as cost, reliability will determine whether or not RLVs can successfully perform military missions.

Political Considerations

International. No RLV capabilities or operations described in either CONOPS A or B would violate international treaties. To some, this may be surprising. Since the dawn of the space age, the popular image of space activities has been that they are peaceful and nonmilitary. This image has been reinforced by governments, including the US government, to help guarantee the use of space for unimpeded reconnaissance. As such, there are international laws and treaties such as the Nuclear Test Ban Treaty (1963), the Outer Space Treaty (1967), and the Treaty on the Limitation of Anti-Ballistic Missile Systems (ABM) Treaty (1972), which restrict military space activities. However, RLV forces can live within these treaties as long as they are not used to carry weapons of mass destruction, conduct ABM testing, deployment, or operations, or interfere with "national technical means (space intelligence systems)" which are being used to verify treaty provisions during peacetime.[120] This is not an exhaustive list of prohibitions, but highlights the main areas of caution for RLV military applications.

Treaties and law are not the only international political concerns related to RLVs. Developing such a dramatic new military weapon capability could appear threatening to other states. It is conceivable that other nations would resent the US's ability to strike from space or within space with little or no warning. They might respond to this threat by developing similar capabilities or by developing ASAT or anti-RLV weapons. If deployment of an RLV force were perceived as an attempt to extend American global hegemony, it could encourage other states to form alliances against the United States. Political scientist, Stephen M. Walt suggests that this sort of balancing mechanism led to a favorable balance of power for the United States during the cold war. The Soviets appeared threatening to other states, which drove them into the US camp.[121] Given its completely commercial operations and more inhibited use in strike operations, CONOPS B might prove less threatening and minimize the appearance of US aggressiveness relative to CONOPS A.

On the plus side, RLVs could be used for conventional strikes with the range and nearly the promptness of ICBMs, but without the nuclear baggage. Assuming the RLV force would never test or employ nuclear weapons, there should be no international concern about the start of a nuclear conflict with the launch of an RLV.

Domestic. Domestically, one concern which must be addressed is the potential US political concern associated with ASAT deployment.[122] While this would certainly not prohibit RLV development and use in space lift, reconnaissance, and transspace operations, it might complicate development of a strike capability. If the prohibition stands, strikes in space will not be possible. Surface strikes might be allowed under the ban, but Congress would have to be convinced that the system would not operate in an ASAT role.

On the executive side of the government, NASA headquarters direction that the RLV must replace the space shuttle comes across loud and clear. While this is understandable, viewing an RLV as a shuttle replacement can be detrimental in three ways. First, it can be detrimental if it limits the designers' and planners' imagination. Second, it could be detrimental if the shuttle replacement paradigm leads simply to swapping RLVs for space shuttle orbiters, but retaining the same dated concept of operations and support facilities. Third, it could be detrimental if it forces the RLV to accommodate the same large payload sizes and weights as the space shuttle without an objective evaluation to consider if there are better options.[123]

Outside of the government, industry requires profit to survive. NASA leaders have experienced frustration in their attempt to get the private sector to fund a significant share of RLV development costs. NASA administrator Daniel Goldin recently criticized the X-33 contractors for their "lack of courage to step up to the plate and make it happen."[124] The two-stage X-34 demonstration vehicle program has already been a casualty in the effort to encourage industry to fund reusable launch vehicle technology. The contractor team of Orbital Sciences and Rockwell International "withdrew from development of the X-34 launch vehicle after determining it would not be commercially viable."[125] The reality of industry's motivation for profit should not be surprising. It indicates that unless government is willing to fund the RLV program completely itself, the design will have to be commercially viable.

It is not likely that the government will completely fund the RLV program. The current budget environment is severely constrained and is expected to remain this way for the foreseeable future. Both the public and the Congress want a frugal government. The NASA budget in particular is on a downward trend. In fiscal year 1995, the programmed NASA budget for the year 2000 was $14.7 billion. The fiscal year 1997 program cut NASA's year 2000 budget down to $11.6 billion.[126] The DOD budget has suffered from the same trend and the future appears to offer little relief.[127] In short, support is not likely to be found for an expensive RLV development effort reminiscent of cold-war-era space and defense programs.

RLVs will have to be developed with industry contributions. Again, commercial viability will dictate development investment and time lines.

Summary of RLV Concepts Analysis

Using capability, cost, operations efficiency and effectiveness, and politics as a framework for analyzing RLV concepts of operations yields several insights. First, capability analysis indicates either RLV can be used as a multirole space superiority weapon. Each CONOPS provides for spacecraft deployment, spacecraft sustainment, reconnaissance of the space realm, and strike within space as well as to the surface—key capabilities for controlling the space environment. CONOPS A may require augmentation with large ELVs given its use of the smaller RLV-A. CONOPS B may provide less flexibility and strike utility given its longer response and turnaround times. CONOPS A may have the advantage in transspace operations given the potential for RLV-A requiring smaller propellant facilities and the accompanying relaxation of mobility requirements.

Second, cost analysis indicates advantages and disadvantages for each CONOPS. CONOPS A will be sensitive to operating base costs, and may require the additional expense of maintaining access to space for heavy payloads using ELVs. It is difficult to imagine either CONOPS providing economically competitive transportation from one point on the earth's surface to another, but there may be some military utility for such missions in the distant future. CONOPS B may suffer in development costs because RLV-B is more likely to push the limits of technology, thus failing to take full advantage of the cost reductions possible through technology maturation. Related to this observation is the final conclusion of cost analysis—funding for propulsion technology development should be increased.

Third, operations efficiency and effectiveness analysis indicates cryogenic propellants will present a challenge to designers and operators. While these propellant systems offer high specific impulse, they do not lend themselves to simplicity and ease of operations. Fortuitously, today's cryogenically propelled systems meet the required time lines for either CONOPS. Deployability will be a challenge as well. Power and propulsion systems for RLV forces will likely be physically large. Efforts to decrease the size and amount of support equipment will ease the deployment burden. Reliability is perhaps the most important attribute within the operations efficiency and effectiveness category. Conclusions drawn from the analysis indicate RLV-A may have the advantage of wider performance margins and greater reliability assuming no major propulsion technology breakthroughs are made.

Fourth, political analysis indicates a tougher environment at home than internationally for RLVs. Neither CONOPS violates international treaties or laws, although it might be in the United States's best interest to soften the RLV's military appearance, perhaps an advantage for CONOPS B. Domestically, the congressional ASAT ban would prohibit the use of RLVs for strike missions in space and complicate attempts to use them for strikes to the surface as well. The domestic fiscal environment poses the greatest difficulty for RLV development. NASA cannot afford to foot the entire bill for an RLV fleet, and industry will only fund what market analysis indicates is a profitable venture. The DOD is also unlikely to fund RLV development independently.

Conclusions

Our safety as a nation may depend upon our achieving "space superiority." Several decades from now the important battles may not be sea battles or air battles, but space battles, and we should be spending a certain fraction of our national resources to insure that we do not lag in obtaining space superiority.

—Maj Gen Bernard A. Schriever

The United States and the Western World has an exciting and vital future in space activities of all kinds, the key to that future, be it in security activities, in scientific exploration or in commercial exploitation, the key is responsive and cost effective space transportation.

—Lt Gen James A. Abrahamson

Maj Gen Bernard A. Schriever, commander, Western Development Division, a powerful force behind early developments in US military missile and space capabilities, was premature in predicting the importance of space battles, in a speech at San Diego, California, February 1957, although the future may prove him correct. Given the increasing importance of space support to recent battles on the land and sea, as well as in the air, his emphasis on achieving space superiority may be more appropriate today. However, it is ironic to read Lt Gen James A. Abrahamson's words when he was director, Strategic Defense Initiative Organization of almost 30 years later (Congressional Testimony, 23 July 1985). These remarks reflect the view of the top leader in development of the largest and most lethal space weapon system ever seriously considered for deployment. Yet he chose to emphasize the need for responsive and cost-effective space transportation, not weapons, as the key to future space activity of any kind. It is also interesting to note that the program which may be credited with inspiring the current pursuit of reusable rockets, the McDonnell Douglas DC-X, was started by General Abrahamson's organization.

Having derived RLV requirements, described RLVs and their attributes, elaborated two concepts of operations, and analyzed those CONOPS, conclusions may now be drawn in attempt to answer the initial research question. These conclusions are followed by recommendations and a summary of the research.

RLVs Have Military Potential. It is clear from both the CONOPS and the subsequent analysis that RLVs have potentially important military applications. In many ways, they can provide a multirole tool to help achieve the space superiority General Schriever discussed almost 40 years ago. An RLV's potential for accomplishing strike missions, especially to the surface, will be higher if turnaround and response times are shorter. Increasing the tempo of operations can make the force appear larger. Military missions also benefit from RLVs with greater cross-range capability allowing the kind of operations described by General Ashy.[128]

Taking full advantage of RLVs' space lift capabilities may require a paradigm shift in spacecraft design, deployment, and sustainment. The launch on demand strategy possible with an

RLV is not in fashion today. Successful implementation of such a strategy will support space superiority but will require spacecraft ready to launch on short notice and ready to operate immediately upon deployment in orbit. These requirements could motivate development of cheaper, single-mission satellites since it may not be feasible to build and store billion-dollar multimission satellites, or to expect them to be operational immediately upon deployment.[129] Capitalizing on the RLV's ability to recover and return payloads, or to service them on orbit, would similarly require satellite design changes.[130]

RLV Design Is a Determinant of CONOPS. The potential impacts of RLV sizing have been addressed throughout this study. There is no unanimity regarding the proper size for an operational RLV. Nevertheless, many argue that the current size identified for a full-scale RLV as part of the NASA-led program involves high technical risk which means high financial risk as well.

This study has suggested that derivatives of current propulsion systems will not deliver the performance levels required, or if they do deliver, there will be no performance margin and reliability will suffer. This assessment may be supported or proven false by further technology development and demonstrations. But due to the limited objectives of the X-33 flight tests, even these demonstrations may fail to give developers and investors the necessary confidence to go full scale. Perhaps the best course of action with respect to this issue is to ensure marketing analysts, developmental engineers, and operators remain in the closest contact to ensure the best RLV size is chosen.

The choice of RLV size must also be informed about the negative consequences and opportunity costs associated with each option. This study suggests that choosing smaller, cheaper RLVs can provide savings to apply towards a larger fleet and more bases. With such a force structure we can accomplish militarily significant activities to an extent not allowed by the choice of a smaller fleet and fewer bases. However, choosing a more militarily useful RLV design and force structure could result in negative consequences for commercial and civil operators. A more militarily useful RLV design might include increased thermal protection system requirements to facilitate the greater cross-range capability needed to take off and land at the same base after one orbit. It may also require the additional weight and cost of onboard self-protection systems. Meeting these requirements will not be cost free. Whether the costs are in dollars, weight, or space, trade-offs will have to be made.

Propulsion Technology Development Is Required. One way to mitigate some of the challenges faced in developing a full-scale RLV is to pursue propulsion technology development more vigorously. Regardless of the size chosen for an operational RLV, advances in thrust-to-weight ratios such as the 100 percent increase sought in the integrated high payoff rocket propulsion technology program can dramatically decrease technical and financial risk.

Such efforts should be funded at the full level recommended by the Moorman Study. An investment of $120 million per year pales in comparison to the potential cost of developing an RLV. A lack of investment in propulsion technology development up front is bound to prove penny-wise and pound-foolish.

Top Priority Must Be Cheap and Responsive Space Access. While RLVs have tremendous potential to perform military missions well beyond simply conducting space lift, an objective evaluation of priorities leads to other conclusions. The US military possesses tremendous strike and reconnaissance capabilities through existing and planned land, sea, and air systems.[131] Space-based reconnaissance has also been conducted since the dawn of the space age. What the US military, and the entire nation, does not possess is cheap and responsive space access.

General Abrahamson's words quoted earlier were prophetic. Less than a year after his address, the United States's space access program literally crashed as a result of poor policy choices and a string of accidents that left the United States with a grounded STS fleet and a limited and unreliable ELV fleet. Talk of achieving space superiority is cheap. We must first have access to the space realm before we can begin to gain superiority.

Recommendations

Three recommendations are offered here. First, the US military, especially the USAF, is already a participant in the RLV program, but it should become more active in this area. If, as this study assumes, today's fiscally constrained environment continues, the US military will not have the luxury of developing an RLV fleet independently. Accordingly, the US military will have to blend its requirements with those of other users to pursue militarily significant applications. The current focus of the RLV program appears to be on NASA and commercial requirements. There is an implicit assumption that whatever is developed will spin on some military capability.[132] If the US military is a passive participant in the RLV program, then the assumed spin-on capabilities may be limited or nonexistent. Military requirements must be defined and stated if the United States is to develop a triple-use, rather than merely dual-use, RLV fleet. If the current fiscally constrained environment does not continue, then active participation in the current program is still warranted. An investment in defining military RLV requirements now will reap dividends should the time come when a military-unique RLV force can be developed.

Second, whether or not the United States develops a dual-use or triple-use RLV fleet or two fleets with one being military-unique, it should not do so before the technology is ready. If this study's assumption that RLV technology will become operationally feasible by 2012 does not prove valid, then their development should not be pursued until the technology matures. The current RLV program appears to include this tenet, but NASA may be tempted to seek high-risk development to acquire a shuttle replacement. For that matter, military space plane advocates may desire a similar approach in pursuit of a seemingly invincible weapon. Both parties undoubtedly have the US's best interests at heart, but could lead us to squander our treasure in pursuit of a dream not yet ready to be realized. Careful evaluation of progress at each step in the program is the prudent course. The earliest opportunity, confidently to assess the merits of developing an operational vehicle, will not come until the turn of the millennium.

Third, regardless of the embryonic state of reusable rocket technology, it is not too early for the US military to think deeply about the implications of RLV operational use. If operational RLVs become a reality, there will be serious implications for war-fighting strategy, force structure planning, training, and doctrine. Concepts of operations should be developed in more depth and breadth than this study could achieve. In this regard, the analytical criteria used in this study may prove to be a useful framework for evaluating new RLV CONOPS. Another way to support preparation for the birth of operational RLVs is to keep military people active in the flight test programs. Today's DC-XA flight tests include uniformed personnel from the USAF's acquisition command.[133] It is not too soon to include operators and maintenance personnel in this activity. One of the often heard objectives of the RLV program is to develop aircraft-like operations. An excellent way to pursue this worthy goal would be to leverage the experience of seasoned military aircraft maintainers. A handful of senior crew chiefs working with RLV developers and test teams may provide helpful advice on how to establish efficient RLV generation and recovery systems and procedures. At the same time, these crew chiefs would also be developing a knowledge base for future military planning and operations.

Final Summary

The US military must be prepared to take advantage of reusable launch vehicles should the NASA-led effort to develop an RLV demonstrator prove successful. The focus of this study was an explanation of how the US military could use RLVs by describing and analyzing two concepts of operations. Four assumptions which guided the research are worthy of mention. First, the estimate that RLV technology will become operationally feasible by 2012 is reasonable. Second, a fiscally constrained environment will continue. Third, the US government will continue to support growth and development of the US commercial space lift industry and encourage dual use, or perhaps triple use, of related facilities and systems. Fourth, the US government's national security strategy will continue to emphasize international leadership and engagement to further its political, economic, and security objectives.

Before developing and analyzing concepts of operations for military use of RLVs, requirements were stated as space lift, reconnaissance, transspace operations, and strike (in and from space). Then, to provide a basis for CONOPS development and analysis, current RLV concepts and attributes were summarized, and hypothetical attributes of a notional RLV for use in military applications were suggested. Following discussion of RLV concepts and attributes, two concepts of operations were presented and subsequently analyzed. The criteria used in the analysis included capability, cost, operations efficiency, operations effectiveness, and political considerations (table 39).

Four major conclusions resulted from the analysis. First, RLVs have military potential. Second, design choices for an operational RLV will have effects on risk, cost, capability, and operations efficiency and effectiveness, the choice of a larger vehicle being accompanied by more risk. Third, increased investment in propulsion technology is warranted. Fourth, the top priority for the RLV program, even from the DOD perspective, should remain cheap and responsive access to space.

Three recommendations were offered. First, the US military should become a more active participant in the RLV program. Second, the United States should not pursue development of operational RLVs before the technology is ready. Third, it is not too early for the US military to think deeply about the implications of operational RLVs for warfighting strategy, force structure planning, training, and doctrine.

The small steps being taken by the DC-X Delta Clipper experimental in the New Mexico desert today may be recognized in coming years as having warmed and strengthened our muscles for the giant leap into an "exciting and vital future in space activities of all kinds."[134] The United States and its military must be prepared for that future.

Table 39

Summary of Analysis

Analytical Criteria	CONOPS A	CONOPS B
Capability		
Spacelift	Responsive, cannot lift all payloads	Responsive, lifts all payloads
Reconnaissance	Capable and responsive	Capable, but less responsive
Strike to Surface	Capable and responsive	Capable, but less responsive
Strike to Space (LEO)	Capable and responsive	Capable, but less responsive
Transspace	Capable—advantage of smaller propellant facilities	Capable—disadvantage of large propellant facilities
Cost		
Operating Base (nonrecurring cost)	At least $350 million	At least $150 million
ELV Augmentation	$26.6 billion through year 2030	None
Transspace	Not economically viable	Not economically viable
Technology Maturation	Decreases development costs—moderate requirement	Decreases development costs—essential
Operation Efficiency and Effectiveness		
Cryogenic Propellants	Complicates operations	Complicates operations
Deployment	Challenging—benefits from smaller propellant facilities	Challenging—suffers from large propellant facilities
Reliability	Good—lower propulsion performance requirement	Poor—lack of performance margin
Politics		
International	Lives within treaties and law—potentially threatening	Lives within treaties and law—less-threatening appearance
Domestic	ASAT ban prohibits space strike Fiscal constraints drive triple use	ASAT ban prohibits space strike Fiscal constraints drive triple use

Notes

1. Maj Michael A. Rampino, personal observations of DC-XA flight test number nine, White Sands Missile Range, N.Mex., 18 May 1996. The DC-XA could hardly be called a spaceship given its limited altitude capability—the ninth flight only went to 800 feet and it never climbed above 10,000 feet. It

also lacks any payload carrying capability. However, it does prove a point about the feasibility of performing reusable rocket operations.

2. For an overview of the program, see the following sources. "RLV Program Overview," Marshall Space Flight Center, Huntsville, Ala., October 1995, n.p., on-line, Internet, available from http://rlv.msfc.nasa.gov/RLV_ HTMLs/RLVOverview.html. "A Draft Cooperative Agreement Notice, X-33 Phase II: Design and Demonstration," Marshall Space Flight Center, 14 December 1995, on-line, Internet, available from http://procure.msfc. nasa.gov/pub/solicit/can_8_3/.

3. Office of Science and Technology Policy, National Space Transportation Policy Fact Sheet (Washington, D.C.: White House Press Release, 5 August 1994), 3–4.

4. Maj Victor Villhard, staff officer, secretary of the Air Force, Space Policy Office, telephone interview with author, 27 November 1995. According to Major Villhard, the USAF and NASA have an ongoing cooperative effort with seven IPTs examining areas for improved efficiency in space operations. One of these teams, the space lift activities IPT, has a panel for the express purpose of interchange and cooperation on the subject of reusable launch vehicles.

5. Col Eric Anderson, director of space lift acquisition, assistant secretary of the Air Force for acquisition (SAF/AQSL), Pentagon, interviewed by author during visit to Marshall Space Flight Center, Huntsville, Ala., 11 December 1995.

6. Integrated Product Team for Space Launch Activities, "Terms of Reference," draft, undated. USAF officers on the IPT encouraged this research effort to help satisfy the very real need to identify DOD requirements.

7. "Air Force Forms Study Group on SSTO," *Military Space* 13, no. 2 (22 January 1996): 4.

8. Michael K. French, "Industry Officials Cautiously Applaud Tax-Break Bill," *Space News* 7, no. 10 (11–17 March 1996): 15. The term *dual use* refers to the idea of government and commercial entities using the same system or facility. The term *triple use* is of more recent origin and has become popular to emphasize that both civil and defense government agencies use a system or facility along with commercial entities. A recent manifestation of this government policy was a proposal to create tax breaks for commercial space ventures.

9. Directorate of Plans, Air Force Space Command, "FY96 MAP Alignments and Definitions," briefing, Air Force Space Command Headquarters, Peterson AFB, Colo., 23 January 1996.

10. Directorate of Plans, Air Force Space Command, *Space Lift Mission Area Plan*, 1 December 1995, 20–23 and 28. Transspace operations as described in the MAP "are those that occur in the boundary regions between the atmosphere and space." These operations involve moving people and material to and through space. See also Glenn A. Kent, *A Framework for Defense Planning*, R-3721-AF/OSD (Santa Monica, Calif.: RAND, 1989), 1. Strategies to task is "a force planning process that focuses on the building

blocks of *operational capability* . . . clearly linking national security objectives to the timely procurement of hardware."

11. *Space Lift Mission Area Plan*, 32. The cost-related deficiencies are the high recurring operations and maintenance costs of launch vehicle infrastructure and the ranges; the operability concerns stemming from space lift's manpower intensiveness and long launch preparation times; long-range turnaround times; the poor supportability resulting from nonstandard ranges; the ranges' inability to support all users; and the difficulty DOD has in identifying and validating the growing commercial sector requirements.

12. *Space Lift Mission Area Plan*, 36 and 85. Space-based range system refers to replacing the ground- and air-based radar, telemetry receivers, tracking, and command systems now used in support of space lift operations with space-based systems. For example, some space lift operations depend on support from aircraft to collect telemetry when the launch vehicle is not in view of a ground station. It may be possible to use a satellite to collect this information. Using a satellite instead of the aircraft may be cheaper, more reliable, and more flexible.

13. *Space Lift Mission Area Plan*, iii.

14. This accident is a theme consistently heard by the author during discussions with USAF officers involved in space lift policy, operations, and acquisition business. At least one author has claimed the USAF's emphasis on expendable launch vehicles has more to do with its organizational essence and intercontinental ballistic missile heritage. If there is any truth to this it would certainly be hard to prove. It would be even harder to prove its relevance given the strong evidence of rational reasons for the USAF to pursue its current course. See Maj William W. Bruner III, "National Security Implications of Inexpensive Space Access" (master's thesis, School of Advanced Airpower Studies, Maxwell AFB, Ala., June 1995).

15. USAF Scientific Advisory Board, *New World Vistas: Air and Space Power for the 21st Century*, Summary Volume (Washington, D.C.: USAF Scientific Advisory Board, 15 December 1995). The USAF Scientific Advisory Board's recently published *New World Vistas* study is one example of revolutionary planning effort. It was specifically designed by the USAF as an external complement to the USAF modernization planning process due to the recognized limitations of a strategies-to-tasks approach to acquisition.

16. Message, 221435Z DEC 95, commander, Air Force Space Command, to vice chief of staff, USAF, commander, Air Force Materiel Command, and commander, Air Combat Command, 22 December 1995.

17. "Latest News, Official Announcements," Marshall Space Flight Center, Huntsville, Ala., n.p., on-line, Internet, 19 February 1996, available from ftp://rlv.msfc.nasa.gov/RLV_HTMLs/ RLVNews.html.

18. "A Draft Cooperative Agreement Notice, X-33 Phase II: Design and Demonstration," Marshall Space Flight Center, Huntsville, Ala., on-line, Internet, 14 December 1995, available from
http://procure.msfc.nasa.gov/ pub/ solicit/can_8_3/, A-2–A-3.

19. Office of Space Systems Development, NASA Headquarters, *Access to Space Study*, January 1994, 61, 69.

20. The data in table 35 came from a number of sources. Data on RLV payload capability, propulsion, mission life, and costs came from the following documents: Report of the Moorman Study, *Space Launch Modernization Plan*, 5 May 1994; National Research Council, *Reusable Launch Vehicle Technology Development and Test Program* (Washington, D.C.: National Academy Press, 1995); "Executive Review, RLV Technology Program," NASA Headquarters, 24 May 1995, n.p., on-line, Internet, available from http://rlv.msfc.nasa.gov/RLV_HTMLs/LIBGen.html. Competing contractors' cost estimates are at the lower end of the spectrum but are not cited as they are proprietary and may also prove to be optimistic. The cost figures also ignore what the unit production costs might be after technology development. Data on RLV takeoff and landing methods, operations base requirements, cross-range capabilities, and turnaround times came from the following telephone interviews: Mike Pitman, Rockwell RLV/X-33 systems engineer, telephone interviews with author, 21 and 26 February 1996; David Urie, former Lockheed RLV/X-33 program manager, telephone interview with author, 22 February 1996; Paul Klevatt, McDonnell Douglas RLV/X-33 program manager, telephone interview with author, 20 February 1996. Data on the Black Horse TAV came from Robert M. Zubrin and Mitchell Burnside Clapp, "Black Horse: Winging It to Orbit," *Ad Astra* 7, no. 2 (March/April 1995): 40–43; "Space Lift: Suborbital, Earth to Orbit, and on Orbit," *Airpower Journal* 9, no. 2 (Summer 1995): 42–64; and Capt Mitchell B. Clapp, USAF Phillips Laboratory, telephone interview with author, 26 February 1996.

21. George P. Sutton, *Rocket Propulsion Elements, An Introduction to the Engineering of Rockets* (New York: John Wiley & Sons, 1986), 21–22. Specific impulse, I_{sp} or I_s, is defined as the total impulse per unit weight of propellant. Total impulse, I_t, is the thrust force, F (which can vary with time), integrated over the burning time t. According to Sutton, I_{sp} has "units of newton-second^3/kilogram-meter. Since a newton is defined as that force which gives a mass of 1 kilogram an acceleration of 1 meter/second^2, the units of I_{sp} can be expressed simply in seconds. However, it is really a thrust per unit weight flow."

22. "Executive Review, RLV Technology Program," n.p.

23. National Research Council, *Reusable Launch Vehicle Technology Development and Test Program* (Washington, D.C.: National Academy Press, 1995), 1–8, 21, and 73. Mass fraction, also known as mass ratio or MR, is defined to be the final mass of a vehicle (after propellants are consumed) divided by the initial mass (before the propellants are consumed); and Sutton, 23.

24. The potential cost impact of overland flight certification was highlighted by Paul Klevatt, McDonnell Douglas's RLV/X-33 program manager during a telephone interview with author, 20 February 1996. The importance of this issue was echoed by Dennis Smith, Marshall Space Flight

Center, RLV program assistant for technology, telephone interview with author, 23 February 1996.

25. There is not complete consensus on this issue, and it is addressed later.

26. Sutton, 63. The linear aerospike engine is a class of plug nozzle engine. A plug nozzle engine has a center body and an annular chamber, unlike the traditional bell-shaped or contour nozzle common on today's expendable rockets and STS. An aerospike nozzle is a plug nozzle where low-velocity gases (e.g., from a separate gas source) are injected in the center and replace the center body. This allows a very short nozzle hardware configuration, which is desirable for a compact vehicle design.

27. Rick Bachtel, RLV program manager, Marshall Space Flight Center, telephone interview with author, 22 March 1996. Both the MDA and the RSSD concepts will most likely use engines derived from the space shuttle main engine or the Russian RD-0120.

28. "Space Lift; Suborbital, Earth to Orbit, and on Orbit," 42–64.

29. To be fair to the Black Horse advocates on the *SPACECAST 2020* team, they did not intend to suggest great payload capability for the Black Horse. Their concept included revolutionary reductions in satellite size and weight and a greater focus on missions other than delivering payload to orbit.

30. Analysis conducted by the Aerospace Corporation and discussed at the 15 February 1996 meeting of the USAF's Military Space Plane Applications Working Group in Colorado Springs, Colo., indicates the Black Horse as described by the *SPACECAST 2020* group is not feasible. With a change from aluminum to composite material structure and a significant increase in size, the vehicle might be able to achieve orbit, but just barely. Phillips Lab's Black Horse experts are in the process of rebutting this Aerospace Corporation analysis.

31. Assistant Secretary of Defense (Economic Security), Department of Defense, *Industrial Assessment for Space Launch Vehicles*, January 1995, ES-4 and ES-12–ES-13.

32. The belief that a smaller vehicle is more feasible was the consensus of participants at a meeting of the Military Space Plane Applications Working Group held in Colorado Springs, Colo., on 15 February 1996. The participants included NASA personnel, one of whom had worked on the agency's *Access to Space Study* and claimed the analysis behind this study supported the conclusion that developing smaller RLVs involves less technical risk. This perspective is also shared by Dr. Len Worlund, director of technology for the RLV program at Marshall Space Flight Center, although he was also quick to point out that a smaller vehicle will not necessarily meet the requirements of all the users, such as NASA. David Urie, recently retired RLV/X-33 program manager for Lockheed Advanced Development Company, telephone interview with author, 22 February 1996, representing a contrary view, suggested a smaller vehicle size was actually less feasible than the full-scale RLV envisioned by NASA. Like Dr. Worlund, he was also

quick to point out that a smaller vehicle would not meet all user requirements.

33. National Research Council, 3.

34. Ibid., 8.

35. Ibid., 73.

36. Ibid.

37. Assistant Secretary of Defense (Economic Security), II–2. Earlier studies reached a similar conclusion. For example, see National Research Council, *From Earth to Orbit, An Assessment of Transportation Options* (Washington, D.C.: National Academy Press, 1992), 4.

38. "French Launcher Design: Evolved Ariane," *Military Space* 13, no. 5 (4 March 1996): 3–5.

39. The first two scheduled *Ariane 5* launches will carry European Space Agency, not commercial, payloads. Craig Covault, "Cluster to Inaugurate Ariane 5 Flights," *Aviation Week & Space Technology* 144, no. 13 (25 March 1996): 48–50; and Craig Covault, "Reentry Flight Test Set for Second Ariane 5," *Aviation Week & Space Technology* 144, no. 13 (25 March 1996): 51–53.

40. "News Briefs, Satellite Launch Review," *Space News* 7, no. 10 (11–17 March 1996): 17. According to the Arianespace consortium, the average weight of a communications satellite will increase from the current 2,400 to 3,200 within the next four years then level off after that. This average does not include the satellite constellations for mobile communications. These weigh less than 1,000 each.

41. Assistant Secretary of Defense (Economic Security), IV–3.

42. Ibid., I–12.

43. Lt Col Jess Sponable, briefing to Military Space Plane Applications Working Group, Colorado Springs, Colo., 15 February 1996. The pop-up maneuver is essentially a nonoptimum staging maneuver in which the payload, with an appropriate upper stage, is deployed only a few thousand feet short of orbit. This maneuver can significantly increase the payload capability to orbit (or into an intercontinental ballistic trajectory).

44. Office of Space Systems Development, NASA Headquarters, 56.

45. Lt Col Jess Sponable, briefing to Military Space Plane Applications Working Group, 15 February 1996.

46. Bruce A. Smith, "Rockwell Completes Design of Key X-33 Components," *Aviation Week & Space Technology* 144, no. 13 (25 March 1996): 56–57.

47. Maj Michael A. Rampino, personal observations during DC-XA flight test number nine, White Sands Missile Range, N. Mex., 18 May 1996. Based on the 18 May 1996 DC-XA flight test, this requirement may change. During this test, McDonnell Douglas used the grate and trench system for the first time with poor results. Instead of relieving thermal stress on the base on the vehicle, it actually focused the DC-XA's exhaust flame back up toward the rocket causing a fire.

48. An RLV's ability maneuver in space is a function of its propulsion system and available propellant. To some extent, an RLV may be able to

trade payload weight carried for fuel increasing its ability to change its orbital path. However, given the mass fractions required of an RLV, trading all the payload capacity may still translate into very little out-of-plane maneuverability.

49. Dr. Len Worlund, RLV program technology director, telephone interview with author, 22 February 1996.

50. Wiley J. Larson and James R. Wertz, eds. *Space Mission Analysis and Design* (Torrance, Calif.: Dordrecht, 1992), 135–36 (equation 6-10). The greater cross-range requirement necessary to support landing at the base of origin after one revolution is driven by the fact that the earth will have rotated some 22.5 degrees by the time the RLV completes one 90-minute orbit. After circling the earth once, the RLV will find its orbital path is west of where it started. $P = 4(360^{o} - \Delta L)$, where P is the orbital period, and ΔL is the change in longitude that the satellite goes through between successive ascending nodes in degrees. For this example, $90 = 4(360^{o} - \Delta L)$, and $\Delta L = 337.5$.

51. David Urie, former Lockheed Advanced Development Company RLV/X-33 program manager, telephone interview with author, 22 February 1996.

52. Joseph C. Anselmo, "NASA Confident of Shuttle Backups," *Aviation Week & Space Technology* 144, no. 15 (8 April 1996): 54. Drawing on analogies with today's space shuttles would not be helpful. The orbiter with the most flights, *Discovery*, has flown only 21 times. One could also assert that depot-level maintenance is required after every flight. If this kind of performance is repeated by RLVs, there is no hope for achieving the necessary efficiencies.

53. For insight into how weather affects current launch operations, see J. T. Madura, "Weather Impacts on Space Operations," July 1992 (Air University Library number M-U 44059, no. 92-02) and 30th and 45th Space Wing, "Space Lift Concept of Employment," draft, 1 October 1994.

54. Urie. According to Urie, Lockheed plans on an RLV crew size between 50 and 60 people, with a total launch organization of 150 people, including administrative and logistics support personnel. NASA's Cooperative Agreement Notice for the X-33 Phase II requires demonstration of "at least three X-33 landing-to-reflight turnarounds with a ground crew (touch labor) of less than or equal to 50 personnel." "A Draft Cooperative Agreement Notice," B-11.

55. Smith, 57. Rockwell plans on this type of arrangement for their RLV concept. There is a reflection of a broader trend in the space industry to adopt common standards. In the space launch industry, DOD is funding an effort to develop standards to benefit the defense, civil, and commercial sectors. See also Jennifer Heronema, "Space Industry Officials Advocate Adopting Standards," *Space News* 7, no. 8 (26 February–3 March 1996): 10.

56. Paul Jackson, ed., *Jane's All The World's Aircraft* (London: Jane's Information Group Limited, 1995), 567–76. A nominal weapons load for an F-16 consists of six 500-pound or two 2,000-pound bombs in addition to

external fuel tanks and air-to-air missiles. The greatest practical bomb load for the F-16 consists of 12 500-pound bombs or four 2,000-pound bombs, assuming no external fuel tanks are carried. The F-117 can carry two 2,000-pound bombs. The former Soviet Union's SS-18, the ICBM with the greatest throw weight, could deliver 16,700 pounds. Max Walmer, *An Illustrated Guide to Strategic Weapons* (New York: Prentice Hall Press, 1988), 12. These figures obviously do not account for enhanced weapons effects due to the method of delivery.

57. The capabilities described in this table are based on the discussion of RLV attributes in the previous chapter. CONOPS A capabilities reflect the desired RLV attributes described in table 36. CONOPS B capabilities reflect a composite of the attributes currently conceived by the three RLV program competitors. In the case of turnaround time, this table actually reflects the shortest time of any of the concepts rather than an average. Table 37 also assumes all RLVs are completely mission capable—none undergoing maintenance, lost in accidents, or lost to enemy activity. Finally, the RLVs will require some form of weapons dispenser. Based on examination of the USAF's most recently developed bomber, the Northrop-Grumman B-2 Spirit, it is estimated that eight 2,000-pound weapons may be carried on one rotary launcher assembly (RLA), and that each RLA weighs 4,000 pounds. Paul Jackson, ed., *Jane's All The World's Aircraft* (London: Jane's Information Group Limited, 1995), 614–17.

58. This sentiment is echoed in the *New World Vistas* study. "The first attempt to apply new concepts is a necessary, but not sufficient step. In military systems, the second step in the development of a radically new concept must be determined after operational deployment. The warfighters will use the system in innovative ways not described in the manuals." USAF Scientific Advisory Board, *New World Vistas: Air and Space Power for the 21st Century,* Summary Volume (Washington, D.C.: USAF Scientific Advisory Board, 15 December 1995), 13.

59. This is a very conservative estimate of the development cost savings possible with the smaller RLV—it could well be twice as cheap. But this conservative estimate adds more balance to the two CONOPS and may help highlight tradeoffs. One premise for the lower cost estimate on development of the smaller RLV is that a major new engine development program is likely to be required to support the full-scale RLV large payload capacity and size. The National Research Council's RLV program review supports this premise. National Research Council, *Reusable Launch Vehicle Technology Development and Test Program* (Washington, D.C.: National Academy Press, 1995).

60. The terms *space superiority, precision employment, global mobility,* and *information dominance* are used to describe four of the five USAF core competencies. They are not used to imply the USAF must own or operate RLVs for military applications, but to illustrate the connections between the capabilities and a larger mission area or strategy. For example, the ability to quickly launch national security spacecraft in response to some contingency in addition to the ability to reconnoiter and strike enemy spacecraft

as necessary can provide the basis for affecting control of the space environment for friendly exploitation while denying the environment to an adversary—the essence of space superiority.

61. In the earliest days of the space age, the USAF proposed satellite interceptor project (SAINT), involved using an orbital vehicle to inspect potentially hostile spacecraft. The USAF also hoped to develop SAINT into an ASAT system. The project was canceled on 3 December 1962 for a number of reasons. It contradicted the US government's desire to emphasize the peaceful nature of its space program, and it experienced technical, conceptual, and financial difficulties. According to historian Paul Stares, by the mid-1960s ground-based systems were capable of a great deal of information gathering without the added expense and potential political problems of an orbital system. Paul B. Stares, *The Militarization of Space* (Ithaca, N.Y.: Cornell University Press, 1985), 112–17.

62. The USAF's Scientific Advisory Board's *New World Vistas* study discussed the utility of these types of weapons for space control. USAF Scientific Advisory Board, 46–47.

63. Joint Publication 1-02, *Department of Defense Dictionary of Military and Associated Terms*, 23 March 1994, 88. A contingency is "an emergency involving military forces caused by natural disasters, terrorists, subversives, or by required military operations. Due to the uncertainty of the situation, contingencies require plans, rapid response, and special procedures to ensure the safety and readiness of personnel, installations, and equipment."

64. At least one theorist writing about future war predicts space will be "a strategic center of gravity in any future war. Both sides will want space control." Col Jeffery R. Barnett, *Future War: An Assessment of Aerospace Campaigns in 2010* (Maxwell AFB, Ala.: Air University Press, 1996), xxv. This prediction is also made in the *New World Vistas* study. USAF Scientific Advisory Board, 11, 46, and 61.

65. Barnett, xxv. Colonel Barnett predicts that in 2010, if the US "chooses to oppose an invasion of an ally, it must do so during the initial stages of the attack. Failure to immediately engage the enemy could prove disastrous." While this passage may contain some hyperbole, it seems intuitively obvious that being able to strike an adversary while his offensive is unfolding can be advantageous. The ability to do this without having to deploy large forces to the theater of conflict would be even more advantageous.

66. This standard should be the same for EELV.

67. Max Walmer, *An Illustrated Guide to Strategic Weapons* (New York: Prentice Hall Press, 1988), 12. These maneuverable reentry vehicles might have a precision strike capability much like the "Precision-Guided Reentry Vehicle (PGRV)."

68. Rockwell has identified two baseline locations for RLV operations, Cape Canaveral, Fla., and Edwards AFB, Calif. They ruled out Vandenberg AFB, Calif., because of concerns about flying over environmentally sensitive areas when launching to the east. While this type of overflight restriction

may be lifted by 2012, its reality today drove the choice of coastal primary operating bases in this study. If overflight restrictions are relaxed or completely lifted in the future, then primary operating bases in the interior of the CONUS may be a better choice to decrease vulnerability. Bruce A. Smith, "Rockwell Completes Design of Key X-33 Components," *Aviation Week & Space Technology* 144, no. 13 (25 March 1996): 57. Vandenberg is included here because of the anticipation that polar orbits may be desirable for some military missions.

69. Giulio Douhet, *The Command of the Air*, trans. Dino Ferrari (1942; new imprint, Washington, D.C.: Office of Air Force History, 1983), 18–19. Douhet's words "there is no practical way to prevent the enemy from attacking us with his air force" are indicative of his belief in the offensive nature of airpower and the ineffectiveness of ground-based defense against air attack.

70. Andrew Wilson, ed., *Jane's Space Directory* (London: Jane's Information Group Limited, 1995), 162–63 and 172–73.

71. It is conceivable that complete autonomy would not be acceptable for strike missions when collateral damage or fratricide concerns are extremely high.

72. Joint Publication 0-2, *Unified Action Armed Forces*, defines COCOM as "the authority of a combatant commander to perform those functions of command over assigned forces involving organizing and employing commands and forces, assigning tasks, designating objectives, and giving authoritative direction over all aspects of military operations, joint training, and logistics necessary to accomplish the missions assigned to the command." Quoted in Armed Forces Staff College (AFSC) Publication 1, *The Joint Staff Officer's Guide 1993*, 2-20–2-21.

73. Lt Col Robert Owen, "The Airlift System," *Airpower Journal* 9, no. 3 (Fall 1995): 16–29, proposes four tenets of airlift. One of these tenets suggests the military component of the US's airlift system should only do what the civilian component can't or won't do. This tenet might well apply to space lift, especially if there is a viable military component as described in this CONOPS.

74. A similar arrangement already exists today with the civil reserve air fleet.

75. Joint Pub 0-2 defines TACON as "the detailed and usually local direction and control of movements or maneuvers necessary to accomplish mission or assigned tasks." Quoted in AFSC Pub 1, 2-22.

76. According to Joint Pub 3-56.1, *Command and Control for Joint Air Operations*, 14 November 1994, II-2–II-3, the joint force air component commanders (JFACC) responsibilities do not include space forces. However, if the JFACC role is to function as supported commander for strategic attack operations, counterair operations, theater airborne reconnaissance and surveillance, and the JFC's overall interdiction effort, as described in Joint Pub 3-56.1, then it may make sense to give the JFACC authority and responsibility for planning, coordination, allocation, and tasking of RLVs used in

support of a theater campaign plan and to include RLV strikes on whatever the air tasking order evolves into by the year 2012.

77. This is obviously not a critical issue for the CONOPS, but the RLV unit headquarters would be best located where most of the activity is likely to be. With the fall of the former Soviet Union, Cape Canaveral has become the busiest launch base in the world.

78. Jacob Neufeld, *Ballistic Missiles* (Washington, D.C.: Office of Air Force History, 1990), 103, 208, 252–53. Interestingly, in 1954 the USAF considered using civilian contractor personnel to operate its first Atlas ICBMs to attain an early "emergency operational capability" by 1958. This option, described as a "PhD-type capability," was never exercised. Instead, the Atlas achieved operational status in September 1959 only after a Strategic Air Command crew completed a successful training launch.

79. The assumption here is that the commercial organization would be an American company. If the operator were to be a multinational corporation, tasking for military missions would be more complicated. At the same time, operations by a multinational corporation could provide a measure of deterrence. Any attack on a multinational RLV might invite a response from other nations as well as the United States.

80. Joint Pub 0-2 defines OPCON as "the authority delegated to a commander to perform those functions of command over subordinate forces involving the composition of subordinate forces, the assignment of tasks, the designation of objectives, and the authoritative direction necessary to accomplish the mission." Quoted in AFSC Pub 1, 2-21–2-22.

81. Report of the Moorman Study, "Space Launch Modernization Plan," 5 May 1994, 12. The minimum time required from call-up to launch for today's expendable launch vehicles is 2–4 months for *Pegasus*, 90 days for *Titan II*, 98 days for *Delta II*, and 180 days for *Titan IV*. The shuttle requires 12–33 months from call-up to launch.

82. Joel W. Powell, "Space Station Hardware," *Spaceflight* 38, no. 2 (February 1996): 54–56.

83. Rick Bachtel, RLV program manager, Marshall Space Flight Center, telephone interview with author, 22 March 1996. Bachtel estimated the crew module will weigh approximately 20,000 pounds and carry a crew of three to four astronauts.

84. Office of Space Systems Development, NASA Headquarters, *Access to Space Study*, January 1994, 40 and 64–65.

85. Powell, 54. The current plans for ISSA deployment have the Russians delivering two of the largest components of ISSA, the functional cargo block at 19,340 kg and the service module at 21,020 kg, from Baykonour cosmodrome using Proton rockets. Former astronaut Buzz Aldrin has suggested using the Russians' largest rocket, the 220,000-pound-to-LEO *Energia*, to lift ISSA payloads. See Darrin Guilbeau, "International Cooperation and Concerns in Space Logistics," *Air Force Journal of Logistics* 19, no. 1 (Winter 1995): 25–31. This lift capability is four times that of the *Titan IV* and could easily handle lifting the largest planned payloads.

86. Cape Canaveral is located at 28.5 degrees (28 degrees, 30 minutes) north latitude, while Baykonour is located at 45.9 degrees (45 degrees, 54 minutes) north latitude. Wiley J. Larson and James R. Wertz, eds., *Space Mission Analysis and Design* (Torrance, Calif.: Dordrecht, 1992), 680. One equation used here as the basis for this estimation can be found in Roger R. Bate, Donald D. Mueller, and Jerry E. White, *Fundamentals of Astrodynamics* (New York: Dover Publications, 1971), 307 (equation 6.4-1). V_o = 1,524 cos L_o , where V_o is the speed of a launch point on the surface of the earth, 1,524 ft/sec is the eastward speed of a point on the equator, and L_o is the latitude of the launch site. For launch out of Baykonour, V_o = 1,524 cos (45.9) = 1,060.6 ft/sec. For launch out of the Cape, V_o = 1,524 cos (28.5) = 1,339 ft/sec. To determine how much of the eastward velocity of the launch site actually may be used to help boost a payload, some basic trigonometry may be used. $V_1 = V_o \cos \varnothing$, where V_1 is the velocity advantage in the direction of the launch azimuth gained from the rotation of the earth, V_o is the speed of a launch point on the surface of the earth, and $\varnothing$ is the angle between the launch azimuth and a line due east of the launch site. For a launch out of Baykonour into the ISSA orbit (51.6 degree inclination), V_1 = 1,060.6 cos (26.8) = 946.7 ft/sec. For a launch out of the Cape into the ISSA orbit, V_1 = 1,339 cos (47.12) = 911.1 ft/sec. Thus, for launches into the ISSA orbit, Baykonour actually benefits more (946.7 – 911.1 = 35.6 ft/sec) from the rotation of the earth. (Launch azimuths to the ISSA orbit from the Cape and Baykonour were provided by the 45th Range Squadron, Cape Canaveral Air Force Station, Fla., and Ed Faudree of the ANSER Corporation, Washington, D.C., respectively.)

87. Craig Covault, "Reentry Flight Test Set for Second Ariane 5," 51–53.

88. The Titan IV can lift 39,100 pounds to LEO. *Ariane 5* data may be found in "French Launcher Design: Evolved Ariane," *Military Space* 13, no. 5 (4 March 1996): 4. Titan IV data may be found in Bretton S. Alexander et al., "1994 Space Launch Activities," ANSER Aerospace Division Note ADN 95-2 (Arlington, Va.: ANSER, 1995).

89. See "National Reconnaissance Office Home Page," National Reconnaissance Office, n.p., on-line, Internet, available from http://www.nro.odci.gov/, for an official introduction to the US government's space-based national security reconnaissance capabilities.

90. There are numerous print and electronic media sources which contain information about US government space-based reconnaissance systems and provide ephemeris data which could be used to predict orbits and overflight times. This author makes no judgment about the accuracy of this publicly available information. However, if there is any truth to it, then even an unsophisticated adversary might thwart US attempts to monitor them using reconnaissance satellites in space.

91. Paul B. Stares, *The Militarization of Space* (Ithaca, N.Y.: Cornell University Press, 1985), 112–17.

92. Paul Jackson, ed., *Jane's All The World's Aircraft* (London: Jane's Information Group Limited, 1995), 617. The B-2 has a maximum weapons

load of 40,000 pounds—the same as the payload weight capacity of RLV-B. However, when delivering weapons such as the AGM 169 advanced cruise missile (ACM), the joint direct attack munition, or the Mk 84 2,000-pound bomb, only 16 of these weapons can be carried due to the weight of the rotary launcher assembly required to carry and dispense them. Obviously, an RLV will not drop these same weapons. Weapons designed specifically for delivery from space will be required.

93. There are many assumptions underlying the numbers in this table. With respect to the RLVs, assumptions include a 100 percent mission-capable rate, no other missions, such as space lift, being accomplished, no losses, RLA carries eight 2,000-pound weapons and weighs 4,000 pounds, mission execution time is 90 minutes to go around the earth and return to launch base while making strike en route (45 minutes into orbit). With respect to the B-2, assumptions include a cost of $2 billion for each aircraft allowing for a cost-equivalent fleet of 10 aircraft (corresponds to RLV development budget of $20 billion), one-hour response time, sixteen 2,000-pound weapons delivered by each aircraft, flying out of the continental US (CONUS) with an 18-hour flight required to reach the target, recovery at the point of origin in the CONUS (Whiteman AFB, Mont.), and three-hour turnaround time. The $2 billion cost figure for each B-2 is a unit program cost—the total program cost divided by the number of aircraft acquired (see Jackson, 615). A lower cost figure for the B-2 could be used if one only counted the current unit production costs, not including program development costs. But then one could use the same method to arrive at a lower cost figure for RLVs as well.

94. At least one writer has suggested using an RLV-like vehicle to deliver US Marines "From Space" in the 2040 time frame. Maj William C. Redmond, "Campaign 2040: Victory From Space" (master's thesis, US Marine Corps School of Advanced Warfighting Studies, 1995), 34.

95. Dave Schweikle, McDonnell Douglas Aerospace DC-X/DC-XA program manager, telephone interview with author, 2 April 1996, and personal observations of the author at White Sands Missile Range, 16–18 May 1996.

96. Ibid., 2 April 1996.

97. Report of the Moorman Study, D-14.

98. Telephone inquiry to package delivery service, 3 April 1996.

99. Daniel McDonald, staff member, Aerial Ports Operations Division, Air Mobility Command, telephone interview with author, 4 April 1996. These cost and time figures averages are based on all shipments given a "999" priority, the highest possible, shipped from Dover to Ramstein during the period of October 1995 through February 1996.

100. For two examples, see Office of Space Systems Development, NASA headquarters, 70, and Report of the Moorman Study, 27.

101. Report of the Moorman Study, D-10–D-11.

102. National Research Council, *Reusable Launch Vehicle Technology Development and Test Program* (Washington, D.C.: National Academy Press, 1995), 13.

103. Ibid., 9.

104. Briefing, OL-AC PL/RKF (USAF Phillips Lab, Fundamental Technologies Division), subject: Integrated High Payoff Rocket Propulsion Technology program, 12 March 1996.

105. Recommendation number eight of the Moorman Study was to "increase funding for a core space launch technology program as an enabler for future investment" (see Report of the Moorman Study, 26). Funding for the IHPRPT program prior to the Moorman Study report was in the $50–60 million range—it has remained at that level since. James Chew, staff member, Directorate of Advanced Technology, director of Defense Research and Engineering, telephone interview with author, 19 March 1996.

106. The Moorman Study recommended "that the space lift core technology program within DoD be increased from its current level to $120M total by FY 96" (see Report of the Moorman Study, C-1–3).

107. Neufeld, 208–19 and 234–35.

108. A cryogenic propellant is a liquefied gas at low temperature. Cryogenics are very high performance propellants (specific impulses as high as 450 seconds may be achieved) but are complicated to handle. Hypergolic propellants are liquid propellants which spontaneously ignite as the oxidizer and fuel come in contact with each other. Hypergolics like those used in the Titan are liquid at ambient temperature that can be stored for long periods in sealed tanks, easing some of the complications encountered with cryogenics, but only achieve specific impulses of 300–340 seconds. Solid propellants are simple, reliable, and relatively low cost, but have more limited performance (specific impulses of 300 seconds or less). George P. Sutton, *Rocket Propulsion Elements, An Introduction to the Engineering of Rockets* (New York: John Wiley & Sons, 1986), 147, 175, and 292. Larson and Wertz, 644–45.

109. Third Space Launch Squadron Operations Checklist, *AC-74 Launch Management Countdown Checklist*, 12 April 1994.

110. Schweikle.

111. Edward Boyle, Matthew Tracy, and Lt Col Donald Smoot, "Rethinking Support Equipment," *Air Force Journal of Logistics* 19, no. 4 (Fall 1995): 28–31.

112. Quoted in Boyle, Tracy, and Smoot, 28.

113. Quoted in Capt Sandra M. Gregory, "Improving Space Lift Reliability through Robust Design," *Air Force Journal of Logistics* 19, no. 1 (Winter 1995): 16–18.

114. Ibid., 16.

115. Report of the Moorman Study, 10.

116. Gregory, 17.

117. Smith, "Rockwell Completes Design of Key X-33 Components," 56–57.

118. Gregory, 17.

119. The McDonnell Douglas Aerospace and Rockwell RLV propulsion concepts plan on using SSME or RD-0120 derivative engines. The Lockheed

RLV concept requires new engine development given its nontraditional linear aerospike engine design. Bachtel, telephone interview, 22 March 1996.

120. Joint Doctrine, Tactics, Techniques, and Procedures (JDTTP) 3-14, "Space Operations," final draft, 15 April 1992, II-18–II-22 and A-1–A-17.

121. Stephen M. Walt, "Alliance Formation and the Balance of World Power," *International Security* 9, no. 4 (Spring 1985): 3–41.

122. Air Force Doctrine Document 4, "Space Doctrine," draft, 1 August 1995, 5.

123. NASA leaders fear losing the shuttle, so they will take the greater technical risk involved with developing a full-scale rocket instead of initially shooting for reusability in a smaller-scale vehicle. This is entirely speculative but may provide interesting insights into organizational behavior related to the RLV program. It may also suggest why the DOD would take a more risk-averse approach. The DOD sees development of an RLV as a potential gain but is confident of its ability to continue using ELVs for space access. Thus, DOD is less motivated to take a high risk in the quest for an operational RLV. DOD pursues the lower-risk approach of incrementally improving its ELV fleet through the EELV program.

124. Quoted in Joseph C. Anselmo, "NASA Issues Wake-up Call to Industry," *Aviation Week & Space Technology* 144, no. 8 (19 February 1996): 20.

125. Ibid., 20.

126. Joseph C. Anselmo, "Clinton Budget Chops NASA Again," *Aviation Week & Space Technology* 144, no. 13 (25 March 1996): 23–24.

127. Lt Gen Howell Estes, director of operations, Joint Staff, "Operationalizing Space," address to the Second Annual C^4I Symposium, USAF Academy, Colo., 28 March 1996. Paul G. Kaminski, undersecretary for acquisition and technology, "Building a Ready Force for the 21st Century," *Defense Issues* 11, no. 6 (17 January 1996): n.p., on-line, Internet, available from http://www.dtic.mil/defenselink/ pubs/di96/di1106.html. Rep. Ronald V. Dellums, "Toward a Post-Transition World: New Strategies for a New Century," *SAIS Review* 25, no. 10 (Winter–Spring 1995): 93–111. On the plus side, the DOD's 1997 budget request for space programs is up 12 percent in spite of the gloomy fiscal environment. Jennifer Heronema, "US Military Has Less, But Allocates More to Space," *Space News* 7, no. 10 (11–17 March 1996): 8.

128. Gen Joseph W. Ashy, CINCSPACE, identified the capabilities to "take-off on demand, overfly any location in the world in approximately one hour and return and land within two hours at the take-off base" as desirable. This was used as one basis for RLV requirements at the outset of this study (see chap. 1). Message, 221435Z Dec 95, commander, Air Force Space Command, to vice chief of staff, USAF, commander, Air Force Materiel Command, and commander, Air Combat Command, 22 December 1995.

129. The USAF Scientific Advisory Board's *New World Vistas* report describes a potential future in space where many low-cost single-function satellites work in cooperative networks to achieve even greater capability

than that possible with a few high cost multifunction satellites. USAF Scientific Advisory Board, *New World Vistas: Air and Space Power for the 21st Century,* Summary Volume (Washington, D.C.: USAF Scientific Advisory Board, 15 December 1995).

130. Wally McCoy, "Sustaining Space Systems for Strategic and Theater Operations," *Air Force Journal of Logistics* 19, no. 1 (Winter 1995): 32–33. The feasibility of on-orbit support has been studied within the military, as recently in 1993 by USSPACECOM, and made a reality by NASA through use of the space shuttle. The Hubble Space Telescope is one example of a spacecraft specifically designed for on-orbit servicing.

131. Report of the Moorman Study, "Space Launch Modernization Plan," 5 May 1994, C-1-2-3. If the ability to take off, overfly a target half-way around the world within an hour, and land back at the base of origin within two hours of takeoff is strongly desired by America's military leaders, then there may be a better way to obtain that capability. Scramjet technology, such as that developed as part of the National Aerospace Plane Program (HYFLITE) might be pursued to achieve TAV-like capability. This might also satisfy the deficiency in transspace operations as defined in the Space lift MAP.

132. John A. Alic et al., *Beyond Spinoff, Military and Commercial Technologies in a Changing World* (Boston: Harvard Business School Press, 1992), 7–8, 73. The term *spin on* refers to reverse spin-off. With spin on, technologies developed entirely in the commercial sector are used, or are adapted for use, by the defense sector.

133. Maj Michael A. Rampino, personal observations at DC-XA flight test number nine, White Sands Missile Range, N.Mex., 16–18 May 1996.

134. Lt Gen James A. Abrahamson, quoted in House, *Assured Access to Space during the 1990s: Joint Hearings before the Subcommittee on Space Science Applications of the Committee on Science and Technology and the Subcommittee on Research and Development of the Committee on Armed Services*, 99th Cong., 1st sess., 1985, 41.

Bibliography

Books

Alexander, Bretton S., et al. *1994 Space Launch Activities*. Arlington, Va.: ANSER, 1995. ANSER Aerospace Division Note ADN 95-2.

Alic, John A., et al. *Beyond Spinoff: Military and Commercial Technologies in a Changing World*. Boston: Harvard Business School Press, 1992.

Barnett, Col Jeffery R. *Future War: An Assessment of Aerospace Campaigns in 2010*. Maxwell AFB, Ala.: Air University Press, 1996.

Bate, Roger R., Donald D. Mueller, and Jerry E. White. *Fundamentals of Astrodynamics*. New York: Dover Publications, 1971.

Douhet, Giulio. *The Command of the Air*. Translated by Dino Ferrari. Washington, D.C.: Office of Air Force History, 1983.

Futrell, Robert Frank. *Ideas, Concepts, Doctrine: Basic Thinking in the United States Air Force, 1907–1960*. Vol. 1. Maxwell AFB, Ala.: Air University Press, December 1989.

Jackson, Paul, ed. *Jane's All the World's Aircraft*. London: Jane's Information Group Limited, 1995.

Kent, Glenn A. *A Framework for Defense Planning*. Santa Monica, Calif.: RAND, 1989.

Larson, Wiley J., and James R. Wertz, eds. *Space Mission Analysis and Design*. Torrance, Calif.: Dordrecht, 1992.

National Research Council. *From Earth to Orbit, An Assessment of Transportation Options*. Washington, D.C.: National Academy Press, 1992.

———. *Reusable Launch Vehicle Technology Development and Test Program*. Washington, D.C.: National Academy Press, 1995.

Neufeld, Jacob. *Ballistic Missiles*. Washington, D.C.: Office of Air Force History, 1990.

Stares, Paul B. *The Militarization of Space: U.S. Policy, 1945–1985*. Ithaca, N.Y.: Cornell University Press, 1985.

Sutton, George P. *Rocket Propulsion Elements: An Introduction to the Engineering of Rockets*. New York: Wiley, 1976.

Walmer, Max. *An Illustrated Guide to Strategic Weapons*. New York: Prentice Hall Press, 1988.

Wilson, Andrew, ed. *Jane's Space Directory*. London: Jane's Information Group, 1995.

Theses and Papers

Bruner, Maj William W., III. "National Security Implications of Inexpensive Space Access." Master's thesis, School of Advanced Airpower Studies, Maxwell AFB, Ala., June 1995.

Redmond, Maj William. "Campaign 2040: Victory from Space." Master's thesis, USMC School of Advanced Warfighting Studies, 1995.

Government Documents

3d Space Launch Squadron, 45th Space Wing, Air Force Space Command. *AC-74 Launch Management Countdown Checklist*. Operations checklist, 12 April 1994.

30th and 45th Space Wings, Air Force Space Command. "Space lift Concept of Employment." Draft, 1 October 1994.

Air Force Doctrine Document 4. "Space Doctrine." Draft, 1 August 1995.

Armed Forces Staff College Publication 1. *The Joint Staff Officer's Guide 1993*, 1993.

Assistant Secretary of Defense (Economic Security), Department of Defense. *Industrial Assessment for Space Launch Vehicles*, January 1995.

Commander, Air Force Space Command. *Implementation of the Conclusions and Recommendations of the Air Force Scientific Advisory Board's New World Vistas Study*.

Directorate of Plans, Air Force Space Command. *Space lift Mission Area Plan*, 1 December 1995.

Integrated Product Team for Space Launch Activities. "Terms of Reference." Draft, undated.

Joint Doctrine, Tactics, Techniques, and Procedures 3-14. *Space Operations*. Final draft, 15 April 1992.

Joint Publication 1-02. *Department of Defense Dictionary of Military and Associated Terms*, 23 March 1994.

Joint Publication 3-56.1. *Command and Control for Joint Air Operations,* 14 November 1994.

Madura, Col J. T. *Weather Impacts on Space Operations,* July 1992. Air University Library no. M-U 44059, no. 92–02.

Message, 221435Z Dec 95, to vice chief of staff, USAF, commander, Air Force Materiel Command, and commander, Air Combat Command, 22 December 1995.

Office of Science and Technology Policy, The White House. *National Space Transportation Policy Fact Sheet,* 5 August 1994.

Office of Space Systems Development, NASA Headquarters. *Access to Space Study,* January 1994.

Report of the Moorman Study. *Space Launch Modernization Plan,* 5 May 1994.

USAF Scientific Advisory Board. *New World Vistas: Air and Space Power for the 21st Century,* Summary Volume. Washington, D.C.: USAF Scientific Advisory Board, 15 December 1995.

US House. *Assured Access to Space during the 1990s: Joint Hearings before the Subcommittee on Space Science Applications of the Committee on Science and Technology and the Subcommittee on Research and Development of the Committee on Armed Services.* 99th Cong., 1st sess., 23–25 July 1985.

Articles

"Air Force Forms Study Group on SSTO." *Military Space* 13, no. 2 (22 January 1996): 4.

Anselmo, Joseph C. "Clinton Budget Chops NASA Again." *Aviation Week & Space Technology* 144, no. 13 (25 March 1996): 23–24.

———. "NASA Confident of Shuttle Backups." *Aviation Week & Space Technology* 144, no. 15 (8 April 1996): 54.

———. "NASA Issues Wake-up Call to Industry." *Aviation Week & Space Technology* 144, no. 8 (19 February 1996): 20–21.

Boyle, Edward, Matthew Tracy, and Lt Col Donald Smoot. "Rethinking Support Equipment." *Air Force Journal of Logistics* 19, no. 4 (Fall 1995): 28–31.

Covault, Craig. "Cluster to Inaugurate Ariane 5 Flights." *Aviation Week & Space Technology* 144, no. 13 (25 March 1996): 48–50.

———. "Reentry Flight Test Set For Second Ariane 5." *Aviation Week & Space Technology* 144, no. 13 (25 March 1996): 51–53.

Dellums, Rep. Ronald V. "Toward a Post-Transition World: New Strategies for a New Century." *SAIS Review* 25, no. 10 (Winter–Spring 1995): 93–111.

"French Launcher Design: Evolved Ariane." *Military Space* 13, no. 5 (4 March 1996): 3–5.

French, Michael K. "Industry Officials Cautiously Applaud Tax-Break Bill." *Space News* 7, no. 10 (11–17 March 1996): 15.

Gregory, Capt Sandra M. "Improving Space lift Reliability through Robust Design." *Air Force Journal of Logistics* 19, no. 1 (Winter 1995): 16–18.

Guilbeau, Darrin. "International Cooperation and Concerns in Space Logistics." *Air Force Journal of Logistics* 19, no. 1 (Winter 1995): 25–31.

Heronema, Jennifer. "Space Industry Officials Advocate Adopting Standards." *Space News* 7, no. 8 (26 February–3 March 1996): 10.

———. "US Military Has Less, But Allocates More to Space." *Space News* 7, no. 10 (11–17 March 1996): 8.

McCoy, Wally. "Sustaining Space Systems for Strategic and Theater Operations." *Air Force Journal of Logistics* 19, no. 1 (Winter 1995): 32–33.

"News Briefs, Satellite Launch Review." *Space News* 7, no. 10 (11–17 March 1996): 17.

Owen, Lt Col Robert C. "The Airlift System: A Primer." *Airpower Journal* 9, no. 3 (Fall 1995): 16–29.

Powell, Joel W. "Space Station Hardware." *Spaceflight* 38, no. 2 (February 1996): 54–56.

Smith, Bruce A. "Rockwell Completes Design of Key X-33 Components." *Aviation Week & Space Technology* 144, no. 13 (25 March 1996): 56–57.

"Space Lift: Suborbital, Earth to Orbit, and on Orbit." *Airpower Journal* 9, no. 2 (Summer 1995): 42–64.

Walt, Stephen M. "Alliance Formation and the Balance of World Power." *International Security* 9, no. 4 (Spring 1985): 3–41.

Zubrin, Robert M., and Capt Mitchell B. Clapp. "Black Horse: Winging It to Orbit." *Ad Astra* 7, no. 2 (March/April 1995): 40–43.

Briefings and Speeches

Directorate of Plans, Air Force Space Command. "FY96 MAP Alignments and Definitions." Briefing, Air Force Space Command Headquarters, Peterson AFB, Colo., 23 January 1996.

Estes, Lt Gen Howell. "Operationalizing Space." Briefing, Second Annual C[4]I Symposium, USAF Academy, Colo., 28 March 1996.

James, Maj Glenn. "IHPRPT Is." Briefing, USAF Phillips Laboratory, Fundamental Technologies Division, Edwards AFB, Calif., 12 March 1996.

Sponable, Lt Col Jess. "Military Spaceplane Technology and Applications." Briefing, Military Spaceplane Applications Working Group, Colorado Springs, Colo., 15 February 1996.

Internet Sources

"A Draft Cooperative Agreement Notice, X-33 Phase II: Design and Demonstration." Marshall Space Flight Center, 14 December 1995. On-line. Internet. Available from ftp://procure.msfc.nasa.gov/pub/solicit/can_8_3/.

"Executive Review, RLV Technology Program." NASA Headquarters, 24 May 1995. On-line. Internet. Available from http://rlv.msfc.nasa.gov/RLV_HTMLs/LIBGen.html.

Kaminski, Paul G. "Building a Ready Force for the 21st Century." *Defense Issues* 11, no. 6 (17 January 1996). On-line. Internet. Available from http://www.dtic.mil/ defenselink/pubs/di96/di1106.html.

"Latest News, Official Announcements." Marshall Space Flight Center, 19 February 1996. On-line. Internet. Available from http://rlv.msfc.nasa.gov/RLV_HTMLs/RLVNews. html.

"National Reconnaissance Office Home Page." National Reconnaissance Office, April 1996. On-line. Internet. Available from http://www.nro.odci.gov/.

"RLV Program Overview." Marshall Space Flight Center, October 1995. On-line. Internet. Available from http:// rlv.msfc.nasa.gov/RLV_HTMLs/RLVOverview.html.

Author's Notes

Rampino, Michael A. Discussions at 15 February 1996 Military Spaceplane Applications Working Group. Colorado Springs, Colo., 15 February 1996.

———. Personal Observations of DC-XA Flight Test Number Nine. White Sands Missile Range, New Mexico, 16–18 May 1996.

Chapter 9

The Inherent Limitations of Space Power: Fact or Fiction?

Gregory Billman

Control of space means control of the world, far more certainly, far more totally than any control that has been achieved by weapons or troops of occupation. Space is the ultimate position, the position of total control over Earth.

—Lyndon Baines Johnson

Now the competition will be for the possession of the unhampered right to traverse and control the most vast, the most important, and the farthest reaching element on the earth, the air, the atmosphere that surrounds us all, that we breathe, live by, and which permeates everything. . . . A new set of rules for the conduct of war will have to be devised and a whole new set of ideas of strategy learned by those charged with the conduct of war.

—Brig Gen William "Billy" Mitchell

Is US space power's current subordinate position to terrestrial military powers—air, land, and sea—due to inherent limitations? Space power today is limited in its ability to accomplish many military missions. Whether those limitations are predominantly inherent to the space environment or are self-imposed by the current US approach to space is the subject of this study.

Following a clear definition of space power, three steps are taken in the analysis process. First, the evolving relative importance of space power, as it is generally regarded, is discussed in relation to the other forms of military power. Histori-

This work was accomplished in partial fulfillment of the master's degree requirements of the School of Advanced Airpower Studies, Air University, Maxwell AFB, Ala., 1996.
Advisor: Col Phillip Meilinger, PhD
Reader: Lt Col Robert Owen, PhD

cal analogy with the accession of airpower from the early twentieth century onward seems particularly appropriate. Terrestrial military theory and space theory are subsequently discussed from a historical context, leading to a discussion of current space doctrine as it relates to space power's current supporting military role. In the application of theory and doctrine, current technologies are considered as they demonstrate space capabilities beyond those presently fielded. Second, the physical attributes of space are examined to establish whether conduct of operations within the medium has inherent physical limitations. Third, beyond physical limitations, the issue of inherent limitations due to a lack of military utility is addressed. Military power characteristics are discussed as they apply to (1) terrestrial power, (2) currently fielded space forces, and (3) space forces which are technologically feasible. The characteristics include strategic agility, ability to demonstrate commitment and credibility, and economic, military and political considerations.

Conclusions and implications are discussed as they apply to the future potential of US space power. Depending on the findings, doctrinal implications exist to properly use space power—either as an adjunct force with terrestrial power, or as an independent military force.

No standard definitions seem to exist for air, land, or sea power. However, all seem to have similar characteristics, and hence space power can be defined in a similar manner. As Lt Col David E. Lupton writes in his work, *On Space Warfare: A Spacepower Doctrine;* "Spacepower is the ability of a nation to exploit the space environment in pursuit of national goals and purposes and includes the entire astronautical capabilities of the nation."[1]

The United States depends on space power. It has a space infrastructure, both civilian and military, and is presently exploiting space for many purposes. As naval forces supply the military component of sea power, and air forces provide the military component of airpower, space forces supply the military component of space power. The following is an examination of how space power has developed as compared to other military powers, specifically airpower, and how this development has affected current space power doctrine.

Military Power Development and Space Power's Relative Importance

The air war of yesterday becomes the space war of tomorrow.

—1960 Democratic Party Policy Statement

Space power is evolving into a mature military entity—much like airpower evolved into a dominating military force. This section reviews space power's development and its relationship to airpower's historical development to demonstrate that space power, like airpower, owes its potential rapid rise as a dominating form of warfare to its unique ability to affect adversaries in ways previously unimaginable.

Many similarities exist between airpower's development and space power; some are cursory, while others are more concrete. Cursory similarities include the difficult conceptual thought required in both cases to develop theories exploiting the military potential of fundamentally new and environmentally hostile mediums, the requirement for a technological knowledge base of current and future developments, and the need for a doctrinal push by military organizations to claim the developmental turf of a new medium. More concrete similarities include the way in which each power's resources were first employed, the evolution and relationship of each power's technologies to new roles, and the organizational development of each power within the military.

Early Employment of Airpower and Space Power

Peter Hays states "the first military use of these two new mediums was for observation and reconnaissance."[2] In actuality, in 1911–12, prior to World War I, the Italians used airpower in all four present-day mission areas against the Turks in Libya (force application, force enhancement, control and support). Lee B. Kennett, in his book *The First Air War, 1914–18,* suggests that it was this experience that caused a young Italian artillery officer, named Giulio Douhet, to remark, "A new weapon has come forth, the sky has become a new battlefield." Though Hays's remark is not completely accurate, his concept is predominantly regarded as valid. Force application did play a considerable, if not a major, role in World War I airpower. The rudiments of counterair weapons began development in the early years, but it wasn't until the latter part of the war that purpose-built antiaircraft weapons appeared on the battlefields. Additionally, lighter-than-air German dirigibles were used early on in both reconnaissance and bombing roles.[3] It is generally accepted that the overwhelming bulk of sorties flown by any side in World War I involved aircraft and airships in tactical observation and reconnaissance roles. Likewise, Operation Desert Storm, fought 77 years later, utilized military space assets in much the same way. Desert Storm has been called "The First Space War," harking to World War I's appellation, "The First Air War."[4]

World War I gave airpower its first large-scale opportunity to contribute, mostly by observing for artillery placement and reconnaissance of enemy troop movements and dispositions. In his book *The First Air War: 1914–1918,* Kennett emphasizes this point. He illustrates airpower's contributions and technical development from both sides, including German airpower's value at the Battle of Tannenberg and Allied airpower's efforts at the Battle of the Marne. He discusses specifically how airpower supported earth-bound forces via communications, positioning, and intelligence and surveillance.

In the early stages of World War I, as the Germans moved swiftly across the European continent, German observation planes were found most compatible with this rapid movement. From 15 August until 9 September 1914, the *Fliegerabteilung* of the German Third Army Corps changed airfields 18 times and during that time was grounded by bad weather only two days.[5] Aircraft became essential to command and control of German forces. Commanders now had much better information to determine where enemy armies were and, consequently, were better prepared to move their troops. Kennett writes, "German observation planes played a significant role in the east, where their reports, coupled with interceptions of Russian radio transmissions, set the stage for the victory at Tannenberg. Field Marshal Paul von Hindenburg acknowledged his debt to the German Air Service: 'Without the airmen no Tannenberg.'"[6]

Positioning of friendly (for navigation and tactical purposes) and enemy forces (for targeting purposes) became exceedingly more precise with the advent of the balloon, and then the airplane. Not only did aircrews directly report positions of friendly and enemy troops, but balloons, which were in use for observation purposes, were used by fixed-wing aviators and ground troops to determine their position relative to friendly lines.

Though, as previously discussed, airpower played other significant roles in World War I, intelligence and surveillance were generally regarded as its raison d'être. Altitude, and the capability to travel well behind enemy lines gave airmen the unique capability to see and determine things never before available to opposing forces. Information on force movements, troop dispositions, deployed weaponry, and enemy resupply capabilities all became available to the commander who was lucky enough to be supported by air machines. In short, the visibility restricted by the "fog of war" became a bit clearer with the introduction of airpower.

There are strong analogies to be made between the emergence of airpower in World War I and the emergence of space power in Operation Desert Storm. While it is true that space power in the Gulf did not contribute to all four mission areas as airpower did in World War I, the Gulf War provided space power with its first large-scale opportunity to demonstrate its capabilities. Similarly, these capabilities were generally limited to force enhancement, reconnaissance, and other command and control-enhancing operations. Space power severely reduced the "fog and friction of war" for supported commanders, while it increased the opportunity for "fog" to cloud the enemy's decision making and "friction" to increase the enemy commander's difficulties. As World War I proved the efficacy of airpower as a valuable tool for future conflict, the Gulf War seems to have proved the efficacy of space power as a viable arm of future military operations. Similar to the dominant role of airpower in World War I, allied space assets in the Gulf were limited to functions which supported the other military arms. In today's terms, space power's main focus in the Gulf was direct support to the war fighter, or force enhancement.

Force enhancement includes space power capabilities that "provide effective operational support to military forces."[7] As airpower multiplied the combat effectiveness of surface-bound forces in World War I, so too did space power multiply the combat effectiveness of terrestrial weapon systems in the Gulf. In fact, a comparison of each power's early functions demonstrates the similarities. Specifically, in the Gulf, force enhancement capabilities included communications, navigation, positioning, intelligence and surveillance (including weather).

Communications in Desert Shield/Desert Storm were accomplished via the Defense Satellite Communications System (DSCS) and Fleetsat spacecraft. The system provided a high data rate, high capacity, worldwide, secure voice communications system for command and control, crises management, and intelligence data transmission between the field units, theater command structure and the National Command Authorities (NCA). As well as supplying direct communications links, DSCS also provided a bridge for terrestrial communications systems with line-of-sight restrictions across the vast expanses of desert. DSCS provided real-time communications between land, air, and sea units, as well as television into and out of theater. As the military communication systems became saturated, a "Civil Reserve Space Fleet" concept (analogous to the Civil Reserve Air Fleet or CRAP concept) was adopted to use commercial communications satellites to relay nonsecure and nonpriority traffic.

Navigation and positioning efforts in the Gulf were carried out by the navigational strategic, tactical relay (NAVSTAR) Global Positioning System (GPS) fleet of satellites. This system provided Coalition forces precise three-dimensional location and time information. The featureless desert terrain posed significant navigational challenges, thereby increasing the benefit of GPS. Additionally, many targeting components of US weapons systems (of all services) interfaced with GPS for highly accurate initial, midcourse, and terminal guidance. Its popularity became so widespread that aircrews flying Vietnam-era systems, which used notoriously untrustworthy analog inertial navigation systems, bought personal, hand-held GPS receivers to augment their onboard systems.[8] Parents of some infantry personnel included GPS receivers in their children's "CARE" packages. At the outbreak of Desert Shield, the NAVSTAR system had not yet reached full operational capability, but it soon became integral to the Coalition effort. In the future, planners believe small, lightweight GPS receivers will become standard kit for every deployed US soldier.

Surveillance in the Gulf was accomplished by the US fleet of spy satellites whose name(s), configuration(s), and specific characteristic(s) is/are classified. Civil and commercial satellites, such as the French *Systeme Probatoire pour l'Observation de la Terre* (SPOT), used for earth observation, were pressed into service to provide additional surveillance for the Coalition. A widely publicized, key capability of US surveillance satellites is multispectral sensing. Satellites over the Gulf provided US commanders and decision makers with optical, radar, and infrared (IR) high-resolution images. Other capabilities included electronics intelligence (ELINT) gathering, though this capability was not as valuable once the war began since Iraqi command and control capability was rapidly degraded early in the conflict. High quality and rapid battle damage assessment (BDA) was another significant advance credited to space systems. Unit complaints about untimely BDA can be attributed to human errors in developing an inefficient and ineffective dissemination system vice technical inadequacies of space systems, though the overclassification of information borne of space capability also contributed. The highly accurate photo and radar images provided by space platforms allowed for intricate interpretation of damage caused by Coalition "smart bombs."

Surveillance assets were also used to assist in targeting Iraqi Scud missile launches. The Defense Support Program (DSP) fleet of spacecraft provided this capability. These satellites sat in their geostationary orbits, constantly looking for telltale Scud IR plumes. Once observed, the system relayed the location, time and trajectory to the United States Space Command (USSPACECOM) operations crews, who then evaluated and assessed the data before relaying, via the DSCS, to Patriot missile crews in Saudi Arabia, Israel, or Turkey, using newly established, refined, and exercised warning-alert communications paths. DSP—a cold war era space resource—saved lives.

Evolution and Relationship of Airpower and Space Power Technologies to New Roles

Simultaneous with the evolution of airpower's operational role was an evolution of air-related technology. Similarly, the evolution of space-related technology has accompanied the evolution of space power's operational role.

As airpower developed rapidly from a technological standpoint, so too did its military potential. Within the span of just a few years, the dominant role of airpower evolved from general support to directly offensive. Space power is now advancing technologically in a very rapid manner. There seems no reason to assume that space power cannot, technologically, mimic airpower's offensive evolution. The most powerful early doctrines developed for both airpower and space power emphasized the war-winning potential of strategic applications of force from these new combat mediums.[9] As space power doctrine evolves, the strategic usefulness of applying force from this medium must be acknowledged, but care must be taken not to fall into early airpower's doctrinal trap of promising too much, too soon.

As airpower advanced technologically after World War I, its primary function became strategic bombardment. Giulio Douhet, impressed by airpower's capabilities and potential that he witnessed in World War I, drafted an offensive airpower theory in *The Command of the Air.* Other great airpower thinkers followed, Gen William "Billy" Mitchell, Alexander de Seversky, Air Vice Marshal Hugh Trenchard, and Gen James "Jimmy" Doolittle, to name a few. All focused on the offensive capabilities of airpower. The lesson which flowed from these air-advocates was clear: an air force's sole concern should be to do the enemy the greatest possible amount of surface damage in the shortest possible time (in consonance with the theater campaign plan).[10]

By the end of World War I, technology allowed airpower to be used separate from surface forces to bomb well beyond the battle front in efforts to affect the enemy infrastructure and its ability to wage war. Government centers, industry, and transportation links were targetable by air though such targets were not at great risk due to the limits of the technology. This changed over time. An analogy can be drawn with the evolutionary advance of space power today. Technologically, space power clearly has access to the battlefield, yet it is limited in what it can do offensively. This, too, may change over time. And when it does, the political and military advantages of being able to rapidly and widely affect an enemy with minimal regard to friendly vulnerabilities will be great.

Organizational Development of Airpower and Space Power within the Military

Part and parcel to the operational and technological development of airpower was its organizational development. Many writers advance the idea that it was actually the wish to identify a need for independent air forces that spawned offensive airpower theories. Due to airpower's evolution during the interwar years, incremental organizational changes took place. In the United States, the changes led from the Air Service to the Air Corps to the Army Air Forces, and finally to an independent Air Force. The emergence of Air Force Space Command (AFSPACECOM) and USSPACECOM may be similar organizational steps towards the eventual creation of an independent space force.[11]

Douhet, Mitchell, and other airpower thinkers pushed for establishment of separate air forces due to their perceptions that airpower proved a decisive form of warfare. Strategic bombardment became the backbone of the air mission. This legacy is apparent in Air Force Manual 1-1, as it contrasts surface forces with "aerospace power," which "can be the decisive force in warfare."[12] It seems with all of the attention paid to strategic attack as the raison d'être of the USAF, the removal of this mission might signal a certain lack of legitimacy in the institution. Without a strategic attack capability, the Air Force would be nothing more than a support arm for surface forces, providing air superiority over the battle space, close air support, and resupply missions. The debate would hark back to the arguments of the early forties. Why have a separate air force if airpower is purely a support function for the surface forces?

This hypothetical situation is somewhat analogous to today's space operations.[13] As mentioned earlier, space forces exist today to support terrestrial forces. However, the realization of space force application and space control (akin to air superiority) capabilities, like airpower's development, would present space power with its own raison d'être, thereby making the establishment of a separate space force a possibility.

Summary of Air and Space Power Development

Space power development seems to be mimicking the development of airpower. Extrapolating the analogy, the importance of space power will rise accordingly. Airpower's rapid rise to dominance as a form of warfare was due to the unique advantages that vertical positioning, speed, and eventually range, gave to the war fighter. From early military uses as observation and reconnaissance platforms, both airpower and space power continue to evolve. Airpower gained its present status as a separate, and some would argue dominant, form of warfare by technologically developing its offensive capability. Similarly, space power could one day achieve such status, given technological innovation as well as political will. The similarities between airpower and space power developments seem to suggest that space power will evolve into a dominant military force in the future.

Physical Attributes of Space and a Comparison to Terrestrial Environments

National security policy makers, planners, programmers, and operators take geography into constant account, because it exerts strong influence on strategies, tactics, logistics, and force postures. Geography, however, excludes most of the Earth-Moon system, which comprises a vast environment loosely known as space.

—John M. Collins
Report to Congress

This section attempts to determine if there are systemic environmental deficiencies which limit space power's potential to realize more independence. It defines space, discusses the physical characteristics of the medium, and compares these to terrestrial environments. It begins with a discussion of space and such celestial phenomena as libration points (also known as Lagrangian points) and the gravity well. It concludes by analyzing differences between the space environment and the terrestrial environments in which other forms of military power operate.

Space Defined

A short discussion on what constitutes "space" provides a better understanding of this thesis (fig. 13). Many sources describe the medium referred to as "space." However, John Collins' work, *Military Space Forces: The Next Fifty Years*, discusses it in militarily significant terms. "The Earth-Moon System circumscribes four discrete regions: Earth and Atmosphere; Circumterrestrial Space; Moon and Environs; and Outer Envelope. Boundaries are blurred and some attributes overlap, but each nevertheless is individualistic. . . . Earth's atmosphere, gravity, and rotation strongly influence transit between that infrastructure and space. Most effects are adverse, but a few are advantageous."[14]

In space, there are areas in which objects theoretically will require little or no energy to maintain position, and from which energy can be used advantageously to affect near-earth space, as well as the earth itself. These are termed the libration points. Collins writes:

> The five so-called libration points are not points at all, but three-dimensional positions in space. Mathematical models and computer simulations indicate that free-floating objects within their respective spheres of influence tend to remain there, because the gravitational fields of the Earth and moon are in balance. Spacecraft could theoretically linger for long periods without expending significant fuel. L1 through L3, on a line with Earth and moon, are considered unstable. Objects at those locations, perturbed by the sun and other forces, will wander farther and farther away, if calculations are correct. L4 and L5, 60 degrees ahead of and behind the moon in its orbit, assertedly are stable. Objects at those locations probably resist drift more vigorously and, if it begins, remain in that general region.[15]

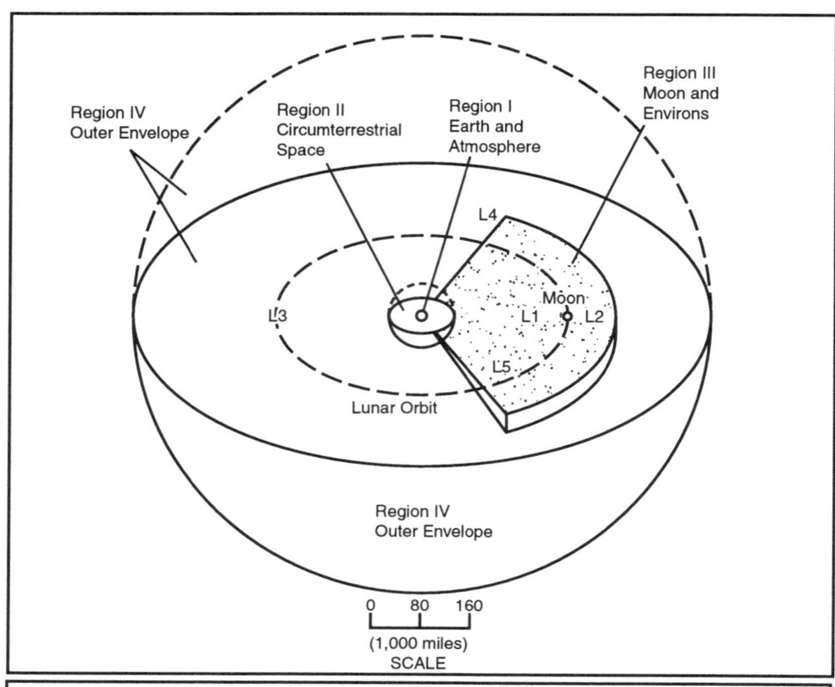

Distance in Miles			
From Earth		From Moon	
Region I	Surface to 60	L1	45,000
II	60 to 50,000	L2	42,000
III	50,000 to 360,000	L3	480,000
IV	240,000 to 480,000	L4	60° ahead of moon
Lunar Orbit	240,000	L5	60° behind moon

Note: Regions I, II, and IV are globe shaped. Region III is like a quarter slice of pie, with little depth in comparison. L1 through L5 are lunar libration points.

Source: John M. Collins, *Military Space Forces: The Next Fifty Years* (Washington, D.C., Pergamon-Brassey's 1989), 7.

Figure 13. Space Regions and Environs

As in air-to-air combat, of "God's G," or converting from a position of relative energy advantage due to high-potential-energy positioning (high to low), is also applicable to space operations (fig.14). Military space forces operating from "low" potential energy states in low or near earth orbit, are disadvantaged from those operating farther away—"at the top of the 'gravity well.'" They also experience less maneuvering room

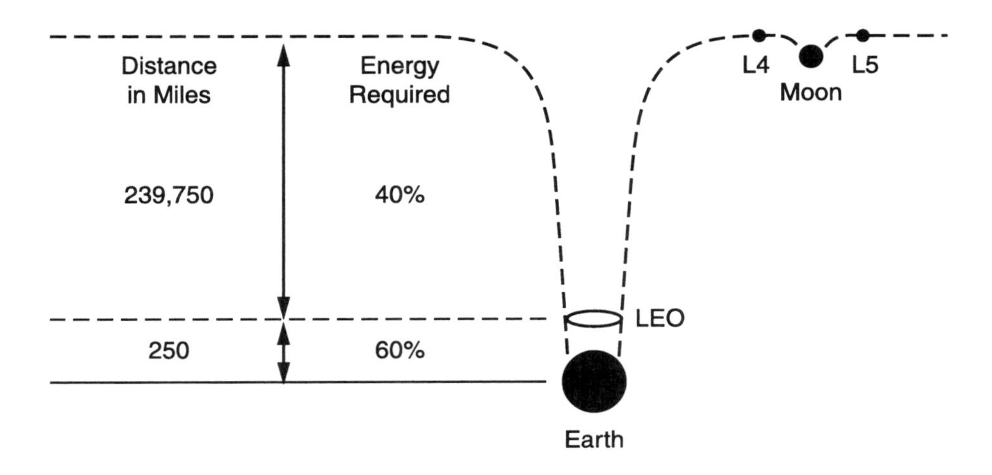

Source: John M. Collins, *Military Space Forces: The Next Fifty Years* (Washington, D.C., Pergamon-Brassey's 1989), 24.

Figure 14. Earth and Moon Gravity Wells

and reaction time. Whereas gravity hinders earth-to-space transit, it helps space-to-earth flight. "Put simply, it takes less energy to drop objects down a well than to cast them out."[16]

Although submariners operate in a seemingly unique environment, they retain standard terrestrial realities as direction and geo-position. Similarly, air forces operate in their own medium, but they too retain many similarities to their sister terrestrial forces, such as direction, geo-position, and constant physical effects of operating within the atmosphere. Obviously, space is a unique operating environment for military forces.

Medium Differences

A concise discussion of medium differences may be helpful in grasping the physical uniqueness of space. In his congressional study, *Military Space Forces: The Next Fifty Years,* John M. Collins provides a complete accounting of medium differences:

> Air, water, weather, climate and vegetation within the Earth-Moon System are exclusively indigenous to this planet. So are populations and industries at present. Land forms and natural resources are restricted to the Earth, moon and asteroids. Cosmic radiation, solar winds, micrometeorites and negligible or neutralized gravity are unique properties of free space. Near vacuum is present everywhere except Earth and vicinity.
>
> Space and oceans are superficially similar, but differences are more remarkable. Continents bound all seven seas, which are liquid and almost opaque. Topographic features conceal ocean bottoms. The Earth's curvature limits visibility to line-of-sight; natural light never illuminates deeply. Water temperature, pressure and salinity anomalies are common.
>
> Space has no north, east, south or west. Right ascension and declination, calculated in different terms than latitude and longitude, designate location and direction. A nonrotating celestial sphere of infinite radius, with its center at Earth's core is the reference frame. Declination, the astronomical analog of latitude, is the angular distance north or south of the celestial equator. Right ascension is the astronomical analog of longitude. The constellation Aries, against which spectators on Earth see the sun when it crosses Earth's equator in spring, defines the prime meridian. Angular positions in space are measured east from that celestial counterpart of Greenwich Observatory.
>
> Distances are meaningful mainly in terms of time. Merchant ships en route from our Pacific coast to the Persian Gulf, for example, take a month to travel 12,000 nautical miles. Apollo 11 made it to the moon—20 times farther—in slightly more than three days.[17]

In short, though it is relatively unique, space is a place—but so is the air, land, and sea.[18] Air and space share some similar advantages, specifically vantage point and speed of access to the surface. The difference between the two on those grounds is simply a matter of degree. Continual operations in both media require countering the force of gravity. In the air, this is generally done by buoyancy, such as lighter-than-air operations, or by lift, via lifting bodies and thrust—but not by speed, because of the frictional drag and heat of the atmosphere. Conversely, in space the effects of gravity are countered by speed and position, the required speed being determined by altitude above the earth's surface.

In the air, typically below 20 miles, the vantage point and speed of access are not as great as they are in space—above 100 miles. But in both media, the vantage point increases as one gains altitude. There is a large difference in the speed of access between air and space—for semiglobal distances, the access time through the air is a fraction of a day, but through space it can be fractions of an hour. As one goes higher in space, one trades speed of access for vantage point—until one reaches the maximum vantage point for specific operations.

The military significance should be obvious: vantage point and access allow observation, communication, navigation, and when developed, force application. The air provides an order-of-magnitude increase in line-of-sight coverage for any place on the earth's surface, while space provides yet another order-of-magnitude increase.

Speed of access is militarily important because it allows operations inside the response times of adversaries. To develop optimal speeds of access, one must operate in space to avoid air friction. RAND Corporation's Carl Builder claims, "If observing, communicating, and navigating [and if possible, applying force] is important to militaries, space is the dominant medium." Builder concludes:

> So space is an important military medium for the same reason that air is an important medium, except multiplied by another order-of-magnitude. Space provides for unprecedented vantage points and speeds of access. Those qualities are not essential to all military activities, and there are significant costs associated with operations in space, so the air, land and sea will remain important media for many operations. But wherever vantage point and speed of access are critical aspects of military operations—space will be the dominant medium.[19]

Summary of the Physical Attributes of Space

Physical attributes of the space medium exhibit nothing that systemically, or inherently, limits it as compared to other war-fighting media. Space is a medium somewhat unique from other war-fighting media. Certain aspects of space allow advantages, though other space aspects tend to be detrimental. Libration points allow for little or no energy expense for station keeping, while operating from atop the "gravity well" allows for high-potential-energy positioning. Differences between the media are noted, as are similarities.

The terrestrial medium most similar to space seems to be the air. Vantage point and speed of access are shared physical characteristics of these media. What, then, makes space special? It is the order-of-magnitude advantage gained over the air in vantage and access—when these qualities are militarily required.

Characteristics of Military Power and Space Forces

We should on all occasions avoid a general Action, or Put anything to Risque, unless compelled by a necessity, into which we ought never be drawn.

—George Washington, 1776

The essential ingredients that lead to an expanded role for space are coming together.

—Gen Thomas Moorman

Previous discussion demonstrated space power's current subordinate position in regard to the other military powers, but it also illustrated space power's potential, given current technologically feasible capabilities. Upon recognizing its subordinate position and superior potential, the next step examined physical differences and similarities between the military operating media. From that background, this section combines these concepts by comparing military power characteristics of terrestrial and space forces.

The general characteristics include strategic agility, ability to demonstrate commitment and credibility, economic considerations, military considerations, and political considerations. To conclude, the relationship between these characteristics and political flexibility is discussed. Ultimately, this comparison demonstrates whether space power is limited by any inherent, systemic inadequacies.

Military forces are often compared in various ways, but it's the comparison of *real politik* characteristics which generally carries the day. Unique equipment, operating mediums, and doctrinal differences are common parameters for comparison, but such analysis is often incomplete, as seams between military forces are typically unclear. For example, airpower connotes abilities of multiple military forces employing various types of equipment, for numerous doctrinal reasons. Herein, forces are compared based on perceived political applicability. At the end of the day, it is the political applicability of a force, not operational dissimilarities from another, which is most meaningful.

Airpower as Part of Terrestrial Military Power

Airpower's characteristics are discussed as part and parcel of terrestrial military power characteristics because airpower's medium is limited in many of the same ways as ground forces and naval forces. Though airpower has one great liberating characteristic from other terrestrial forces—elevation—due to atmospheric (drag, gravity, etc.) and geopolitical limitations (overflight restrictions, basing concerns, etc.), it is similarly limited.

Speed and elevation allow airpower to rapidly mass large quantities of power anywhere in the world, treaty limitations notwithstanding. It can attack strategic targets that surface forces cannot. However, it remains terrestrially limited, as compared to space power, by footprint size, geopolitical concerns, and persistence. Another observer has noted: "In addition, just like surface forces, political restrictions could determine where aircraft flew, when, and for what purpose." He points out that, 75 years later, airpower remains similarly limited. His paper discusses airpower advocacy pitfalls, as well as the evolutionary, or revolutionary, questions senior leaders dealt with in determining the viability of a separate aviation service. The important conclusion, however, is that separateness does not equal singularity. "Wars are fought in many ways with many weapons. Seldom is one service used to wage a campaign or war, although one service may be dominant in them. The nature of the enemy and the war, the objectives to be achieved, and the price to be paid by the people will determine what military instruments will be employed and in what proportion."[20]

Characteristics of Military Power

Regardless of the type of military power considered, they all share common characteristics which represent, in the author's view, diverse considerations capturing the essence of military power. They are diverse yet interrelated, which reflects the association between military power and political will. It is this relationship which, in the end, determines the usefulness of any form of military power in any given situation.

The characteristics of military power include strategic agility, ability to demonstrate commitment and credibility, economic considerations, military considerations, and political considerations. The application of these general characteristics changes as the status of forces changes from being home based to deploying, to engaging in combat.* All of these characteristics are essential to determine the political flexibility of applying the military element of power. Basic definitions of these characteristics follow.

Strategic Agility. *Strategic agility* refers to the ability to respond rapidly, over global distances, with appropriate capabilities to carry out operations in support of US international interests.[21] This concept takes on even greater import as US forces are restructured and decreased, while US global interests, and possible trouble spots, increase. Various "futures" studies have noted the probability of multiple conflicts in various stages of resolution, occurring in areas around the world vital to US national interests. Hence, the ability to respond rapidly anywhere in the world with appropriate force is a basic requirement for effective military response, and is therefore within the US national interest.

Commitment and Credibility. The terms *commitment* and *credibility* go hand-in-hand. *Commitment* refers to the state of being bound emotionally, or intellectually, to a course of action or ideal. The dictionary refers to it as a pledge to act, while *credibility* takes this concept another step by making

*Home basing does not imply US basing. Rather, it denotes a force located in its primary position, with all of its required logistics for permanent operations and sustainability. A fighter wing is home based if it is at its primary location, e.g., Lakenheath, United Kingdom, whereas a carrier battle group is home based if it is totally integrated and sustainable—theoretically, this could be "on station."

this commitment, or pledge to act, plausible. For instance, the perceived capability of US assets makes the actor the United States wishes to influence believe the United States will act on a notion of international interest.[22]

In the past, these terms have been closely identified with the concept of deterrence. Thomas Schelling, in his work *Arms and Influence*, talks at length about what he terms "The Art of Commitment." He frames his argument in terms of the cold war, and posits that an adversary must be communicated with effectively if one is to realize one's strategy. If a country has gone to great lengths to influence an adversary, but has not communicated its commitment or credibility to act, then it has failed—the adversary remains uninfluenced. Interestingly, in the cold war paradigm, Schelling suggests that to effectively communicate to an adversary its commitment and credibility, the country must physically, or morally, put itself into a situation from which its only rational response is to act. In his words, "Just saying so won't do it. What we have to do is to get ourselves into a position where we cannot fail to react as we said we would—where we just cannot help it—or where we would be obliged by some overwhelming cost of not reacting in the manner we had declared. Often we must maneuver into a position where we no longer have much choice left. Thus is the old business of burning bridges."[23] The paradigm of conventional terrestrial force commitment and credibility has always included the notion of putting forces at risk to make a point. This approach remains valid today.

However, this thesis suggests not an alternative solution, but a unique application of these concepts as applied to an adjunct force. Such a force could demonstrate commitment and credibility for less Machievellian reasons. If a force were "easy" to use—economically, militarily, and politically—it would be engendered with commitment and credibility. The adversary need not consider that US personnel and equipment are at risk to prove credibility and commitment; rather, these concepts would exist by US capability to apply force with little regard to risk of any kind. This virtual lack of risk, then, becomes the mechanism to convince adversarial leadership of US ability and willingness to act. This "third wave" concept is the antithesis of the industrial warfare paradigm of proving commitment and credibility through putting one's forces in harm's way.[24]

Economic Considerations. A discussion of the myriad of issues involved in the fiscal realities of military forces is beyond the scope of this thesis. However, a more narrow focus for this paper is akin to a USAF perspective: the USAF seems to believe the basic economic consideration for military forces is the ability to efficiently allocate resources required to deploy and employ capabilities.[25] Military forces are expensive and, generally, their size and capability demonstrate the vastness, or lack thereof, of a country's treasure and international stature. One need only refer to present-day media to discern the immense amount of fiscal resources involved in fielding a credible and able fighting force. As the United States downsizes its military and takes advantage of the "peace dividend," the susceptibility of US forces to physical loss or damage, or increasing expense involved in deployment and operations, weighs heavily into political decision making. When one is comparing forms of military power by economic considerations, many variables exist. Susceptibility of forces to loss or damage, research and development costs, acquisition costs, sunk costs, operational costs, and associated costs (manning, infrastructure, etc.) are all considerations.

However, when considering the economics of military force, one must realize forces are bought and exist for two basic purposes—as diplomatic tools and to provide national security. If national security is at risk, or serious diplomatic endeavors are in jeopardy, many of these cost issues may become insignificant. For example, if forces were used in the interest of a close ally, or for operations upon which monumental national economic priorities exist, economic arguments against using such force may be mute. This said, though, if the same effects could be rendered by an adjunct force with fewer risks, regardless of diplomatic or national security priorities, this would seem advantageous.

Military Considerations. The concept of military considerations is closely associated with economic as well as political considerations. The susceptibility of a force to degradation or destruction is the measure of its military vulnerability. As the USAF defines it, survivability is the key, that is, the ability to limit risks.[26] For a deployed force, this plays heavily into command planning functions.[27] Other considerations include training, replacements, loss rates, family considerations, media relations, unit cohesiveness, and coalition dynamics, to name a few.

Aside from these "negative" aspects, military forces are built and maintained with one mission in mind—war fighting.[28] As discussed previously, this mission relates to two objectives—diplomatic utility and national security. Sufficient numbers are planned for attrition, and advancing technology is offered to increase force effectiveness, though fiscal realities make such planning increasingly problematic. Quality and effectiveness are hallmarks of US military forces, though certain contingents cast counter dispersions. Many quarters prior to the Gulf War were doubting the effectiveness of high-cost US weapon systems. Such contingents were noticeably quiet after the war ended and US technological superiority was widely recognized. The US military has generally had quality training and equipment to meet most contingencies—but such assets cannot make up for fallacious policy.

Political Considerations. The effect of the above considerations rests firmly on the political fulcrum. As economic and military considerations ebb and flow, so too do a nation's political considerations. A nation's political fortunes are closely tied to its economic and military robustness. Hence in the end, the susceptibility of a nation's economic health and military power to degradation affect the nation's political viability. This interrelationship is one of the most critical and absorbing problems of statesmanship—it involves the security of the nation and, in large measure, determines the extent to which the individual may enjoy life, liberty, prosperity, and happiness.[29] Other valid political considerations include media relations, public relations, and world geopolitical dynamics (alliances, coalitions, neutral, gray, third party states, and enemy states).

Regarding the counter argument, a successful military operation generally results in great political benefits, thereby mitigating any negative considerations that may have existed. For example, President George Bush was inundated with cautious overtones from many political quarters prior to the beginning of hostilities in the Gulf. Many deemed the political considerations of such an operation too costly. However, after successfully engaging his forces, the same man was regaled from the same quarters, and more, for his astute statesmanship and political guts. The president's polls were the highest of any president in recent memory—some showing approval ratings as high as 90 percent. Politicians can ride the wave of popularity following successful military operations, or can be swept up in the despair of a nation which uses its forces less than effectively.

With these basic definitions, this work considers how each characteristic applies to forces on a continuum of deployment. Terrestrial force characteristics are discussed as the force moves from home base to deployment and engagement. The work then discusses how these same characteristics apply to two variations of space forces—current fielded forces and current technologically feasible forces. The first analysis is of a terrestrial force located at its home base (fig. 15).

Terrestrial Home-Based Forces and Strategic Agility. Generally, a terrestrial force enjoys its maximum responsive capability when based at home. Upon receiving mobilization orders, and generating to deployable status, the force is ready to be deployed anywhere in the world at varying rates. Obviously, ground forces must be either airlifted or sealifted to an area of concern, and this process takes time—given a relatively large force commitment, this period could be weeks to months. Wings or squadrons of fighter, attack, bomber, or reconnaissance aircraft can be in place with large amounts of firepower within a theater of operations in probably the shortest time for terrestrial forces—within hours to days. Naval assets, depending where they are located when the decision is made to deploy them, are available anywhere from within hours to days to weeks. Generally, to generate a large enough force to be decisive in any contingency, the deployment time will be days to weeks for naval forces.

Terrestrial Force Characteristics

Home-Based Forces:

Terrestrial Power Characteristics

↑ Strategic Agility

↓ Commitment/Credibility

- Economic Considerations
- Military Considerations
- Political Considerations

Home **Deployed** **Engaged** →

Note: The use of arrows in Figures 15 through 18 are meant to indicate the significance of the issues associated with them. While ↑ Strategic Agility means *increasing* strategic agility and ↑ Commitment means *increasing* commitment, ↑ Economic, Military, or Political Considerations means that the considerations discussed are of *heightened importance*. The length of the vectors denotes relative magnitude.

Figure 15. Continuum of Operations and Characteristics of Home-Based Terrestrial Forces

"At home" for a naval carrier task force could in reality be "deployed" if such a force is integrated and sustainable in its location. If this force is not located at exactly the proper spot, relocating such force could take days to weeks. Sustained operations require arrival of more combat and support forces. However, naval forces themselves have theoretically indefinite sustainment capability, given effective replenishment.

Terrestrial Home-Based Forces and Commitment/Credibility. On the other hand, while terrestrial forces remain at home, the commitment of the nation to respond to crises and its credibility with its alliances/coalitions, and its adversary, is at an ebb. The nation's potential adversary may remain unimpressed and affected by only whatever diplomatic rhetoric is exchanged. Even if the rhetoric includes outright or veiled threats of military response, the adversary may not perceive the intention of the communication.

A force still at home may demonstrate an unwillingness to react militarily. (Nuclear alert forces are the exception, although the threat of their use in most regional contingencies is regarded by adversaries as low.) The reasons for nonreaction could be many, most of which could actually be valid domestic, national, or international concerns. However, the perception by the concerned parties is the same—a fundamental lack of commitment and credibility to react with sufficient force to stem the tide of an international event.

Terrestrial Home-Based Forces and Economic Considerations. Economic considerations for home-based forces tend to be neutral. Generally, the cheapest basing mode for a terrestrial military force is at home. Units at the home base subsist on a system that is integrated, streamlined, and reasonably efficient. Units train effectively and efficiently based on many years of experience. Historically, accidents resulting in dead and injured personnel, as well as destroyed and damaged equipment, are lower while a unit is home based.[30]

The personnel and equipment are maintained most efficiently in this mode as well. The result is that national treasure remains relatively unaffected. The force is maintained within budgetary constraints mandated by government, and no "surge" funding is required to meet unanticipated needs. The force is most easily maintained combat-ready at home.

The exceptions to this concept include possible funding of US operations by another party, though few operations involving US forces have been sufficiently funded by another party to negate a loss to US budgets. At times, training experience can be better at deployed locations, for example Red Flag, though records indicate loss rates are generally higher. However, higher loss rates do not necessarily negate the added value of such training. Another exception is that home-based forces could be attacked either by another nation's forces or by terrorists. If attacks on home-based forces are broad, well targeted, and successful, the economic impact on the nation could be quite immense. However, if such an attack were to escalate into a war with the nation's survival at risk, the impact on the nation's economy by an attack on its home-based forces would be a relatively minuscule concern.

Terrestrial Home-Based Forces and Military Considerations

Military considerations for home-based forces tend to be neutral as well. Obviously, the susceptibility of a terrestrial force to damage or defeat is almost nonexistent when it remains at home and the country is not at war. On the other hand, the force is usually not operationally viable in such a position. (Exceptions include carrier task forces and certain airpower capabilities.)

Military assets, both personnel and equipment, are in their least susceptible state when based at home. The exception to this concept would be a nation that experiences attacks on its military forces within its own borders. In this instance, the fact that forces remain in an undeployed state actually makes them more liable to degradation from an attack, and the chance that the force will be degraded or destroyed is higher than if the forces were deployed. The military impact of such an occurrence could have drastic consequences. If the attack was but a prelude to a full-scale war with the nation's survival at stake, the impact of the susceptibility of the nation's forces to such an attack would be great indeed.

Terrestrial Home-Based Forces and Political Considerations. As economic and military considerations remain neutral for home-based terrestrial forces, so too do the political considerations. The susceptibility of the nation's political realm to domestic as well as international outcry and indignation is low—so too is the chance for great political windfall given a successful military operation. As a result, the political leadership of the nation remains relatively flexible in use of terrestrial armed force. For example, there is often the little public outcry as home-based military forces are mobilized to help with natural and man-made disasters.

Considering the exception, as noted in the above discussion, if the home-based force were to suffer degradation or destruction from an attack, political considerations could skyrocket. Again, if the nation's existence were at stake, this would be of little regard. However, if the attack was but a nuisance operation to demonstrate resolve or capability, the leadership of the nation could experience great disrepute. The dangerous consequence might be an inappropriate use of force in reaction to the humiliation, thereby escalating opportunities for drastic occurrences. This paper now turns its attention to terrestrial forces in a deployed state (fig. 16).

Terrestrial Deployed Forces and Strategic Agility. As terrestrial forces leave their home base and are committed to a theater of operations, their capability of redeploying to yet another theater decreases. There are many aspects to this dilemma. First, if a force is committed to a theater to carry out the lavishes of its political leadership, the wish to relocate it supposes a severe reason to do so. Doing so may result in unachieved objectives in the original theater. Second, as US forces are drawn down, the amount of force required to effect desired outcomes becomes critical. Theoretically, if just the right amount of force exists in a theater, moving any of it to another theater could result in either defeat or unwarranted losses in both theaters. Third, once a force is deployed and in place, it becomes physically difficult to relocate it to another area by virtue of logistics requirements.

Terrestrial Deployed Forces and Commitment/Credibility. As US forces are deployed into harm's way, commitment and credibility of US resolve increases. Friends and enemies alike realize the significance of US leadership deciding to jeopardize personnel, equipment, national treasure, and domestic and

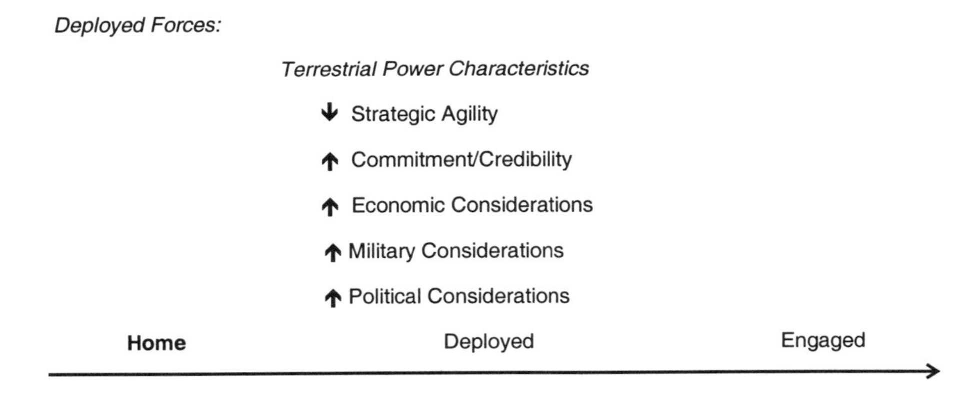

Figure 16. Continuum of Operations and Characteristics of Deployed Terrestrial Forces

international goodwill by sending forces into a theater of operations.

The deployment of forces heralds the increased capability of the United States to react. It is the perception of this increased capability that is the bulwark of demonstrating commitment and credibility. The perception of US resolve, both by friends and adversaries, is greatly increased by the deployment of forces.

Terrestrial Deployed Forces and Economic Considerations. As forces are deployed, economic costs of all kinds tend to increase. No longer is the force sustained by a system whose efficiency has been honed through years of use. Field conditions demand additional housing, food, water, transportation, medical care, maintenance, and other things as these functions must now be afforded apart from an established base. The force cannot necessarily be maintained within its legal budgetary constraints. The possibility of additional funding to ensure adequate operations increases dramatically.

Historically, the accident rate for military forces increases as they deploy into unfamiliar territory. Emotions run high, units tend to train more "realistically," crews are not operating in territory or under conditions they are used to. In such a state, equipment tends to break down and personnel tend to be injured or killed more than when forces are home based.

Maintenance of equipment becomes more expensive and problematic at deployed locations. Major and minor maintenance on equipment becomes more difficult. Depots exist half a world away, and industry technical representatives are not always immediately accessible. Equipment is fixed with what was brought with the force. If the proper tool or part is not available, the entire system remains unusable. Inefficient transportation practices are put into use to field important parts and ensure rapid repairs. Cost increases as the distance from the home base increases for terrestrial forces. These costs are varied and span the spectrum of military requirements.

It can be argued that economic considerations would be of no import if the act of deploying forces deterred war. This would be true given an electorate fully cognizant of essential facts and politicians willing to risk such an act. Given a situation where deployment of forces successfully deterred war, even though costs were high in lives, equipment, and treasure, it would be difficult to prove that it was simply the act of deploying such force which resulted in peace. In fact, such a situation seems much too simple. Even with deployed force, it remains the dynamic of diplomacy which results in peace. With US media coverage, the loss of treasure, lives, and equipment would be on the minds of Americans, even though peace may be at hand. However, if one considers military forces exist to fight, and fighting generally connotes deployment of US forces, and deployments generally connote economic losses, which are therefore accepted, this argument falls apart. In other words, a nation will tolerate the loss of national treasure if forces are properly deployed and engaged with successful results.

Terrestrial Deployed Forces and Military Considerations. The susceptibility of US terrestrial forces to attack increases as the force deploys to a theater of operations. Forces are outside of the protective boundary afforded by US airspace and sea buffers and are nearer the enemy forces' capability to strike. Terrorist or unconventional warfare forces can also attack with greater ease once US forces arrive in theater. As with economic considerations, as the distance increases away from the home base, military considerations of US forces increases. Both of these considerations affect US political considerations as well.

However, this susceptibility may be a must consideration if one considers that military forces exist to fight, and to fight US forces are generally deployed, and deployments tend to risk degradation and destruction. With such a notion, the fact that military forces may be degraded or destroyed has little or no significant impact on the decision to use them—for they exist to be used. The risk of their destruction or degradation is of little significance, the argument goes, because such problems have been planned for in force structure and effectiveness decisions. This argument is more convincing on paper than in reality.

Terrestrial Deployed Forces and Political Considerations. If deployed US forces demonstrate a heightened importance of economic and military considerations more than when they are at home, then so too do US political leaders' considerations increase in significance. If US national wealth becomes susceptible to increased diminution, and US personnel are put at increased physical risk, US political leaders begin to walk a fine line as they carry out national policy by using terrestrial armed forces.

It can be argued that this point is only true if the political objectives for deploying forces are not achieved. If they are achieved, it is argued, the deployment of forces actually allows for the potential of enormous political gains. The question that must be answered for this point to be valid is, "Are politicians willing to take this risk, based on recent deployment track records?" If US public opinion still had events like Vietnam, Desert One, Somalia, or Bosnia on its mind, this line of reasoning seems debatable. Events such as Haiti and Grenada could be looked at either way, while the initial public reaction to the result of Desert Storm could support this line of reasoning.

Terrestrial Engaged Forces and Strategic Agility. Engaged force characteristics tend to mimic deployed force characteristics, but with greater impact (fig. 17). The concept of decreased strategic agility that held for deployed terrestrial forces holds true for engaged terrestrial forces as well. However, the extent to which strategic agility is decreased is much larger. It becomes much more difficult to move terrestrial forces to another theater once they are engaged. The problems of just disengaging the forces are so immense—politically, militarily, and logistically—as to prohibit the thought of redeploying them elsewhere.

Terrestrial Engaged Forces and Commitment/Credibility. Commitment and credibility rapidly rise exponentially when forces are engaged in combat. The adversary and the alliance partners become strong believers in US resolve. The ultimate expression of resolve is to put national resources, such as lives and equipment, into direct contact with the enemy.

Terrestrial Engaged Forces and Economic Considerations. Unfortunately, as commitment and credibility rapidly increase as forces come into contact, so too does the susceptibility to losing vast amounts of national wealth. Present and future weapon systems are exceedingly expensive, and operators of

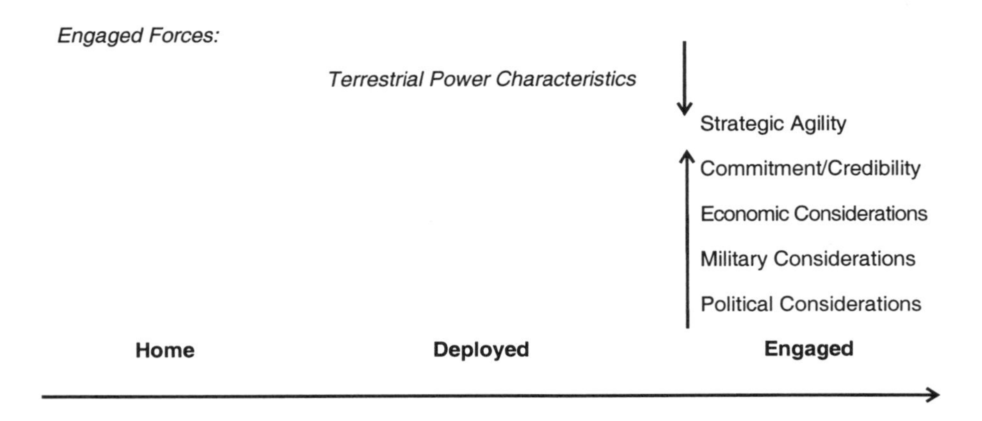

Figure 17. Continuum of Operations and Characteristics of Engaged Terrestrial Forces

these systems are ever more highly trained. The economic impact of their losses is great, and the chances that this impact will be felt rises rapidly as terrestrial forces meet the enemy. War is always costly, but it continues to be waged because costs of not waging it are perceived to be high. The other counter arguments remain the same as for deployed forces. The point is, however, that if these considerations and costs can be mitigated, they should be.

Terrestrial Engaged Forces and Military Considerations. As with economic considerations, the susceptibility of losing military forces is exponential once rounds begin to be exchanged in the terrestrial battle space. As weapon systems become ever more complex and expensive, their numbers dwindle. Therefore, each one becomes more militarily valuable. The loss of each system, and/or the highly trained operator, is that much more militarily significant. On the other hand, such systems and soldiers are meant to fight, therefore their loss is generally accepted as attrition, and properly planned for in force structure debates. Hopefully, this is true and the planning is accurate.

Terrestrial Engaged Forces and Political Considerations. As with the previous discussion, the political vulnerability associated with forces engaged in combat is quite high. The moral and economic impact of one's forces engaged in combat bring with it a high susceptibility of political leadership to ridicule and blame. As American mothers' sons and daughters are injured and killed, and media coverage of civilian casualties—even large amounts of enemy troops—is broadcast into American homes, domestic moral outrage could be quite high. Even without a protracted conflict to drain US coffers, domestic opinion of losing large amounts of high-cost weapon systems will begin to emerge. The sum of this discontent will fall squarely on the shoulders of US elected officials—most of whom are interested in continued employment and prestige.

Obviously, this applies only to perceived losers. President George Bush's 90 percent popularity rating in the wake of the Gulf War is evidence of the enormous political boon "winning" entails. The question remains, however, are most politicians willing to take such a risk? The United States was considered the out-and-out winner in Kuwait, but what were public perceptions regarding Korea, Vietnam, El Salvador, Desert One, Panama, Somalia, and Bosnia? It seems that wildly successful campaigns are far outnumbered by perceived questionable or outright poor results.

Comparative Space Power Characteristics

With all of this said about terrestrial force characteristics across the spectrum of deployment, how are space power's characteristics affected on this same continuum? Where does space power fit on this continuum? Certain space forces constantly exist somewhere between the deployed and engaged states. This assumes the asset is successfully launched and placed into proper orbit.[31] In such a location, the asset is deployed. From such a deployed location, the asset can engage.

A case could be made, however, that space assets are always engaged. This concept stems from the idea that space power, due to its position, is constantly present in the mind of allies and potential adversaries. Force can be immediately, or relatively rapidly, employed in concert with or against an actor—either virtually or really (much like terrestrial nuclear alert forces). This concept is termed *presence*.

The question for space power today is: What forces can be brought to bear? What capabilities are "present" in the mind of the actor? As previously discussed, space power is limited today in what "force" it can provide. This limitation, however, is not due to technological limitations as much as political considerations. Regardless, the limits are real. Such limitations, however, do not negate the applicability of this analysis. To the contrary, this analysis may demonstrate the advantages of fielding such capabilities. To demonstrate this, military force characteristics will be discussed as they apply to space power both in current capabilities and in technologically feasible, projected capabilities. The delta between these variations could demonstrate the advisability of pursuing current, technologically feasible capabilities from economic, military, and political standpoints.

Space Forces and Strategic Agility. Being forward deployed, and due to physical capabilities associated with the medium, space power entails a responsive capability. Satellites in geosynchronous orbit can maintain a constant presence over a specified area for years at a time. Even in low earth orbit, constellations of satellites could work in unison to effectively influence areas separated by vast overland distances. Satellites can also be moved. Though today this process is slow and expensive in fuel requirements, technological developments in solar energy collection, conversion, and storage offer new possibilities. The size of satellites is also being reduced, correspondingly reducing the energy requirements to make them maneuverable. Additionally, concepts of directed energy transfer and reusable launch vehicle resupply are on the drawing board or in development. The vantage of space allows a broad footprint that continues to grow and be more maneuverable. Even if moved only degrees per day, that footprint casts a large effective area.

Given today's standing space force agility capabilities described above, as well as emerging capabilities, no longer would the US military and policy makers be restrained from engaging elsewhere when their forces are deployed or engaged in one area of the globe. US space forces retain strategic agility to affect virtually any area, any time. This agility is conditional, however. If the resource is self-reliant, or not supporting a terrestrial system, it maintains its maximum agility. If the system supports a terrestrial system, for example cueing sensors, it then is limited by the terrestrial system's agility, unless it retains a capability to support multiple, geographically separated terrestrial systems.

Except for current systems already on orbit, much of this argument is mute if the United States does not pursue technologies now in development to ensure rapid, responsive, affordable space lift. Without a capability to place forces into proper position rapidly and affordably, be it orbital or suborbital, space power's strategic agility is limited to present on-orbit assets. Shuttle missions to repair satellites or place assets in orbit are prohibitively expensive and time consuming, thereby driving up economic, military, and political considerations.

Space Forces and Commitment/Credibility. The paradigm of putting forces at risk is replaced with the notion that exactly because forces are not at risk, the plausibility of use of such force increases, thereby increasing the notion that US policy makers will use it—commitment and credibility. On orbit, space forces can be thought of as always deployed, or in certain instances, even engaged. Adversaries no longer need question US commitment. No longer do costly deployments of personnel and equipment need be carried out in a show of force. With space power, the force exists on station, all the time, or at least can get on station very rapidly—depending on space force basing modes. Given space assets which are technologically feasible today, the commitment and credibility of such a force is inherent.

The degree of commitment and credibility, however, is limited by today's actual space forces. The lack of an autonomous force application capability to directly influence an actor mitigates the forces' ability to demonstrate commitment and credibility. (This capability need not even require kinetic or directed energy weapons; information warfare systems would be sufficient, perhaps even superior. In this information age, such systems could influence technologically advanced adversaries just as well, if not better, than more conventional weapons.) In other words, though certain space forces are considered by an actor (reconnaissance platforms, for example), their lack of ability to influence directly requires the old paradigm of putting terrestrial forces into harm's way to demonstrate US commitment and credibility.

Space Forces and Economic Considerations. Today's actual space forces, as well as those technologically feasible, do not require escalating support and operational costs upon deployment and engagement (as do terrestrial forces). The majority of space power costs are those incurred as sunk costs, that is, paid at and prior to acquisition. Maintenance costs and life cycle costs can be drastically reduced with a lift capability allowing either on-orbit replenishment, or rapid, contingency-oriented delivery capability, such as a transatmospheric vehicle or a reusable single-stage-to-orbit system.[32]

The cost of attack on home-based space power resources depends on the systems' basing modes. Orbital systems, obviously, are least affected by such an attack, unless the systems are singularly tied to, and reliant upon, a ground-based station. Reusable systems are most vulnerable to this situation and efforts are required to minimize this chance. Today's fielded technology presently requires widely dispersed ground stations, some well outside of the protective boundaries of the United States. This presents a significant security problem for today's US space assets. Considering the presently available technology, such bases could be maintained well inside US territory, allowing worldwide control via constellation interconnectivity, providing maximum security.[33]

Problems with space forces include their extremely high initial cost. The loss of one such asset would be felt much deeper than the loss of multiple terrestrial force resources. This fact calls for the early establishment of a space control capability to ward off such possibilities. As with airpower, superiority of the medium is crucial to the ability to operate from the medium. It also calls for rapid realization of cheap, responsive lift.

Space Forces and Military Considerations. Like home-based terrestrial forces, the susceptibility of space power assets to damage or defeat is relatively lower than deployed or engaged terrestrial forces. This implies a US space control capability to negate any space-borne, or surface-based capability against US space systems. If the United States would attempt to influence a space-capable actor, that actor's possession of an antisatellite weapon could negate the concept of US space power's lower susceptibility to degradation and destruction. Such an antisatellite system is technologically feasible today, though unclassified sources indicate that none have been fielded. Additionally, as previously discussed, minimizing foreign-based ground stations can negate security problems. Technologies presently exist to minimize this risk, allowing ground stations to be located within the contiguous United States, relaying data along constellations.

Tied directly to economic considerations, relatively fewer space assets can be deployed as compared to terrestrial assets. This is due to high cost, as well as the multiple-capability characteristic of space assets. Both of these issues could cause the loss of just a few space assets to adversely affect military operations—much more so than the loss of similar numbers of terrestrial assets. Again, such a fact harkens the need for early space control capability and rapid, reliable lift capability.

Space Forces and Political Considerations. Due to the relatively low economic and military considerations space power resources enjoy, as compared to deployed and engaged terrestrial forces, the political repercussions of utilizing such assets is correspondingly lower. Whereas policy makers have to contend with possible loss of troops' lives when considering deploying or engaging terrestrial forces, the use of space forces carries no such political liabilities when considering unmanned assets, and little chance of political liability when considering manned assets. As Maj Gen Roger G. DeKok, Air Force Space Command's director of Operations and Plans, remarked, "Satellites have no mothers."[34]

Given today's technologically feasible capabilities, as well as today's fielded systems, the inherent lack of political problems with using space power is instrumental in making it an extremely flexible political tool of national power. It can be used with little regard to political ramifications at home in many situations previously deemed as too politically sensitive. National policy decisions no longer need to be restricted by visions of the media displaying dead Marines being dragged through the streets of a foreign land. Space power can complement and support the other elements of power while not increasing chances of early US withdrawal due to loss of life or equipment. This fact makes it more plausible in the mind of the enemy that the US will act, and that equates to deterrence.

On the other hand, as with other considerations, due to economic and military implications of losing just a few space assets, political ramifications of such a loss are high, though remote. Space control remains a high economic, military, and political priority if deploying a space capability. Today, without this space control capability, political, economic, and military ramifications of losing space power advantages to a space-capable adversary could be high. Considering the high degree of space support terrestrial systems have come to rely upon, loss of such capabilities could be disastrous. Additionally, this degree of reliance is increasing.

There is a clear difference as the characteristics of space power apply to today's actual space force and to today's technologically feasible space force. That difference demonstrates the need to pursue space control capabilities that are technologically feasible. Today's actual space forces have a balance of advantages and disadvantages when compared characteristically to terrestrial forces. Conversely, the space force that is technologically feasible today and in the near future seems to demonstrate many characteristic advantages and few disadvantages when compared to terrestrial forces (fig. 18).

Risk management is the hallmark of military characteristics as they apply to space power. Space power, due to its inherent characteristics of nonprovocativeness of position, decreased economic, military, and political considerations, coupled with increased strategic agility and demonstration of commitment and credibility, can act to influence entities with decreased risk as compared to terrestrial forces. To a certain extent this is true given today's fielded space systems, though limitations previously discussed, such as a lack of force application capabilities, mitigate the realization of some benefits. The benefits that space power brings to the diplomatic forum seem to be

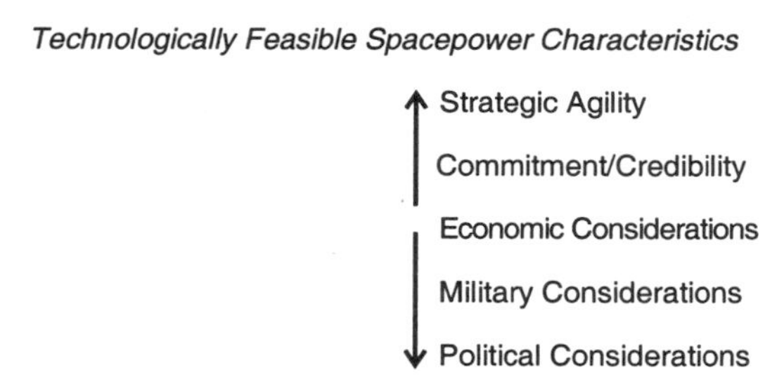

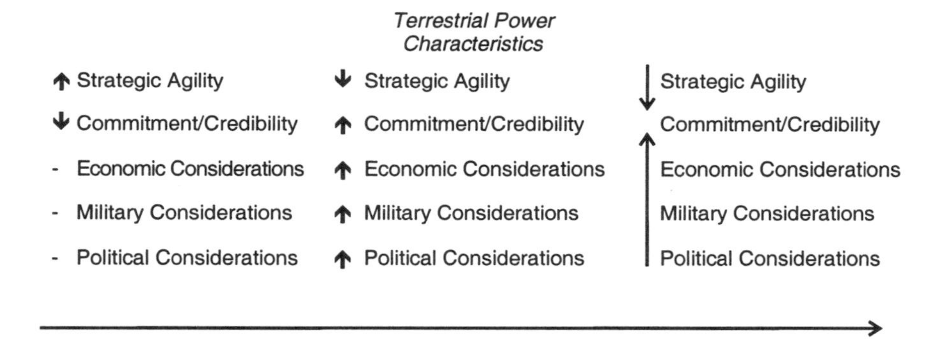

Figure 18. Continuum of Operations and Characteristics of Space Power versus Terrestrial Forces

great as compared to terrestrial forces by the characteristics of strategic agility, ability to demonstrate commitment and credibility, and economic, military, and political considerations.

Many of the considerations discussed with regard to terrestrial forces were moderated by views that military forces exist to deploy and fight. Such views hold that since this is so, the forces' economic, military, and political considerations need not regard their degradation or loss as a primary limiting factor in their use. While this notion has credence, it remains true, it seems, that the ability of an adjunct force to affect an actor in a similar way, but without risking such loss or degradation, has great advantages. Though such a force does not totally exist today, due to space power's lack of ability to apply force, it does not follow that such a capability should not be sought. In fact, from an analysis of military characteristics as they apply to terrestrial forces, today's actual space force, and to a space force technologically feasible today, it seems such a force would be beneficial and should be sought.

Summary: Political Flexibility

It is said that the military is the extended arm of diplomacy. Inherent in this concept is political flexibility to use military force. If the domestic or international political ramifications of using military force are too great, the likelihood that government will resort to it seems low. This notion is modified, however, when considering a fight for national survival or in operations of similarly great import. In other, more routine operations, if left without this sometimes last recourse, government could be left impotent to influence events, and may be forced to stand by and observe events which are counter to national interests.

This political inflexibility results from many factors. Considering recent crises the United States has been embroiled in, however, it seems economic, military, and political considerations are paramount. The problems, as well as advantages, inherent with terrestrial forces and these factors have been discussed. As the probability of actual military confrontation increases, so too do the significance of the considerations. Hence, the political flexibility to use the military instrument tends to decrease. However, due to some of the advantages of space power discussed above (tempered by today's limitations, and bolstered by today's technological capabilities), these considerations can be drastically reduced across the spectrum of military action. This decrease allows much more political flexibility, thereby allowing the government another realistic diplomatic tool with which to ensure US national interests are met.

Conclusion and Implications

Victory smiles upon those who anticipate the changes in the character of war, not upon those who wait to adapt themselves after the changes occur.

—Giulio Douhet

Now is the time to take longer strides—time for a great new American enterprise—time for this nation to take a clearly leading role in space achievement, which in many ways holds the key to our future on Earth.

—John F. Kennedy

The introduction of new military capabilities often involves a rethinking, a mental jump to entirely new concepts. It is not a question of doing something better, it is a question of doing something different. Not everyone can make this mental jump.

—Gen Merrill McPeak

This thesis demonstrates that space power is not inherently limited. Space power has the potential to be a fully functional arm of national military power. However, to realize such benefits, the United States must develop doctrine to realize advancing space technologies, thereby allowing full space access and exploitive ability across the mission spectrum. This last section deals with some implications requiring further thought by US leadership. A basic change of thinking is needed regarding future US space capabilities, both in how we think militarily about space and how we think fiscally about space. US space doctrine, currently reflecting space power's subordinate role, needs to be more forward-reaching. Billy Mitchell once remarked about short-sighted doctrine: "National safety would be endangered by an air force whose doctrine and techniques are tied solely to the equipment and processes of the moment. Present equipment is but a step in progress, and any air force which does not keep its doctrine ahead of its equipment, and its vision far into the future, can only delude the nation into a false sense of security."[35]

Space Power Is Not Inherently Limited

Space power has been evolving much like airpower. Airpower has evolved into a military power capable of the independent application of influential force, while retaining its advantages as an integrated part of the overall US force structure. Space power is a viable force today as part of this structure, in that it is used to support all of the terrestrial military arms—and this support is increasing. However, it seems presently stymied as purely a supporting force, with no aggressive trend toward realizing greater independent military potential. While early airpower doctrine generally seemed to consider ever greater capabilities than were presently available, today's space power doctrine seems to reflect its stymied position. Past and current technological projects, however, seemingly demonstrate greater available space power potential.

Space is a physically unique medium as compared to terrestrial mediums. Its physical attributes seem to demonstrate its ability to affect all other war-fighting mediums. Its encompassing nature ensures access to all other mediums, while its ability to exploit gravity, vice fight it, gives it a natural energy advantage over other mediums. Its lack of atmosphere—while limited in certain respects due to heat, radiation, cold, and so forth—requires less energy to be spent for maintaining operational positioning. Airpower's advantages over the other terrestrial mediums include vantage and speed of access. Space power realizes these over the other mediums, and over the air as well, by orders-of-magnitude. Space power's physical attributes, as they compare to other war-fighting mediums, belie nothing that systemically or inherently limits its ability to be a military force able to fully function across the mission spectrum.

A comparison of how military characteristics apply to today's terrestrial forces, today's fielded space forces, and today's technologically feasible space forces illustrates a relatively large difference in limitations and advantages realized by one form of space force as opposed to another. Today's fielded space forces demonstrate certain advantages when compared to terrestrial forces. On the other hand, a comparison of terrestrial forces to today's technologically feasible space forces illustrates an even greater number of advantages. The delta between these two comparisons seems to demonstrate that space power is not inherently limited when compared, by military force characteristics, with terrestrial forces. In fact, it seems space power can actually be more politically useful in most situations—though it is acknowledged that certain missions will always require the application of other military forces. In other words, though space power has the potential to be a leading independent, as well as integrated, element of military power, terrestrial forces will continue to retain their own unique advantages and applications.

Space power's current, relatively subordinate position as a military power is not due to inherent limitations. Why has its potential not been realized? If, in many ways the medium is physically more capable than terrestrial war-fighting mediums, and space power technology exists allowing it to be in many ways more useful than terrestrial forces, then there must be something "artificial"—not systemic or inherent—which is limiting space power development.

Policy may be the limiting factor. Military forces exist at the direction of policy. Policy is generated within the services, the Department of Defense (DOD), and on Capitol Hill. The ramifications of this reasoning go well beyond the pretenses of this paper. In fact, such a line of reasoning seems worthy of its own study. Suffice to say, however, that if space power allows military and political flexibility as described here, it seems worthy of a supportive policy which would be to the long-term advantage of the United States.

Given supportive policy, a force structure should be created that allows both maximum political flexibility and maximum military flexibility—a fully mission-capable space force, coupled with an integrated, well-proportioned, terrestrial force. With such a force, the possibility may exist for long-term fiscal savings through decreased terrestrial force infrastructure, and long-term manpower and equipment sustainment cost savings.

A major policy change such as this seems a long-term solution, if it is even probable. However, three major things can be done now to start US space power down this road. The first two concepts could turn the tide of thinking about space power as a purely supportive force; the latter would allow a more economical transition to a fully functional space force. The first is a required change of thinking within the military about space power: It can be used in its conventional, supportive sense, as well as in more unconventional, independent ways. Some information warfare missions seem ripe for such applications. Part and parcel to this first idea, space should be considered its own area of responsibility. Unity of command is essential for the proper planning and conduct of operations within the medium. The third concept is that fiscal realities require a closer military—civilian space industry reliance.

Required Change of Thinking

When research originally began on this thesis, the author believed military space personnel were inexorably committed to pursuing space capabilities, even those far into the future, which were merely supportive in nature. Perception was that USSPACECOM, as well as each of the services' space commands, were vectoring efforts toward space capabilities that would only support terrestrial military operations. It seemed possible that such an approach could result in a loss of future military capability, national technical ability and prestige, and possibly, national security itself.

After conducting research at USSPACECOM, AFSPACECOM, the Space Warfare Center, Sandia National Laboratory, and Phillips Laboratory, it became evident that developments to support the type of space infrastructure requisite to realize space power's advantages were possible. In fact, such enabling technology as a rapid, responsive, economical, and reusable space lift capability may not be far off.[36] However, throughout the course of research for this thesis, there were those, some with vast amounts of military space expertise, who claimed space power would never attain the requisite capabilities to fully exploit the medium. Some within service space commands, especially outside of the Air Force, seemed intransigent on this position. Varied reasons were given for these views, but most included political, as well as technological and fiscal, concerns.

One need only refer to the historical wrangling airpower experienced in its relative infancy to discern the same arguments. There were those in World War I who continued to disregard airpower's capability as they put gas masks on their ever trusty cavalry horses to ride them into battle. In the thirties, as the strategic ability of airpower became more widely accepted, there was much political discourse about limiting airpower's capabilities for its perceived inherent political and military instability. Fortunately, there existed professional military airmen whose vision outreached those of the naysayers. Though some suffered humiliating career consequences, they aptly demonstrated the effective and efficient ability of airpower to project presence relatively rapidly as compared to other forces of the day.[37]

It seems most of the professional military space cadre realizes the intrinsic value of operating in and from space, just as Mitchell and his ilk realized similar advantages in their day of operating from the air. Personnel within the United States and AFSPACECOM (including the Space Warfare Center), as well as the other forces' space commands, seem to be moving with the momentum of forward thinking. However, some barriers exist based on the old space-as-support-only paradigm. Though we still face political and parochial barriers to realize the military and political advantages of fully integrated space power, at least we are exploring the science, technology, and operational concepts necessary to accomplish it. The "progressives" in the "system" seem to be overtaking the sedimentation of the "status quos."

Due to fiscal realities of today, and tomorrow, it seems the technological breakthroughs are being, and probably will be, achieved mostly in the private sector. Space is to commercial enterprise today, and more so tomorrow, what the airlines, both cargo and passenger, were to yesterday and today. However, space offers so much more in terms of communications, weather, transportation, and other areas that commercial concerns are rapidly outpacing military research in the field.

The United StatesS must realize the advantages military space power offers. Hopefully, this thesis at least touches some salient concepts that demonstrate space power's advantages. With these advantages realized, the United StatesS can fully integrate its technological and operational biases with space power as a dominant factor. Professional military officers have a duty to articulate any particular concept that displays increased military, and therefore political, advantage to their civilian superiors. It is then the politicians' responsibility to ensure US forces are structured in an optimum manner. Such structuring intimates the need for unity of command regarding space.

Space as an Area of Responsibility

High strategic agility for space forces assumes a high degree of command and control. As air forces are centrally controlled, for matters of understanding unique strategic utility and capability, and decentrally executed, for matters of tactical expertise at the unit level, so too should space forces be controlled. The pervasive capability of limited numbers of space resources in high demand sets up the same logical structure for centralized command and control of space. Subjugating high-valued, far-reaching, but limited forces to one commander responsible for a certain theater of operations has been tried and proven, in most instances, to be an inefficient means of command and control.

Based on arguments presented in this thesis, a case can be made that space is its own area of responsibility. In fact, the Russian military considers space a distinct *teatr voyennykh deystviy (TVD)*, or theater of operations.[38] If this is accepted, unity of command demands the appointment of a single commander to this area. Requirements of such command generally include expertise, a fully functional and expert staff, and control of proper equipment and infrastructure. Only one such commander fully fits this requirement—commander in chief, United States Space Command.

As with airpower employment, employing space power requires special knowledge and conceptual internalization. The far-reaching, sometimes global, aspects of airpower employment have demonstrated a requirement for leaders, staffs, and operators trained to think in such terms. Similarly, assets with space power's worldwide capabilities need to be controlled by leaders, staffs, and operators trained to think with such vision.

With unity of command regarding all facets of space, coordination of requirements becomes easier. These requirements span the spectrum from operational to developmental. Developmentally, as well as operationally, as the use of space becomes more commonplace, commercial enterprise will have to be coordinated with military requirements.

Fiscal Realities and Military-Civilian Space Reliance

The future dictates a close relationship between military space requirements and civil space resources, including both operational and research and development realities.[39] Many military space functions closely parallel civil functions. Where these are evident, they should be exploited to save costs to both sectors. Certain functions will continue to be the sole purview of military space. Joint military-civilian space functions include weather, navigation, communications, earth resources, lift, orbit transfer, and tracking and control systems. Integrating many aspects of these systems to serve both military and civil customers could realize massive savings in fiscal requirements to both sectors. Near-term examples of probable and possible joint projects follow.

Space weather capabilities should become more economical as Defense Meteorological Satellite Program (DMSP) and the National Oceanic and Atmospheric Administration (NOAA) combines. The turnover of DMSP responsibilities to NOAA will decrease military investment in weather reconnaissance. Additional savings can be realized by replacing the purchase of next-generation military weather satellites with purchasing such data from commercial sources. Commercial market competition could allow purchase of what you need only when you need it. Care must be taken to ensure on-demand military capability.

Space navigation systems can be streamlined as well. GPS could be assigned to the Department of Commerce or Transportation, since demand for such data is well beyond the purview of strictly DOD functions. Alternatively, current GPS systems could be sold to corporations on a cash plus percentage basis, thereby raising cash for additional space resources or developments. Military users could purchase required services as needed. Military priority and accuracy would need to be protected. Additionally, large constellations such as Intelsat and Iridium could repeat navigation signals for redundant world wide coverage.

Space communications systems seem to be proliferating rapidly. Microsoft Corporation's Bill Gates plans to exponentially expand such capability with his 840 Teledesic satellite constellation. Teledesic's goal is to bring the information superhighway in all its glory to even the most remote reaches of the globe by the end of the century.[40] AFSPACECOM sources expect realization of this within a decade. Commercial enterprises will offer complete, competitive, fast response global coverage, thereby decreasing DOD demands to build and field such systems. Additionally, research for such capabilities is being increasingly funded by the commercial sector due to potential profit.

Lift and orbit transfer may be solved commercially, driven by commercial needs to access space. If such a robust system develops, there would be no need to maintain the military's satellite-booster-operator system. Such a commercial system could make launch-on-demand more realistic due to launch quantity and competitive price forces.

Tracking and control of commercial satellites could be done commercially, with intercorporation commonality and cost sharing decreasing commercial risk. DOD and NASA could follow corporate footsteps for military satellites, with USSPACECOM controlling all military assets. Alternatively, both corporate and military satellites could be controlled by an integrated national tracking and control system, thereby sharing costs among all users.

The bottom line of this approach to joint military-civil space exploitation is that the huge commercial market would likely dwarf the military needs in space, thereby driving down DOD space costs.[41] However, the military would need to maintain certain realistic standards across the marketplace to ensure its ability to use the systems.[42]

Care must be taken to ensure a capability to closely control these functions in the interest of national security. There are two aspects of this concern. First, the military must have unobstructed and complete access capability in the event of a national emergency, much like the current Civil Reserve Air Fleet concept. For example, contracts with civil communication satellite companies to enable daily dual use, and emergency complete use, of the companies' orbiting resources would be required. Second, the United States can increase its security by increasing foreign customer dependence on US-provided systems. For example, the United States could provide GPS data on a day-to-day basis at a price that would monopolize the world wide satellite navigation market, thereby ensuring control of access to or denial of such data in the event of a national emergency.

The space functions that will continue to be the sole purview of the US military include certain surveillance and reconnaissance capabilities, missile warning and defense, most secure communications capabilities, resource protection, command and control warfare, attack, and space system negation capabilities. Certain near-term requirements for a robust space force follow.

Regarding surveillance and reconnaissance, ELINT and imagery intelligence (IMINT) tactical satellites (TACSATS) for earth observation is needed, with improved responsiveness and previous systems provided by the National Reconnaissance Office (NRO). Real-time data fusion of multiple sensor inputs is presently being worked. Accurate geolocation of threats in time and space is needed for prompt preemptive military action.

Space surveillance requirements include providing space traffic control to allow knowledge and control of all space resources, including civil. Resources can be saved by allowing commercial, university, and technical center feeds into a military space traffic control data base to decrease overall collection requirements. Missile warning and defense require surveillance, tip-off, and queuing functions to remain within the military domain for purposes of speed, accuracy, and preemptive capabilities.

Regarding most secure communications capabilities, commercial sources will have corporate secure capabilities. This seems acceptable. In fact, more “routine” military secure transmission requirements could be met more cheaply this way. However, the military must retain a most secure capability for NCA and CINCs’ communications, highest priority national security communications, and data links for lethal national assets.

Resource protection remains a military consideration. Hardening of sensors, receivers, and transmitters is required to maintain the information edge on future threats (close hold data) and to realize the extent of proliferation of high threats (RF, HPM, lasers).

Another singular military requirement is space maneuverability for coverage, evasion, mission responsiveness, and flexibility. Such a capability may be on board a satellite, or may use Site Transition Team (STT) or transatmospheric vehicles (TAV) technology. However, it is not currently relevant to commercial users, so they will not fund such research and development.

Attack and space system negation issues include kinetic energy (KE) and directed energy (DE) force application capabilities, as well as advanced weapons for permanent or temporary, lethal or nonlethal effects. In the age of information warfare, such capability could give the United States a selective attack option on enemy or third-party information suppliers. with such a capability, space power could realize its maximum political and military flexibility.

Space force application capabilities could include KE as well as DE kill or degradation mechanisms. Such projects as the Tactical Reentry Impacting Munition program, Impact Technology program, Discriminating Attack Capability programs, Defense Suppression Vehicle, and Global Prompt Response Capability programs all have demonstrated, or discussed, KE kill technologies. The Sandia Winged Energetic Reentry Vehicle Experimental program and the Hypersonic Glide Vehicle program, both have illustrated high-explosives kill technologies. DE kill technologies have been discussed in the Beam Experiments Aboard Rocket program.

The feasibility of all of these technologies, and more, was demonstrated by Phillips Lab's 1991 Force Applications Study. The study concluded such technologies could be used to satisfy USAF operational requirements to

- reach out and touch anybody, anywhere, anytime;
- operate in the fiscal and geostrategic environment of the 1990s and beyond;
- complement traditional airpower by providing a number of very accurate, very long range, and very responsive weapons; and

- actively support Global Reach-Global Power by lessening reliance on forward deployment and foreign basing, as well as supporting aerospace power objectives of flexibility, range, responsiveness, and lethality.[43]

Due to the ever-increasing importance of information technology, other future force application capabilities should include capabilities to exploit, disrupt, or destroy adversaries' information systems. US systems should be able to control adversaries' knowledge-support computer infrastructures to effectively circumvent enemy leadership's decision processes. Such systems should also be able to exploit and affect enemy industry, electricity, transportation, and computer support networks. These types of capabilities would allow the disruption of such systems without the destruction and inherent risks of strategic air attacks.

Summary

Space is not systemically, or inherently, limited. Its physical attributes and application to military characteristics belie no reason for its present relative position vis-à-vis terrestrial forces. In fact, the advantages of space offer great military and political flexibility. It presently exists subordinate to terrestrial powers primarily due to purely "artificial" reasons. Policy seems to be at the root of those reasons. Though a full study of this notion is beyond the scope of this paper, certain things can be done now to enhance space power's chances of one day realizing full operational capability across the military mission spectrum.

Space, as a medium to be exploited, is still waiting for a user to get its act together, to determine how and why it can be exploited (a theory), and how to organize, train, and equip itself to do so (a doctrine). It must improve on what is already good in its space capability, and fix what is broken. To do this requires a new way of thinking about space and its role in the present, as well as future, world order. Using a comprehensive space-power theory, the United States can organize, train, and equip itself better to exploit space. "Better" implies more efficiently, faster and cheaper, via streamlined requirements and joint military-civil capabilities. The product of this change will be a national ability to defend US worldwide interests rapidly, with decisive force, and at decreased costs in lives, treasure, and natural resources.

Notes

1. Lt Col David Lupton, USAF, Retired, *On Space Warfare—A Spacepower Doctrine* (Maxwell AFB, Ala.: Air University Press, 1988), 7. In his well-written and informative book, Lupton is more thorough than any other author in dealing with the question of space power's usefulness. He articulates four distinct space doctrines, each one based on clearly defined belief structures and historical perspectives. Many of his views are shared by this author and are included in this thesis.

2. Peter L. Hays, "Struggling Towards Space Doctrine: US Military Space Plans, Programs and Perspectives During the Cold War" (PhD diss., Fletcher School of Law and Diplomacy, 1994). Hays compares the development of airpower to space power in an attempt to determine three critical steps in airpower's development that might explain space power's future development.

3. Richard Ernest Dupuy and Trevor Dupuy, *The Encyclopedia of Military History,* 2d rev. ed. (New York: HERO Books Partnership, 1986), 934.

4. Sir Peter Anson and Capt Dennis Cummings, RAF, "The First Space War: The Contributions of Satellites to the Gulf War," *RUSl Journal* 136, no. 4 (Winter 1991): 45; and Lee B. Kennett, *The First Air War* (New York: Free Press, 1991), title page.

5. Kennett, 31.

6. Ibid.

7. Joint Doctrine, Tactics, Techniques and Procedures (JDTTP) 3-14, *Space Operations*, 15 April 1992, I-15.

8. Author's personal experience flying F-111Es in Desert Shield and Desert Storm.

9. Hays, 27.

10. Ibid., 59.

11. Ibid., 27.

12. Air Force Manual 1-1, *Basic Aerospace Doctrine of the United States Air Force,* 1-3.

13. Dana Johnson, "The Evolution of US Military Space Doctrine" (PhD diss., University of Southern California, 1987). Johnson presents an analysis of space power's developmental similarities and differences to the USAF and the Navy. She discusses the impact of the varying requirements placed on the space force by all of the services and concludes that space is a truly joint arena, and should be managed to support all services to ensure minimum duplication of effort. She points out that space leaders could learn from a study of airpower and sea power developments.

14. John M. Collins, *Military Space Forces: The Next Fifty Years* (Washington, D.C.: Pergamon-Brassey's, 1989), 6–8.

15. Ibid., 21–22.

16. Ibid., 23.

17. Ibid., 6.

18. Much of this discussion comes from ideas generated by Carl Builder of the RAND Corporation in notes to Col Richard Szafranski, Air War College, national security chair.

19. Ibid.

20. Phillip Meilinger, Col, USAF, "Ten Propositions Regarding Airpower" (Maxwell AFB, Ala.: School of Advanced Airpower Studies, August 1994), 3.

21. Department of the Air Force, *Global Presence 1995* (Washington, D.C.: Government Printing Office, 1995), 13.

22. Thomas Schelling, *Arms and Influence* (London: Yale University Press, 1966), 36.

23. Ibid., chap. 2.

24. See Alvin and Heidi Toffler, *War and Anti-War* (Boston: Little, Brown and Co., 1993) for a thought-provoking treatise on the evolution of warfare and technology.

25. *Global Presence 1995,* 11.

26. Ibid.

27. As was the case when Checkmate initially briefed Gen Charles A. Horner on the strategic air campaign against Iraq in early August 1990. When the general determined they had no defensive plan, he reacted with dismay.

28. One might argue the strategic difference between war fighting and peacekeeping, but the infantryman taking fire or the airman reacting to antiaircraft fires does not recognize this difference.

29. Edward M. Earle, "Adam Smith, Alexander Hamilton, Freidrich List: The Economic Foundations of Military Power," in *Makers of Modern Strategy* (Princeton, N. J.: Princeton University Press, 1986), 217.

30. This is most evident in the famous "Sunday Briefing" given the first day of all Red Flag deployments. This is a mandatory pre-exercise flight safety briefing and offers statistics showing a relatively higher accident rate. The point of the briefing is not to become a statistic. The USAF Flight Safety School also teaches that accident rates at deployed locations tend to be higher, thereby requiring increased command and safety vigilance.

31. See a later section for a discussion of lift developments or refer to fig. 13 for associated system developments.

32. Proof-of-concept of such systems is occurring now. See Durnheim, "DCX Proving Initial Operating Concepts."

33. Iridium will have a constellation interconnectivity capability.

34. Maj Gen Roger G. DeKok, interview with the author, Peterson AFB, Colo., 13 March 1995.

35. *The War Reports of Marshall, Arnold and King* (New York: Lippincott, 1947), 455.

36. Work on lift and delivery vehicles such as Blackhorse, DCX, and TAVs is rapidly progressing. Some estimates put such capabilities, developed primarily by private contractors, only a decade or so away. Transatmospheric vehicles would allow rapid crisis reaction (approximately 30 minutes to deliver "influence," that is, ordnance, surveillance, supplies, communication, etc. anywhere in the world), flexible targeting (via "atmospheric skipping" the vehicle reenters the atmosphere to reorient its vector to deploy anywhere over the planet), long-duration presence capability (air refuelablility is a hallmark of TAVs), resupply, rearm, refuel (TAVs will return to earth, or maybe a space station in the distant future, to refresh), low vulnerability (possibly a zero vulnerability capability given an unmanned vehicle), and relatively high economic value (projected costs for pound-into-orbit run in the hundreds of dollars, versus thousands of dollars realized by today's inefficient launch means).

37. Brig Gen William "Billy" Mitchell was court-martialed for his outspoken support of a fully integrated airpower architecture. In the foreword of Mitchell's book, *Winged Defense* (New York: Putnam, 1925), the publisher writes, "In June 1925, Mitchell was returned to his permanent rank of Colonel and was sent to Texas on account of his outspoken criticism of our military policy in general and our aeronautical policy in particular. Mitchell has always been a pioneer, and in aeronautics a good deal of a prophet. Prophets, one recalls, aren't always highly regarded at home. At all events, Mitchell, for his outspoken criticisms of things as he sees them in the army and navy, has been pretty well belabored by his official opponents, of whom there are many, however widespread the approval given him and his views by the country at large."

38. Jacob Kipp et al., *Soviet Views on Military Operations in Space* (College Station, Tex.: Center for Strategic Technology, 1986), 223–49. Kipp analyzes Soviet military writings to demonstrate they viewed space as an "independent theater, pursuing independent missions under the direction of a TVD headquarters, the direct representative of the *Stavka*, or Supreme High Command. He suggests that "once an independent space theater becomes feasible, space should then become the main *TVD*."

39. Much of this discussion is gleaned from a presentation the author received from Lt Col Mike Kauthold, SWC/XR, titled, "The Reinvention of Space."

40. S. Faber, "Global Ambitions," *Discover*, January 1995, 3.

41. For a complete discussion of "dual-use," "spin-off," "spin-on," and economic technology interface see such works as John A. Alic et al., *Beyond Spin-Off: Military and Commercial Technologies in a Changing World* (Boston: Harvard Business School Press, 1992); Anna Slomovic, *An Analysis of Military and Commercial Microelectronics: Has DOD's R&D Funding Had the Desired Effect?* (Santa Monica, Calif.: RAND, 1991); and Victor Utgoff, *The American Military in the Twenty-First Century* (New York: St. Martin's Press). After a study of such works, the author believes that "dual-use" technologies, or those that can be used by both military and commercial sectors, but

funded by DOD, and "spin-off" technologies, or those that are developed by DOD for DOD use, but which have certain civil application, do not hold as much fiscal promise, and do not reflect the reality of today's commercial technologic revolution, as "spin-on" technologies do. "Spin-on" technologies refer to those which are generally developed and funded by commercial concerns, but which can be adapted for military use. Overall, it is the free market competition that will generate both quality products and affordability applicable to both the military and civil sectors. As Slomovic writes, "If the costs of weapon systems are to be contained, the electronics must be produced by firms which have incentives and opportunities to reduce costs. As this study demonstrates, in the majority of cases the DOD is not getting more advanced components for the higher prices it pays." DOD R&D funds must be spent on special military requirements, i.e., those which have no application in the civil market. However, certain criteria such as reliability, temperature tolerance, and radiation tolerance, once thought to be within the military's unique interest, are now being designed into civil components. The military must take advantage of these increasing capabilities by applying realistic test criteria and requirements.

42. Alic notes that during heyday of the spin-off paradigm, that is, when commercial requirements were being met with defense research and development dollars, "military requirements distorted priorities toward (overly) complex, high-performance objectives with limited commercial applicability." As the spin-off paradigm is left for a more realistic, contemporary, market-place driven, commercial-military interface, the problem should be partially "self-correcting," that is, defense will decline as a fraction of national technical effort. However, care must be taken by the corporations, so comfortable with the spin-off paradigm of the past, to not adopt civilianized versions of "defense technology paradigms." International competitiveness, as well as national economic and military superiority, would suffer." American business, accustomed to letting DOD carry much of the burden, has been slow in responding to aggressive technological investments by Japanese firms, even as the latter outdistanced them first in process and then in product engineering. The cost to Americans of carrying around the wrong mental image of how the technological system works will be paid in terms of lost markets, overpriced weapons, and wasted resources."

43. "Force Applications Study," final report briefing, unclassified (Kirtland AFB, N.Mex.: Phillips Laboratory, 13 June 1991).

Bibliography

Published Sources

Advanced Technology Warfare. New York: Harmony Books, 1985.

"Air Force Space Doctrine." *Defense Science 2002+.* February 1984, 43–44.

Aldridge, Edward. "Aerial Advantage—Myths About Space Militarization." *Officer,* November 1985, 16–19.

Alic, John, Lewis Branscomb, Brooks Harvey, Ashton Carter, and Gerald Epstein, *Beyond Spinoff: Military and Commercial Technologies in a Changing World.* Boston: Harvard Business School Press, 1992.

American Council on Education. *Space, America's New Competitive Frontier.* Washington, D.C.: Business-Higher Education Forum, 1986.

American Institute of Aeronautics and Astronautics. *Space: A Resource for Earth.* New York, N.Y.: American Institute of Aeronautics and Astronautics, 1977.

Andrews, Walter. "Space-based Defense Called 'Morally Right.'" *Current News,* 31 October 1984, 1–2.

Anson, Sir Peter, and Capt Dennis Cummings. "First Space War: The Contribution of Satellites to the Gulf War." *RUSI Journal,* Winter 1991, 45–53.

Atwood, Donald. "Preparing to Meet the Future in Space." *Defense Issues,* 11 April 1991, 1–2.

Battelle. "The Strategic Implications of Modifying the Space Environment." *Journal of Defense Research,* Fall-Winter 1983, 135-46.

Baum, Michael E. "Defiling the Altar: The Weaponization of Space." *Airpower Journal,* Spring 1994, 61.

Bell, Trudy, and Karl Esch. "The United States in Space." *Spectrum,* August 1991, 18–20, 45–51.

Benko, Marietta. *Space Law in the United Nations.* Boston, Mass.: M. Nijhoff Publishers, 1985.

Berkowitz, Marc. "Future U.S. Security Hinges on Dominant Role in Space." *Signal,* May 1992, 71–73.

Berry, Adrian. "The Next Ten Thousand Years: A Vision of Man's Future in the Universe." New York, N.Y.: *Saturday Review Press,* 1974.

Blau, Thomas, and Daniel Goure. "Military Uses and Implications of Space." *Society*, January–February 1984, 13–17.

Blechman, Barry. *The American Military in the Twenty-First Century*. New York: St. Martin's Press, 1993.

Blow, Thomas. *Defending Against a Space Blockade.* Maxwell AFB, Ala.: Air University Press, 1989.

Bono, Phillip, and Kenneth Gatland. *Frontiers of Space.* London: Blandford Press, 1969.

Bousher, Brig Gen Homer A. "Blueprints for Space." *Air University Quarterly Review,* Spring 1959, 18.

Bova, Ben. *The High Road.* Boston, Mass.: Houghton Mifflin, 1981.

Brzezinski, Zbignew, Robert Jastrow, and Max Kempelman. "Defense in Space Is Not 'Star Wars.'" *New York Times Magazine,* 27 January 1985, 28–29.

Builder, Carl. *The Icarus Syndrome.* New Brunswick, N.J.: Transaction Publishers, RAND, 1993.

———. *The Masks of War.* New Brunswick, N.J.: Transaction Publishers, RAND, 1992.

Building a Consensus Toward Space. Proceedings of the Air War College 1988 Space Issues Symposium. Maxwell AFB, Ala.: Air University Press, 1990.

Burke, William. "Active Space Experiments Affect Treaty Obligations." *Signal,* June 1990, 73–75.

Burrows, William. "Ballistic Missile Defense: The Illusion of Security." *Foreign Affairs,* Spring 1984, 843–56.

———. "Skywalking with Reagan." *Harper's,* January 1984, 50–52.

Chisholm, Robert. *On Space Warfare: Military Strategy for Space Operations.* Maxwell AFB, Ala.: Air University Press, June 1984.

Christol, Carl. *The Modern International Law of Outer Space.* New York: Pergamon Press, 1982.

Cole, Dandridge. *The Next Fifty Years in Space: Man and Maturity.* Philadelphia, Pa.: General Electric Company, 1963.

Collins, John M. *Military Space Forces: The Next Fifty Years.* Washington, D.C.: Pergamon-Brassey's, 1989.

Courter, Jim. "Military Space Policy: The Critical Importance of New Launch Technology." *Strategic Review*, Summer 1994, 14-23.

Department of the Air Force. *Global Presence 1995*. Washington, D.C.: Government Printing Office (GPO), 1995.

Dougherty, Gen Russell, et al. "Facing Up to Space." *Air Force Magazine*, January 1995, 50-54.

Dougherty, William. "Storm From Space." *Proceedings*, August 1992, 48–52.

Douhet, Giulio. *The Command of the Air*. Translated by Dino Ferrari. Washington, D.C.: Office of Air Force History, 1983.

Durch, William. *National Interests and the Military Use of Space*. Cambridge, Mass.: Ballinger, 1984.

Durnheim, Michael A. "DCX Proving Initial Operating Concepts." *Aviation Week & Space Technology*, 8 March 1993, 49.

Dutton, Lyn, ed. *Military Space*. Washington, D.C.: Brassey's, 1990.

Earle, Edward M. "Adam Smith, Alexander Hamilton, Freidrich List: The Economic Foundations of Military Power," in *Makers of Modern Strategy*. Princeton, N.J.: Princeton University Press, 1986.

Faber, Scott. "Global Ambitions." *Discover*, January 1995, 3.

Fradkin, Elvira. *Air Menace: The Answer*. New York: Macmillan Press. 1934.

Friedenstein, Charles. "The Uniqueness of Space Doctrine." *Air University Review*, November–December 1985, 13–23.

Graham, Gen Daniel. *High Frontier: A New National Strategy*. New York: Tom Doherty Associates, 1983.

———. "High Frontier and Arms Control." *Journal of Defense and Diplomacy*, November 1984, 25–28.

Gray, Colin. *American Military Space Policy*. Cambridge, Mass.: Abt Books. 1982.

Hall, Cargill. "The Origins of US Space Policy." *Colloquy: Security Affairs Support Association*, December 1993, 5–24.

Hartinger, Gen James. "The Air Force Space Command: An Update." *Air Force Engineering and Services Quarterly*, Summer 1984, 4–9.

———. "Strategic Space Systems Require a Unified Command." *Defense Systems Review*, February 1984, 19-22.

"High Frontier Can Reduce Defense Budget." *High Frontier Newsletter*, January 1985, 1.

Horner, Gen Charles. "Space Systems: Pivotal to Modern Warfare." *Defense* 94, 1994, 20–29.

———. "Unpredictable World Makes US Space Capabilities Critical." *Defense Issues*, 1994, 1–7.

———. "Space Seen as Challenge, Military's Final Frontier." *Defense Issues*, 1993, 1–10.

Hudson, Richard D. *Infrared Systems Engineering*. New York: John Wiley and Sons, 1969.

"Introduction of Space Weapons Treaty Resolution." *Congressional Record*, 24 April 1985, 1738–39.

Johnson, Nicholas. *Soviet Military Strategy in Space*. London: Jane's Publishing, 1987.

Kennett, Lee. *The First Air War*. New York: Free Press, 1991.

Kingwell, Jeff. "The Militarization of Space." *Space Policy*, May 1990, 107–11.

Kipp, Jacob, et al. *Soviet Views on Military Operations In Space*. College Station, Tex.: Center for Strategic Technology, 1986.

Kolcum, Edward H. "Pratt and Whitney Assessing Family of Engines for Upcoming Space Missions." *Aviation Week & Space Technology*, 6 January 1992, 56.

Kuskuvelis, Ilias. "Satellites for War and Peace." *Proceedings of the Thirty-fourth Colloquium on the Law of Outer Space*, October 1991, 227–32.

Kutyna, Gen Donald. "SPACECOM: We Lead Today, But What About Tomorrow?" *Defense* 91, July–August 1991, 20–9.

———. "The State of Space." *Defense Issues*, 23 April 1991, 1–8.

Lorenzini, Dino. "Space Power Doctrine." *Air University Review*, July–August 1982, 16–21.

Los Alamos National Laboratory. *United States Space Policy: Review and Assessment*. Los Alamos, N. Mex.: Los Alamos National Laboratory, 1988.

Luongo, Kenneth, and Thomas Wander. *The Search for Security in Space.* Ithaca, N.Y.: Cornell University Press, 1989.

Lupton, David. *On Space Warfare: A Space Power Doctrine.* Maxwell AFB, Ala.: Air University Press, 1988.

Manno, Jack. *Arming the Heavens: The Hidden Military Agenda for Space, 1945–1995.* New York: Dodd, Mead and Company, 1984.

Marks, Hans. "War and Peace in Space." *Journal of International Affairs,* Summer 1985, 1–21.

McLean, Alasdair. *The Military Utility of Space.* Aberdeen, Scotland: Centre for Defence Studies, 1991.

McDougall, Walter. "Sputnik, the Space Race, and the Cold War." *Bulletin of the Atomic Scientist,* May 1985, 20–25.

———. *The Heavens and the Earth: A Political History of the Space Age.* New York: Basic Books, 1985.

Mitchell, Brig Gen William. *Winged Defense.* New York: G. P. Putnam's, 1925.

Moore, George, Vic Budura, and Joan Johnson-Freese. "Joint Space Doctrine: Catapulting into the Future." *Joint Force Quarterly,* Summer 1994, 71–76.

Moorman, Lt Gen Thomas. "The 'Space' Component of 'Aerospace'" *Comparative Strategies,* July–September 1993, 251–55.

———. "The Future of USAF Space Operations." *Vital Speeches,* March 1994, 325–29.

Muolo, Michael, ed. *Space Handbook: A Warfighter's Guide to Space,* vol. 1. Maxwell AFB, Ala.: Air University Press, 1993.

Norton, Oliver. *The Attack and Defense of Little Round Top, Gettysburg, July 2, 1863.* New York: Neale Publishing Company, 1913.

Odom, Lt Gen William. "Aerospace Requirements for U.S. Security." *Comparative Strategies,* vol. 12, 257–61.

Osman, Tony. *Space History.* New York: St. Martin's, 1983.

Petrie, W. "Military Activity in Space—Is There a Choice?" *Canadian Defense Quarterly,* Winter 1985–1986, 31–36.

Power, John. "Space Control in the Post-Cold War Era." *Airpower Journal,* Winter 1990, 24–33.

Richardson, Gen Robert. "Technology, Bureaucracy and Defense: The Prospects for the U.S. 'High Frontier' Program." *Journal of Social, Political and Economic Studies*, Fall 1983, 293–99.

Rosenberg, Maj Gen Robert. "The Air Force and Its Military Role in Space." *Air University Review*, November–December 1985, 52–57.

Sadov, Y. "Washington's Space Tricks." *Contemporary Review*, January 1985, 7–8.

Salkeld, Robert. "The Changing Perception of Space: Vehicles, Treaties, Purposes." *Defense Science 2001*, June 1983, 24–29.

Scoville, Herbert. *Can Space Remain a Peaceful Environment?* Muscatine, Iowa.: Stanley Foundation, 1978.

Schelling, Thomas. *Arms and Influence*. London: Yale University Press, 1966.

Sgrosso, Gabriella. "Demilitarisation of Outer Space." *Proceedings of the Thirty-fifth Colloquium on the Law of Outer Space*, August–September 1992, 325–34.

Sonneberg, Steven. "The Ultimate High Ground." *Marine Corps Gazette*, May 1990, 58–65.

"Soviet Propaganda About 'Militarization of Space.'" *Defense Daily*, 4 March 1985, 15.

Spaulding, Oliver L. *Ahriman: A Study in Air Bombardment*. Boston, Mass.: World Peace Foundation, 1939.

SPACECAST 2020 Final Report, vol. 1. Prepared by the students and faculty of Air University, Maxwell AFB, Ala.: Air University, 1994.

Stares, Paul B. "Space and U.S. National Security." *Journal of Strategic Studies*, December 1983, 31–48.

———. *The Militarization of Space: U.S. Policy, 1945–1984*. Ithaca, N.Y.: Cornell University Press, 1985.

Stine, Harry. *Confrontation in Space*. Englewood Cliffs, N.J.: Prentice-Hall, 1981.

Szafranski, Col Richard. *GEO, LEO and the Future*. Maxwell AFB, Ala.: Air University Press, 1991.

The War Reports of Marshall, Arnold and King. New York: Lippincott, 1947.

Toffler, Alvin, and Heidi. *War and Anti-War: Survival at the Dawn of the 21st Century*. Boston, Mass.: Little, Brown, and Company, 1993.

Translations of Two Soviet Articles on Law and Order in Outer Space. Santa Monica, Calif.: RAND Corporation, 1958.

United Nations General Assembly, Committee on Peaceful Uses of Outer Space. *Report from the Committee on the Peaceful Uses of Outer Space*. New York, N.Y.: United Nations, 1959.

United States Air Force Scientific Advisory Board. *Report on Space Power Technology*. Washington, D.C.: GPO, 1991.

United States Joint Chiefs of Staff. *Joint Doctrine; Tactics, Techniques and Procedures for Space Operations*. Washington D.C.: GPO, April 1992.

Vereshchetin, V. S. *Outer Space, Politics and Law*. Moscow: Progress Publishing. 1987.

Verplaetse, Julien. *International Law in Vertical Space: Air, Outer Space, Ether*. South Hackensack, N.J.: F. B. Rothman, 1960.

Wassenbergh, H.A. *Principles of Outer Space Law in Hindsight*. Boston, Mass.: M. Hijhoff Publishers. 1991.

Welling, William. "Policy and Strategy Options for the Next Century." *Defense Science 2003*, June–July 1985, 58–63.

Westwood, James. "Military Strategy and Space Warfare." *Journal of Defense and Diplomacy*, November 1984, 17–21.

Whittington, Mark R. "Stifled By Political Correctness." *Space News*, 25 April–1 May 1994, 15.

Wolf, James. "Toward Operational-Level Doctrine for Space—A Progress Report." *Airpower Journal*, Summer 1991, 28–40.

Worden, Simon, and Bruce Jackson. "Space, Power, and Strategy." *The National Interest*, Fall 1988, 43–52.

Worden, Simon. *SDI and the Alternatives*. Washington, D.C.: National Defense University Press, 1991.

Unpublished Sources

Abrahamson, James. "Progress and Policy Paradigms." Paper presented to the 31st *AIAA Aerospace Sciences Meeting and Exhibit*. Reno, Nev., January 1993.

Albert, David. "Interdependence in Space?" Paper submitted to the *Journal of Conflict Resolution*. Maxwell AFB, Ala., March 1988.

Botte, David. "The United States and National Sovereignty in Outer Space." Armed Forces Staff College Study. Norfolk, Va., May 1962.

Carron, Brian E. "The Value of Space Control and How We Can Achieve It." Air War College paper. Maxwell AFB, Ala., April 1993.

Coursey, Michael. "Evolution of the Space Command from National Space Policy." Air Command and Staff College paper. Maxwell AFB, Ala., March 1984.

Cox, Dyson. "A Comparison of U.S. and U.S.S.R. Views on Space Law." Air War College paper. Maxwell AFB, Ala., January 1963.

Curtis, Edward. "Space Exploration and International Problems in the Use and Control of Outer Space." Thesis, San Francisco State College. San Francisco, Calif., 1963.

Davenport, Richard. "The Birth of Spacepower: A Doctrine for the 21st Century." Air War College paper. Maxwell AFB, Ala., October 1993.

DeSaussure, Hamilton. "The Two Sides of the Law of Outer Space." Paper presented to the *AIAA Colloquium on the Law of Outer Space*. Dresden, Germany, October 1990.

Dunning, Stephen. "U.S. Military Space Strategy." Naval War College paper. Newport, R.I., 14 May 1990.

"Force Applications Study Final Report." Unclassified briefing slides from Phillips Laboratory project. 1991.

Ford, James. "Space Force: Organizing for Effective Military Use of Space." Air Command and Staff College paper. Maxwell AFB, Ala., April 1985.

Hays, Peter. "Struggling Towards Space Doctrine: US Military Space Plans." Doctoral thesis, Fletcher School of Law and Diplomacy, 1994.

Hoch, Karl. "Legal Aspects of Military Operations in Outer Space." Paper presented to the Air University Airpower Symposium. Maxwell AFB, Ala., 1981.

Howard, William, and Robert Rosenberg. "Future Military Space Systems and the Principles of War." Dahlgren, Va.: Naval Space Command, 1990.

Howarth, Thomas. "The Impact of Space on Future Wars (or: Will World War III Be Waged in Space?)." Student thesis, Naval War College. Newport, R.I., February 1989.

Johnson, Dana. "The Evolution of U.S. Military Space Doctrine: Precedents, Prospects and Challenges." Doctoral thesis, University of Southern California, 1987.

Lentini, Joseph. "Will Congress Support a Space Control Policy?" Air Command and Staff College paper. Maxwell AFB, Ala., April 1984.

Lorenzini, Dino. "2001: A U.S. Space Force." Paper presented to the Air University Airpower Symposium. Maxwell AFB, Ala., 1981.

Lyall, Francis. "Space Law—What Law or Which Law?" Paper presented to the Aerospace Industries Association of America (AIAA) Colloquium on the Law of Outer Space. Dresden, Germany, October 1991.

McFarland, R. S. "The Impact of Space Systems on Future Warfare—A Warrior's Perspective." A Naval War College paper. Newport, R.I. 1980.

Meilinger, Col Phillip. "Ten Propositions Regarding Airpower." Paper written in his position as dean, School of Advanced Airpower Studies, Maxwell AFB, Ala., August 1994, 3.

Piotrowski, John, Gen, USAF. "Space Warfare and the Principles of War." United States Space Command. Colorado Springs, Colo., 1989.

Smith, William. "Potential Legitimate Use of Space Weapons as Part of the United States' Strategic Forces." Air Command and Staff College paper. Maxwell AFB, Ala., May 1980.

Sponable, Jess. "Single Stage Rocket Technology Program Review of Future Systems and Applications." Unclassified briefing viewgraphs, Ballistic Missile Defense Organization, January 1993.

Weston, Craig. "The Essence of Spacepower: Important Influences on the Evolution of National Spacepower." Air War College paper, Maxwell AFB, Ala., May 1989.

Index

☆ U.S. GOVERNMENT PRINTING OFFICE: 1999 – 737 - 536